D0152567

André-Marie Ampère

André-Marie Ampère

JAMES R. HOFMANN

CAMBRIDGE
UNIVERSITY PRESS

Published by the Press Syndicate of the University of Cambridge
The Pitt Building, Trumpington Street, Cambridge CB2 1RP
40 West 20th Street, New York, NY 10011-4211, USA
10 Stamford Road, Oakleigh, Melbourne 3166, Australia

First published 1995 by Blackwell Publishers Oxford
Reissued by Cambridge University Press 1996

Printed in Great Britain by Biddles Ltd., Guildford & King's Lynn

British Library Cataloguing in Publication Data available

Library of Congress Cataloging-in-Publication Data available

ISBN 0 521 56220 1 hardback

In commemoration of
Edward Patrick McKenna
(1921–1992)

Contents

List of Illustrations   ix
General Editor's Preface   xi
Acknowledgements   xiv
Introduction   1

PART I   Coming of Age (1775–1804)   5

1   Idyllic Youth   7
2   Marriage and Provincial Life (1800–1804)   24

PART II   Paris (1804–1820)   97

3   Laplacian Physics   99
4   Ampère in Paris   123
5   Metaphysics: Ampère, Kant, and Maine de Biran   144
6   Mathematics, Chemistry, and Physics (1804–1820)   165

PART III   Electrodynamics   225

7   Ampère's Response to the Discovery of
    Electromagnetism (1820)   227
8   Ampère's Electrodynamics (1821–1822)   265
9   Defense and Elaboration of the Theory   309

CONTENTS

PART IV   Closing Years   351

10   The Final Synopsis   353

Notes   368
Bibliography   384
Index   400

Illustrations

1	Portrait of Ampère	12
2	Lavoisier's Table of Elements	190
3	Ampère's Table of Elements	209
4	Ampère's demonstration apparatus for electrodynamic forces between two parallel conductors	239
5	Ampère's angles for the mutual orientation of two current elements: α, β, and γ	241
6	One of Ampère's axially compensated helices	243
7	Ampère's apparatus for two conductors at variable orientations	245
8	Ampère's sketch of the apparatus to be discussed at the 26 December, 1820, Académie meeting	255
9	The final design for Ampère's first equilibrium apparatus	257
10	Ampère's induction apparatus	286
11	Ampère's initial rotation apparatus	294
12	Ampère's improved rotation apparatus (March, 1822)	295
13	Ampère's oscillatory apparatus (1822–1823)	296
14	Ampère's mobile rectangular conductor	298
15	Ampère's early drawing of the "third" equilibrium demonstration	301
16	Ampère's 1822 "third" equilibrium demonstration	304
17	Ampère's floating wire experiment	317

18 Figure 1 from Ampère's 1824 memoir on
    electrodynamic phenomena   329
19 Figure 1 from Ampère's 1827 Brussels memoir   334

General Editor's Preface

Our society depends upon science, and yet to many of us what scientists do is a mystery. The sciences are not just collections of facts but are ordered by theory, which is why Einstein could say that science was a free creation of the human mind. Though it is sometimes presented dispassionately and impersonally, science is a fully human activity, and the personalities of those who practice it are important in its progress and often interesting to us. Looking at the lives of scientists is a way of bringing science to life.

Those scientists who appear in this series have been chosen for their eminence, but the aim of the biographers is to place them in their context. The books are long enough for authors to write about the times as well as the life of their subjects. Science has not long been a profession, and for many eminent practitioners of the past it was very much a part-time activity: their lives will therefore show them practicing medicine or law, fighting wars, looking after estates or parishes, not simply focusing upon their hours in the laboratory. How somebody earned a living, made a career, got on with family and friends is an essential part of a biography, though in this series it is the subjects' commitment to science that has made them eligible and must be always at the back of the biographer's mind.

The name of André-Marie Ampère and his major field of work should be familiar to most people because the unit of electric current is named after him, but it might well be that we do not think of him even when we change a 3-amp fuse. Apart from that, very little is generally known about him, especially in the English-speaking world, and this biography is therefore especially timely. James Hofmann has a fascinating story to tell, and he succeeds in bringing Ampère to life. It was not a very happy life: his father was executed in the French Revolution, his marriages were unhappy for different reasons, his son was a disappointment to him, and his last years were spent in poverty, doing university inspecting to make ends meet – he died during a tour of this duty. But he had had his times of intellectual triumph, which must have brought him great satisfaction.

An idyllic childhood and youth was violently brought to an end in the Terror in Lyon, destroying feelings of security. Then came happy marriage and provincial life, but the death of his wife in 1803 was a terrible blow. For serious work in science, it was at that time necessary to go to Paris, and there Ampère went to make his career. He was very interested in philosophy, particularly that of Immanuel Kant and Maine de Biran; and his work ranged across mathematics, chemistry, and physics. Chemists will recall his passing iodine to Humphry Davy, who speedily recognized it as an analogue of chlorine, and his independent postulation of what we call Avogadro's Hypothesis.

It was in 1820 that his big break came, when Hans-Christian Oersted's experiment proving the magnetic effects of electricity was repeated at the Académie des Sciences. Many others all over Europe took up this exciting research, but it was Ampère who became the founder of electrodynamics. Combining the French tradition of mathematical physics with great experimental skill, he made this a central part of physics in a very few years. Hofmann skillfully leads us through these researches, which mark Ampère out as one of the great scientists. They led, for example, to his election as a Foreign Member of the Royal Society of London in 1827.

Hofmann emphasizes that Ampère was never a representative figure, making a career as some of his contemporaries did. Prone to introspective brooding, and a devout Catholic, he kept up connections not only with Lyon but outside physical science. He was interested in the controversies over classification and evolution between Georges Cuvier and Geoffroy Saint-Hilaire, and had an encyclopedic urge to classify first the chemical elements, then the whole of knowledge. He sought a natural method, rather than an artificial system like that of Linnaeus, and his last and rather lonely years were spent in trying to perfect it.

The Ampère to whom Hofmann introduces us is thus primarily concerned with science, to which he made contributions of fundamental importance, but he also comes across as an individual full of quirks.

He makes an excellent subject for a biography, working as he did in Paris when it was the world's centre of excellence in the sciences, so that his life and times were momentous. Hofmann shows how a good biography can be an excellent way into understanding the science of a past epoch, and also cast light on how the scientific mind, or scientific minds, work in a particular case. I warmly commend it.

David Knight
University of Durham

Acknowledgements

I have accrued enormous debts in writing this book. Over a decade ago, L. Pearce Williams of Cornell University graciously gave me access to his hard-won collection of Ampère microfilms and reprints. His generosity and advice greatly facilitated my own subsequent investigation of the voluminous Ampère archives. Invigorating conversations with Kenneth Caneva also were an important factor during that early period. More recently, Judith Grabiner has generously shared her knowledge of nineteenth-century mathematics. Nancy Caudill has worked wonders as the tireless Supervisor of the Interlibrary Loan Office at California State University in Fullerton. Archivist Christiane Demeulenaere-Douyere and her staff cheerfully helped me negotiate the Ampère archives at the Académie des Sciences in Paris. I am also grateful to the Académie for permission to photograph documents and to use some of them as illustrations. Financial support has been provided by a General Faculty Research Grant from California State University at Fullerton and by a Research Grant from the National Endowment for the Humanities, which I gratefully acknowledge. Portions of chapters 7, 8, and 9 are revisions of earlier publications in the *British Journal for the History of Science* and *Osiris*. Unless otherwise noted, all translations from French are my own.

James R. Hofmann
California State University
Fullerton, July 1994

Introduction

Biographies are shaped as much by the interests and competence of biographers as by their subjects. I find Ampère intriguing in his rare synthesis of scientific brilliance, philosophical reflection, and spiritual sensitivity. While my perspective may be appropriate for a scientific biography, quite different books will be written about Ampère in the future. The few that presently exist do not attempt to convey the burden of Ampère's science. For example, although Louis de Launay (1925) did appreciate the complexity of Ampère's intellectual and emotional life, he provided no detailed analysis of scientific achievements. On the other hand, Valson (1886) and Lewandowski (1936) went to great lengths to herald Ampère as an example of a scientist who ultimately assigned as much, if not more, importance to his religious faith as he did to the pursuit of scientific knowledge. Unfortunately, this portrayal elucidates neither Ampère's science nor the fascination it held for him. It is true that Ampère's scientific career evolved in conjunction with philosophical and religious concerns about the limitations of scientific knowledge. But in this respect his spirituality both haunted and inspired him. With due respect for the complexity of Ampère's character, the present biography places his scientific work at center stage.

As these comments suggest, Ampère was prone to introspective brooding. His letters to friends and relatives are predominantly recitals of his own suffering, boredom, or enthusiasm, together with appeals for support in his daily struggles. This is not surprising in light

1

of the tragic events that punctuated his personal life. His character took on a permanently melancholic cast following the execution of his father by guillotine during the Reign of Terror and the early death of his beloved first wife. Perhaps neither Balzac nor Dickens more forcefully depicted the dismal consequences of misplaced emotion as does the tale of Ampère's disastrous second marriage. On the other hand, Ampère could be exuberant and even boisterous in conversation; he could captivate audiences with the sheer energy of his discourse. This was particularly the case in the company of close personal friends, a context in which he felt encouraged to cross disciplinary boundaries in a manner not tolerated by his professional colleagues. It was also a milieu in which he usually held the center of the stage and did not have to acknowledge alternative viewpoints.

Ampère is by no stretch of the imagination representative of his era. He was idiosyncratic in ways that made him unique. Nor was he a figure who had significant personal impact on the society in which he survived as a gifted but essentially exotic presence. In this respect he stands in contrast to Einstein, Newton, or Darwin who in such large measure helped define the centuries in which they worked. Instead, Ampère's life was a series of intersections, a commingling of the intellectual and social currents he explored. With no formal education, his protective family circle encouraged him to adopt both the optimistic scientific outlook of the Enlightenment and a devotion to the Catholic faith. This combination of intellectual expectations and emotional spirituality produced a tension that became his most definitive characteristic. As a scientist, he made original contributions to mathematics and chemistry as well as electrodynamics, the discipline he virtually created. As a member of the prestigious Académie des Sciences and a professor at the École Polytechnique, he was one of the elite few who fully participated in the highly centralized French scientific community. On the other hand, he made valiant philosophical efforts to locate scientific knowledge within a broader intellectual context; he was an active member of a philosophical circle that drew much of its agenda from Maine de Biran and the study of Kant. While Ampère is generally acknowledged as the man who created the science of electrodynamics, relatively little attention has been devoted to the relationship of this achievement to his broader philosophical interests, a central concern of this biography.

Were we to search for a foil to serve as a contrast to Ampère's multifaceted personality, several men immediately come to mind. Among his contemporaries, Biot, Poisson, and Gay-Lussac all participated in the Parisian scientific environment in ways that bring Ampère's character into sharp relief. These were rigorously trained career men who knew how to be appropriately aggressive in the pursuit of their own interests, particularly in formal institutional settings. Ampère preferred the more relaxed context of casual conversation where exploratory speculations were encouraged. Furthermore, unlike Fresnel or Berthollet, Ampère did not contribute significantly to "applied science." His invention of the galvanometer was an exception to his predominantly theoretical and highly mathematical approach to physics.

With provincial roots in Lyon, Ampère never lost touch with the eclectic circle of friends he established there prior to his departure for Paris in 1804. Of a predominantly spiritual and literary bent, this group included Bredin, Ballanche, and Degérando. But Ampère was atypical in this context as well. Although he participated in the romantic Catholic revival of the Restoration, he was far better informed about contemporary science than the majority of romantic thinkers. For example, while romantics were apt to denounce atomic theory as an expression of materialist philosophy, Ampère's own atomic theory was thoroughly embedded in an anti-materialistic metaphysics. Ampère's intellectual life ultimately came full circle. As a child he studied the quintessential expression of the scientific mentality of the Enlightenment, the *Encyclopédie* edited by Diderot and d'Alembert; his last years were devoted to his own attempt to construct a "natural" classification scheme for all disciplines of human knowledge.

Part I

Coming of Age (1775–1804)

1
Idyllic Youth

Family Roots

In the earliest extensive biographical study of Ampère, Louis de Launay traced Ampère's paternal lineage back five generations to the middle of the seventeenth century. By that point the family had established itself in Lyon. Ampère's great-grandfather, Jean-Joseph Ampère, originally followed his father's trade as a stonemason, but eventually made a transition to the merchant class. These bourgeois roots were strengthened as Ampère's grandfather plied the flourishing silk trade in Lyon and married Anne Berthois, daughter of an *avocat au Parlement*. Their four sons all entered business in Lyon and it was there that the fourth son, Jean-Jacques Ampère, married Jeanne-Antoinette Desutières-Sarcey on 16 July, 1771. At the time of her marriage, Jeanne Sarcey was a 22-year-old orphaned daughter from another Lyon silk-merchant family. Her father had been Claude-Joseph Desutierres-Sarcey of J. Sarcey and Company. Following his death, Jeanne lived in Lyon with his only other child, her sister Antoinette. One of Jeanne Sarcey's uncles provided her with an endowment of 25,000 livres shortly after her marriage. Both sides of Ampère's family tree thus enjoyed a moderate prosperity prior to the upheavals of the Revolution.

Before his marriage at age 28, Jean-Jacques Ampère resided on rue Merciere, centrally located in the heart of the old commercial center of Lyon between the Saône and Rhône rivers. Michelet described this area as a "teeming ant heap nestled among rocks and rivers, crowded

7

into dark streets that slope downward in the rain and the eternal fog."[1] It was common for prosperous bourgeois families to maintain country homes and spend the summer months outside the city. Three weeks prior to his marriage, Jean-Jacques Ampère's financial status allowed him to purchase a country estate at a price of 20,000 livres. Located adjacent to the village of Poleymieux, about ten kilometers northwest of Lyon, the new property provided a summer retreat during the initial years of marriage; the Ampère household would take up permanent residence there following Jean-Jacques' early retirement in 1782.

Lyon and Poleymieux

It was in the midst of the bustling commercial activity of Lyon, however, that the Ampères' first two children were born: a daughter, Antoinette, in 1772 and a son, André-Marie, on 20 January, 1775. Both were baptized in the nearby parish church of Saint-Nizier. Little direct information has survived concerning Ampère's early years in Lyon, the capital of the department of Rhône in southwestern France. The cliffs and valley created by the confluence of the Saône and Rhône rivers provided a serenely beautiful setting. With a relatively low population of only 149,000 as late as 1821,[2] the city was a commercial and cultural center for the primarily rural populations of the surrounding departments. The silk trade's dependance on far-flung markets gave the city a broader perspective than might be expected of a provincial town. Its proximity to Geneva ensured an international aspect intellectually as well. Noted for both an aggressive business climate and a mystically inclined religious intensity, the city later made an eclectic setting for Ampère's early and permanent friendships.

During the first seven years of Ampère's life, Lyon provided his initial encounter with concentrated urban life. During the summers of that same period, however, he also experienced exhilarating rural life at Poleymieux. The combination of Lyon and Poleymieux was to leave an impression which later would haunt Ampère with a longing for a return to the happy scenes of his youth. Separated from Lyon by the short distance of only ten kilometers, Poleymieux is reached by following the Saône valley north and then veering to the west by one of several possible routes so as to reach a small enchanting valley within the Mont d'Or mountain range. Dotted by small hamlets, the valley comes to a head in the west where it is dominated by the peaks of Mount Verdun to the north and Mount Thou to the south, both over

8

600 meters in altitude. The Ruisseau du Thou has its source midway between them and follows a major fault line northeastward until, after a quick descent of 350 meters, it flows into the Saône at Neuville. The approach to Poleymiex from Neuville thus requires an invigorating climb as the valley gradually widens into an amphitheater where the village clings to the steep northern slope. Sun, rich soil, and ample rain support the foliage of a lush and complex forest that directly borders the cleared acreage. In Ampère's time local roads were primitive, particularly in rainy weather. The terrain is somewhat rugged but manageable; from the heights, the views of the distant Saône valley are inspiring, but the overall impression is that of an oasis rather than a citadel. Perhaps the mood Poleymieux inspired in Ampère is best captured by this passage from Chateaubriand's *René*:[3]

> In our endless agitation we Europeans are obliged to erect lonely retreats for ourselves. The greater the turmoil and din in our hearts, the more we are drawn to calmness and silence. These shelters in my country are always open to the sad and the weak. Often they are hidden in little valleys, which seem to harbor in their bosom a vague feeling of sorrow and a hope for a future refuge. Sometimes, too, they are found in high places where the religious soul, like some mountain plant, seems to rise toward heaven, offering up its perfumes.

During Ampère's lifetime, the permanent population of Poleymieux ranged between 400 and 500. In 1788, for example, 406 inhabitants were recorded, a population that was hard pressed financially under the obligations of the Old Regime.[4] The village center was located about a kilometer further up the slope from Jean-Jacques' house; it clustered near the old church and the ancient *château*, a fortified remnant of the Roman Empire. There was no school until 1833, long after Ampère's departure. For centuries the peasant population had coexisted with a small contingent of Lyon bourgeoisie who found the valley an ideal setting for summer estates or permanent retirement. During Ampère's childhood there were five of these families in residence.

Jean-Jacques Ampère's house was itself something of a model representation of bourgeois taste. Of solid stone construction, it was cut into the northern slope of the valley. It had three main rooms, including a reception area, on the ground floor and five rooms upstairs. Southern windows gave splendid views into the valley, as did the southern terrace, shaded by lindens. About 70 acres of land made up the surrounding estate, part of which was rented to local farmers. An assemblage of farm animals completed the scene. The nearest neighboring

9

structure was the *maison de la dîme*, a small storage shed where the peasants delivered the sporadic but burdensome tithe of the Old Regime. The property remained in the possession of the Ampère family until 1818, when André-Marie's financial requirements in Paris forced him gradually to sell parcels of land and eventually the house itself in 1819. The house survived the subsequent century and in 1928 it was purchased by two Americans, the brothers Hernand and Sosthènes Behn. They were directed to the purchase by Paul Janet of the Académie des Sciences, and they donated the home to the Société francaise des Electriciens. The Société des Amis d'André-Marie Ampère was initiated in 1930, and during the following year it assumed the preservation of the Ampère property as a museum.

In 1782 Jean-Jacques Ampère moved his family to Poleymieux; thereafter he allotted only short periods of time during the winters for business in Lyon. About to turn 50 years of age, he placed more importance on the education and nurturing of his children than on the advancement of a business career. In 1785 a second daughter was born and christened Joséphine. Ten years younger than André-Marie, she became a close companion and eventually would be his housekeeper in Paris. In 1786 Marie-Aimé Guillin Dumontet took up residence in the *château* as the new *seigneur* of Poleymieux. A retired Governor of Senegal with a distinguished military record, he soon established a reputation in Poleymieux for highhandedness and intransigence. Nevertheless, Jean-Jacques Ampère was willing to take an administrative and judicial position as Dumontet's *procureur fiscal*.

What little is known about André-Marie's experiences during these years is due primarily to an autobiographical summary he composed many years later. The following passage contains his memories of his early education. Ampère refers to himself in the third person.[5]

His father, who had never ceased to cultivate Latin and French literature, as well as several branches of science, raised him himself in the country near the city where he was born. He never required him to study anything, but he knew how to inspire in him a desire to know. Before being able to read, the young Ampère's greatest pleasure was to listen to passages from Buffon's natural history. He constantly requested to have read to him the history of animals and birds, the names of which he had long since learned while amusing himself by looking at the pictures. The liberty he was allowed to study only when it pleased him to do so was the cause of the fact that, although he had known how to spell for a long time, he did not yet read, and it was only by making an effort on his own to understand the history of birds that he finally learned how to read fluently. Soon the reading of history books and theater pieces that he found in his father's library attracted him as much as Buffon. He became enthusiastic about the Athenians and the Carthaginians and scorned the Lacedaemonians and the Romans when he saw them subjugate or destroy peoples he was fond of. He took a singular pleasure in learning entire scenes from the tragedies of Racine and Voltaire and in reciting them while walking alone. The feelings that this reading developed in him were amplified by the fact that he heard recounted events from the war that England and France were engaged in on the subject of the independence of the United States.

This intriguing summary is open to a variety of interpretations. Although Jean-Jacques Ampère was guided to some extent by Rousseau's educational philosophy, he left no direct evidence about how he put it into practice. Clearly he was selective. We can only speculate as to whether his move to Poleymieux was partially motivated by Rousseau's insistence on the advantages of a rural setting. Surely André-Marie was given ample opportunity to explore the terrain of Poleymieux and to practice Rousseau's admonition both to "learn from things" and to do so according to spontaneous interest. Nor was the boy constrained by the structure of a formal institutional education. On the other hand, contrary to Rousseau's warnings, André-Marie was given early access to his father's library. Nor was he in any manner discouraged from exploring and applying the power of words at an early age, practices strongly discouraged by Rousseau. As Ampère mentioned in his autobiographical sketch, he delighted in memorizing entire passages, particularly dramatic ones. Years later, separated from his own four-year-old son following the death of his wife, Ampère advised his sister Joséphine concerning the reading habits he wished the boy to adopt, habits that he recalled from his own childhood.[6]

> Kiss him well for me and try to inspire in him, if not a taste for reading, at least one for listening to reading. I am sending you some books for this and I beg you to try to make him understand what you read to him by explaining it to him while showing him the corresponding engravings. One must try to make him connect the written ideas with the engraved ideas, either with respect to animals or little stories.

In addition to unrestricted early reading, Jean-Jacques also ignored Rousseau's advice to teach his pupil music and the physical practice of a manual trade. André-Marie would grow up to be rather maladroit physically; an early accident was later said to be responsible for his childlike but very legible handwriting. Similarly, in the religious sphere the young Ampère was not restricted to the precepts of the "natural

11

1   Portrait of Ampère, probably commissioned from an unknown artist near the time of his marriage in August, 1799, when he was 24. In the mid-1930s it was given to the *Musée Ampère* in the subject's old home at Poleymieux, where it remains. (Photograph by Studio Basset, 159 rue Pasteur, 69300 Caluire.)

religion" advocated by Rousseau. The Catholic Ampère family emphasized devotion rather than dogma and *The Imitation of Christ* was soon committed to Ampère's voracious memory. Musically, Ampère allegedly remained entirely indifferent until his young adult years in Paris. Aside from the physics of their material production, musical melodies apparently had no effect upon him up to that point. Then, while listening to a performance with his usual impatience, he was suddenly overcome by a newly discovered sense of beauty and burst into tears at the sensation.

Ironically, Rousseau's description of the results to be expected from an education sheltered from the complexities of social interactions accurately applies to Ampère's later experiences in Paris:[7]

> Whoever has spent his whole youth far from polite society brings to it for the rest of his life an awkward and constrained bearing, conversation that is always off key, and clumsy and maladroit manners which the habit of living in society can no longer undo and which are only made doubly ridiculous by the effort to improve them.

Buffon's natural history and Rousseau's popular essays on botany made a strong impression on the young and cheerfully sheltered Ampère. In particular, he developed an early and abiding interest in classification schemes. Buffon and Rousseau shared an aversion for "artificial" classification criteria such as those of Linnaeus. Whether this attitude had any early influence on Ampère is difficult to say since he never made comments to this effect. Nevertheless, whatever his initial methodological orientation, Ampère shared the eighteenth-century fascination with "herborizing," the collection and classification of plants. Perhaps his interest was channeled by a lack of perception at more than short-range distances. His friend Ballanche later reported that as a young adult Ampère casually tried on a pair of spectacles during a hike in the country and was overjoyed by his new visual impressions of the panorama.

Similarly, it is difficult to specify what articles from the *Encyclopédie* were most influential. François Arago reported that even late in life Ampère could recite entire articles from the *Encyclopédie*, including those on such abstruse topics as heraldry and falconry.[8] Ampère's lifelong respect for Catholicism is an obvious indication that he was not overly impressed by the thrust of anti-religious barbs hurled in such articles as Holbach's "Priests." His son, Jean-Jacques Ampère, later recalled that his father's first communion was one of the most significant events of his youth.[9]

I have several times heard him say that the three events that had the most influence on him were his first communion, which, carried out with the greatest fervor, permanently attached him to the faith of his fathers, the eulogy of Descartes by Thomas, which transported him with a love for the sciences, and the fall of the Bastille, which, only reaching his mountains from a distance as an explosion of liberty, decided the political sentiments of his entire life.

This emphasis on truth as a religious, scientific, and political goal was shared by a significant contingent of introspective, conservative, and romantic thinkers of Ampère's generation. Chateaubriand spoke for many of them when he cited "the three truths which are the foundations of social order: religious truth, philosophic truth and political truth or liberty."[10]

Arago was quite impressed by Ampère's claim that his initial reading of the *Encyclopédie* took place in strict alphabetical order through each volume. With due allowance for exaggeration, this accomplishment does give some insight into Ampère's intellectual character. Gifted with an amazing power of concentration, he could quickly assimilate and utilize vast amounts of detailed information. He also exercised this ability with a rare enthusiasm that later energized both his research and his conversation. At the same time, even as a young adult Ampère was inclined to shift his attention rapidly from one subject to another, a characteristic that is not surprising in light of the literally "encyclopedic" nature of his early reading habits.

We have noted how Ampère's son claimed that his father attributed an early interest in science to reading Antoine Leonard Thomas' *Éloge de René Descartes*. The eulogy had been praised by Voltaire and deemed prize-worthy by the Académie des Sciences in 1765.[11] Characteristically, Ampère left no comments as to why Thomas made such a vivid impression on him, but some aspects of the essay do stand out in light of what we know about Ampère's subsequent career and character. Written in a grand academic style, Thomas portrayed Descartes as a selfless seeker after truth, one who was systematically misunderstood and denigrated by petty and shortsighted contemporaries. With an aplomb typical of the Enlightenment, Thomas dismissed the medieval period as a dark age for science and emphasized sixteenth-century discoveries in astronomy as having an essential role in setting the stage for Descartes. According to Thomas, a "passion for truth" and a "desire to be useful to men"[12] generated the young Descartes' omnivorous reading, a search anchored in the experience of mathematical certainty. The most striking theme advanced by the eulogy is Descartes'

14

heroic contribution to the triumphant progress of knowledge. The overall emotional impact of the piece was probably more significant for the young Ampère than specific details. Later he would write and converse with a similar enthusiasm about the advancement of knowledge, and he would do so with an equally breathless style.

Thomas made no references to electricity or magnetism other than a few brief phrases that attributed phenomena of that type to the "magnetic virtue" or "the electric fluid" that flows in bodies and is rendered active by friction.[13] Some of the more important subordinate themes are worth mentioning. Thomas attributed Descartes' success to an early refusal to allow assertions to stand unchallenged on the basis of tradition or authority. The famous exercise of Cartesian doubt thus became the starting point for Descartes' attempt to ground philosophy on a new foundation of certainty. The rigor of mathematical reasoning was particularly compelling, and the "geometric spirit" became the model for subsequent philosophical inquiry. Ampère probably did not dwell upon Thomas' reference to Descartes' admonition to "climb by analysis and descend by synthesis"; Thomas did not clarify the distinction, nor did it play any significant role in his discussion of Descartes. Nevertheless, this was one of Ampère's early contacts with a topic that would reappear repeatedly during the early years of his career and would eventually become a major key to his research in electrodynamics.

Thomas also emphasized the Cartesian effort to unite the sciences through the consistent application of rigorous method. Typical of this enterprise is the attempt to merge physiology with psychology on the basis of both introspection and experimental biology. Thus Ampère was confronted by a thinker who had attempted to draw connecting links among the disparate disciplines Ampère had encountered more haphazardly in the *Encyclopédie*. Thomas attributed the limitations of the Cartesian project to an excessive distrust of sensory input and a resulting lack of reliance on experimental evidence. The Cartesian emphasis on introspection resulted in overly "systematic" reasoning from alleged causes to effects rather than from trustworthy observations to probable causes. "The spirit of system took the place of truth."[14] Descartes' Catholicism was presented as an essential aspect of his character. His faithful observation of ritual and his reverence for revealed truth and scripture were asserted by Thomas as facts only denied by those who sought underhanded means to attack Descartes. Regardless of the inaccuracy of this account, Ampère was given a portrait in which the scientific and religious pursuit of truth were complementary. Descartes' final years were presented with all the

15

emotion proper to a martyrdom in the face of envy and petty intrigue. Thomas' prose rose to new heights:

> Men of genius, of whatever country you may be, here is your fate. Misfortunes, persecutions, injustices, the scorn of courts, the indifference of people, the calumnies of your rivals, or those who believe themselves to be, indigence, exile, and perhaps an obscure death five hundred leagues from your homeland, that is what I foretell for you. Because of that must you renounce the enlightenment of men? Surely not, . . . are you not born to think as the sun is to radiate its light? . . . What are all your enemies in comparison to the truth? . . . Remember that your soul is immortal and that your name will be as well.[15]

The passage captures both the rational and the romantic facets of the intellectual ambition that would drive Ampère. Confident not only that there was objective truth to be known, but also that he could command recognition of it, Ampère eagerly adopted the role of a long-suffering servant to enlightenment. Given his innate ability and the circumstances in which he found himself, this must have struck him as his natural calling.

Whether directly inspired by Thomas' essay or not, Ampère's interest in mathematics began at an early age. By his own autobiographical account, this happened at 13 when he began to study elementary texts. According to the autobiographical note we have been citing, Ampère soon felt confident enough to start developing his own mathematical ideas.[16] "Not knowing anyone who had the least knowledge of mathematics, he began to compose a treatise on conic sections with the materials he found in these works and some demonstrations that he imagined and which he believed new."

It is possible that one of these early "demonstrations" was written up into a memoir submitted to the Académie de Lyon in 1788 and entitled "Rectification d'un arc quelconque de cercle plus petit que la demi-circonférence." Louis de Launay studied two copies of the memoir; both are written in a hand that is not Ampère's, but one copy is certified by the *secrétaire perpétuel* of the Académie de Lyon to be a 1788 submission by Ampère at age 13.[17] There seems to have been some confusion at the Académie about the topic of the paper because its title suggested the disreputable subject of circle squaring. According to the records of the Académie, Ampère submitted an essay on the squaring of the circle along with a letter requesting an assessment. Since as early as 1783 the Académie had established a policy of not responding to submissions on that subject, no further official action was taken. Nevertheless, the Abbé Claude-Antoine Roux agreed to offer a private opinion. The *procès-verbal* of the Académie also records the submission of a second memoir on the rectification of a circular arc. It is not clear whether Ampère actually did submit two different memoirs or whether his single one was initially misinterpreted as being about squaring the circle. One of the two copies of the rectification memoir does conclude with some comments on circle squaring and this may have generated the initial confusion. At any rate, only rectification manuscripts have survived and neither copy studied by de Launay is in Ampère's hand.

Bibliographical mysteries aside, the rectification paper, if it was indeed written by Ampère, indicates that he took an early interest in the manipulation of infinitesimal magnitudes. The problem he addresses is the geometric construction of a linear magnitude equal in length to an arbitrarily chosen arc of a semicircle. He solves the problem by constructing a curve adjacent to the semicircle and separated from any point on it by a distance proportional to a magnitude determined by how far along the semicircle the point lies. Choosing an arbitrary point on the semicircle, he constructs a triangle based on the segment joining the point to the adjacent curve and with its other two sides determined by drawing tangent lines to the semicircle and the relevant point on the adjacent curve. He then constructs a similar triangle at a point infinitesimally displaced along the curve and uses the resulting proportionalities to show that the length of the semicircle arc is equal to one side of the first triangle. In the process he uses the notation dx and dy to represent infinitesimal quantities and manipulates them algebraically.

At approximately the time at which this essay was written, Ampère had come in contact with references to the differential calculus in his reading of the *Encyclopédie*, including the articles by d'Alembert. Jean-Jacques Ampère was no mathematician and he responded to his son's questions by taking him to Lyon to see the Abbé Daburon, a professor of theology at the Collège de la Trinité. He had an interest in mathematics and Ampère later recalled that[18] "he had the kindness to give him [Ampère] some lessons in differential calculus and integral calculus and in this way ironed out the difficulties that had stopped him." According to one oft-cited account, the boy's initial request for volumes by Bernoulli and Euler prompted Daburon to ask whether he understood the Latin in which those volumes were written. Although forced to answer that he could not as yet read Latin fluently, father and son withdrew to Poleymieux where André-Marie soon mastered the language sufficiently to return to Lyon for the desired volumes.[19] The truth of the story is dubious since father and son had long since shared the

17

reading of Latin poets. At any rate, Ampère's early felicity with Latin and Italian stands in strange contrast to his later failure to learn English or German, languages which would have facilitated his philosophical reading and his communication with other scientists. The study of Euler and Bernoulli was soon followed by the 1788 edition of Lagrange's *Méchanique analitique*. During winter months in Lyon, Jean-Jacques also took his son to some physics lectures at the Collège de Lyon by Joseph Mollet. According to Ampère's recollections:[20] "Upon returning to the country, the young Ampère read some works on physics and, some time thereafter, the reading of Rousseau's letters on botany having inspired a great ardor for the study of this science, he divided his time between herborizations and calculations."

As the examples cited indicate, Ampère's subsequent memories of these years were uniformly positive and included no references to family or social tensions. There is no particular reason to question the accuracy of his description of an emotionally contented and intellectually invigorating childhood. Unburdened by the schedules and taunts of the classroom, Ampère seems to have received ample emotional and intellectual support from both parents. His mother is generally credited with his Catholic religious orientation. Although we have no direct evidence about what practices or beliefs were expected of him, the emphasis was on devotion rather than doctrine. His first communion was remembered as an important event and Thomas à Kempis' *Imitation of Christ* became a lifelong companion. The doctrine of the eternal suffering of the damned would become a source of skepticism in later years, but not during this early period. His father's willingness to introduce him to the scholarly community in Lyon and to provide books and some rudimentary geometrical and scientific equipment are indicative of close attention and a shared intellectual curiosity.

One of André's few friends during this period was a young man named Couppier who lived a few miles to the north in Claveisolles, Beaujolais. He visited Lyon and Poleymieux periodically, and by 1793 the two friends were exchanging lengthy letters. Couppier had a general interest in mathematics and physics, but had not as yet studied calculus. Ampère began tutoring him and enjoyed teaching on an individual basis. In later years his academic career would become increasingly dreary as its formality and repetitious routine replaced the personal interaction he enjoyed most about teaching.

The status of the Ampère family within the pre-revolutionary class structure of the Old Regime was not a major source of concern for the growing boy. He left no record of his impressions as he observed peasants depositing their feudal tithes at the *dîme*, the storage building located just down the slope from the Ampère home. In later life he expressed typically ambivalent attitudes toward the Revolution. In common with all his close associates and friends, he was sympathetic to the ideals of liberty and equality but repulsed by the actual course of revolutionary violence, violence which includes the death of his own father.

The Revolution and the Siege of Lyon

The initial two years of the Revolution, between mid-1789 and 1791, were not particularly violent or dramatic in Poleymieux. A poorly armed National Guard of about 50 men was organized, and some community and national property was redistributed or sold. Jean-Jacques Ampère commemorated the initial stage of the Revolution, particularly the fall of the Bastille, by composing a dramatic piece entitled *Artaxerxe ou le Roi constitutionnel*. As his plot unfolds, the tyrannical Persian King Xerxes is assassinated by the father of a persecuted general, Arbace. The new King, Artaxerxe, defends Arbace from accusations of the crime when he refuses to defend himself for fear of incriminating his father. Artaxerxe's sister loves Arbace but worries that he might be implicated in the death of her father. The ensuing rather formal dialogue explores the theme of the relationship between law and justice. Retrospectively, this narrative took on considerable irony and pathos; Jean-Jacques Ampère would soon face trial and execution at the hands of a revolutionary court. The extreme reaction of his son to that event was thus preceded by lengthy exposure to themes of justice, retribution, and sacrifice. Ampère also inherited his father's appetite for tragic drama and verse, an inclination he would pass on to his own son.

In August of 1789, Jean-Jacques Ampère lost his administrative position in Poleymieux. He was actually fortunate to distance himself from Guillin Dumontet in 1789. The Poleymieux *seigneur* had generated considerable ill will through his calculated response to the decrees issued by the National Assembly during August 1789. The Assembly had abolished peasant dues stemming from "personal" obligations, but it also required compensation for the remission of obligations hallowed by feudal law. Guillin attempted to manipulate the ambiguity of this distinction by revising his charters and rental documents. He had already alienated the peasants by enlarging his manorial property at their expense.

The response to these developments by the peasants in the

Poleymieux area was not as immediate as the reactions that took place in many other French villages. Elsewhere in 1789 there was considerable violence and destruction of manorial châteaux and the feudal documents they contained. Generous consumption of wine often contributed to the excessive violence of this pillaging, including brutal torturing and murder of many a local *seigneur*. In Poleymieux the peasants expressed their discontent through an inquiry into Guillin's legal and financial machinations. He was reprimanded and assessed a tax of 1,000 livres as his *contribution patriotique*. In December of 1790 Guillin's brother, Antoine Guillin Dumontet, was implicated in a counterrevolutionary royalist plot in Lyon. The local citizenry ransacked the Poleymieux château in an unsuccessful search for armaments, but there was no serious damage. The Ampère home was also searched.

Following the abortive flight of Louis XVI in June of 1791, tension again increased in the provinces; on 26 June a house search of Guillin's château was ordered by the Jacobin section from the nearby village of Chasselay. A detailed account of the ensuing events was published on 20 August in the *Mercure de France*, with Guillin's young wife cited as an eyewitness.[21] Guillin initially objected to the search when it was demanded without a legal warrant by the National Guard of Chasselay, Quincieux, and Poleymieux. Fighting broke out and the sounding of the tocsin attracted a larger crowd from a considerable area. Eventually Guillin did allow them to enter but once again nothing was found. As the crowd grew disorder increased, and Madame Guillin escaped with her children. Suspicion, resentment, and wine soon conquered the patience of the mob. Guillin's efforts to defend himself led to indiscriminate violence. The château was pillaged and burned while Guillin himself was decapitated and dismembered. Following the ensuing drunken revelry, charges were filed in Chasselay when two men were apprehended parading through the city with Guillin's arms displayed at the ends of their sabres, a spectacle they embellished by gnawing on his flesh and spitting pieces into the air.

We do not know how the Ampère family reacted to all this. Their own status as bourgeois landowners does not seem to have been immediately threatened. Jean-Jacques would not have been a likely victim of peasant violence since he did not participate in either the grain trade or the manipulation of sharecroppers, lucrative practices typically pursued by large landowners. In the fall of 1791 he rented additional quarters in Lyon where he had accepted new responsibilities as justice of the peace and the presiding legal functionary for the police tribunal. Unfortunately, the latter position was destined to

20

embroil him in the accelerating violence of the revolutionary power struggle.

In the meantime, the first in a sequence of tragic deaths struck the Ampère household. André's elder sister Antoinette died at the age of 20 on 2 March, 1792. She had been the constant companion of his youth and she would always remain a major component of his memories of Poleymieux. Many years later he would continue to write poetry in her memory.[22] But additional disaster awaited Ampère's father in Lyon. Although the Parisian Jacobins came to power by August of 1792, their influence in Lyon was not as strong as it was in many other parts of France. Indeed, one of the major themes of Ampère's life is the tension between his Lyon origins and the demands placed upon his self-identity by his career in Paris. Traditionally resistant to Parisian directives, Lyon retained strong monarchist sympathies throughout the Revolution. But since the Lyon Jacobins also had a strong contingent, the result was a particularly severe episode of civil war during the Reign of Terror in 1792 and 1793. The most extreme Lyon Jacobins were led by Marie Joseph Chalier; he imposed increasingly oppressive security procedures in Lyon during the fall and winter of 1792. He was particularly strident in his suppression of Catholic clergy, a policy that met with stiff resistance from the strong religious tradition in Lyon. Driven to desperation by arbitrary arrests and executions, the Girondin opposition sacked the Jacobin headquarters and took over the municipal government on 29 May, 1792.

Not surprisingly, the provisional Girondin government issued warrants for the arrest of prominent Jacobins, including Chalier. Jean-Jacques Ampère was thus confronted by a difficult political situation. As a civil employee of the city of Lyon, his duty was to carry out the functions assigned to him. These included the certification of arrest warrants and the initiation of inquest proceedings. This was the procedure he followed. On the other hand, 29 May was also a turning point for the suppression of Girondins by Paris Jacobins. In July the Jacobin dominated Convention declared Lyon to be in revolt and its administrators to be traitors whose property was subject to confiscation. Ampère found himself retroactively declared guilty of treason for carrying out what he had considered to be the responsibilities of his office. He later justified his actions by pointing out that he had advocated conformity to legal procedures in the face of demands for immediate vengeance on the part of prominent rebels. He also cited this reason for his effort to delay the proceedings so as to provide time for passions and tempers to cool. Unfortunately, in Lyon the defiant response to the orders from Paris was to guillotine Chalier on 16 July. The Convention reacted by ordering the destruction of the city; the siege of Lyon began with an artillery bombardment on 10 August.

It is impossible to determine what options were considered viable by Jean-Jacques at this point. If he considered abandoning his post, we have no records or suggestions to that effect. He took up residence in Lyon with an elderly relative and left instructions with his wife not to inform their children of the situation, instructions which apparently were carried out successfully. He remained in Lyon during the subsequent two months of the siege. On 9 October the city was captured, owing to collaboration by Lyon Jacobins. Jean-Jacques was among those immediately imprisoned; he remained in the Roanne prison for six weeks awaiting his trial.

Two of Jean-Jacques' letters to his wife during his imprisonment have survived; they indicate a remarkably collected state of mind. In the letter of 17 October he is primarily concerned that Madame Ampère take the necessary steps to retain possession of the property at Poleymieux after his death. With no doubts about the outcome of his trial, he suggested that she rely upon their marriage contract to argue that the property should not be confiscated. This strategy proved successful and the property remained in the family's possession. Jean-Jacques Ampère's trial took place on 23 November, one of the last dates on which some semblance of legal formality was preserved in the face of increasingly haphazard administration of revolutionary justice. As he had expected, Ampère was found guilty as "the justice of the peace who launched the arrest warrant against Chalier." Before ascending the scaffold of the guillotine, he was allowed to return to his cell to write a final letter to Madame Ampère. The result was a remarkably poignant expression of compassion and resignation in the face of injustice.[23]

> I desire my death to be the seal of a general reconciliation between all our brothers; I pardon those who rejoice in it, those who provoked it, and those who ordered it . . . If, from the eternal sojourn where our dear daughter has preceded me, it should be given to me to occupy myself with things down here, you, as well as my dear children, will be the object of my care and kindness. May they enjoy a better fate than their father and always have before their eyes the fear of God, that salutary fear that effects innocence and justice in us in spite of the fragility of our nature . . . As for my son, there is nothing that I do not expect of him.

The religious sentiment expressed in Jean-Jacques' acceptance of a harsh fate indicates that he shared the devotional attitude generally attributed to Madame Ampère; that is, in addition to the intellectual guidance he contributed to André's development, he also provided an example of that rare combination of courage and humility which can be inspired by a religious devotion and faith relatively unconcerned about dogma and ritual. André carried with him the memory of his father's composure and tranquility fostered by a secure faith, a memory that would haunt him during his frequent periods of religious crisis. The siege of Lyon was a shared experience that acted as a strong bond within the circle of close friendships André initiated soon thereafter. Ballanche, Bredin, and Degérando all came through this adventure with a distinctly Lyonnaise perspective. As was the case with Jean-Jacques Ampère, they distinguished the revolutionary ideals to which they remained faithful from the excessive use of violence and extreme anti-religious attitudes they rejected.

The news of the death of his father was broken to André at Poleymieux. The effect was devastating. According to his autobiographical retrospection, he underwent an almost total mental and physical withdrawal.[24]

> M. Ampère left his family in the country and served a duty to not abandon his fellow citizens. He refused to leave the besieged city, and when it succumbed he was one of the first victims of the revolutionary tribunal. According to his orders, the young Ampère was retained in the country where he had left him. He was beguiled by the vain hope that his father was going to be returned to him, and the study of mathematics occupied him more than ever because shortly before the siege of Lyon someone had the care to procure for him the *Mécanique analytique*, the reading of which had animated him with a new ardor. He repeated all the calculations in it, and he was still applying himself to that work at the instant when the fate of his father was revealed to him. For more than a year, abandoned to a sorrow that absorbed him solely, there was no longer any question for him of any study.

The severity of this reaction speaks for itself to some extent. The bond between father and son had been extraordinarily close. Jean-Jacques had been the only significant male figure in André's experience; he had also been almost the sole means through which André made contact with society outside Poleymieux. The anxiety provoked by his father's death was slowly dissipated, but it contributed to the permanently melancholy cast of his adult temperament.

2

Marriage and Provincial Life (1800–1804)

Courtship and Marriage

Ampère's state of mental and physical paralysis lasted for approximately 18 months. He later recalled that two factors were particularly influential during his return to full awareness: a renewed interest in botany, and a volume of Roman poets entitled *Corpus poetarum latinorum*.[1] Prior to his father's death, Ampère was enchanted by Rousseau's *Lettres sur la Botanique*, a set of letters written by Rousseau to Madame Delessert with instructions on how to introduce her daughter to the practice of herborizing. Despising the "artificiality" of the Linnaean system, Rousseau provided painstaking descriptions of his own five-family taxonomy, together with detailed instructions on how to examine and classify specimens. Memories of Rousseau's engaging style may have rekindled Ampère's interest in the lush countryside surrounding him. It was deeply characteristic of Ampère to combine direct experiences of nature with an effort to classify and discern orderly patterns. His experience with natural phenomena was by far most direct in the case of botany; his later work in physics was primarily an effort to discover mathematical laws that govern highly artificial electrodynamic phenomena produced in the laboratory. Botany gave him far more direct access to nature, and his most active

24

botanizing coincided with the emotionally tumultuous period of his youth.

It is possible that an analogous appreciation for order attracted Ampère to the disciplined cadences produced by Latin grammar. Certainly he found Roman poetry conducive to botanizing; he later recalled romping over the hills near Poleymieux reciting Horace, Virgil, and Lucan. Perhaps the poetic combination of stoic fortitude and romantic sensitivity helped reconcile him to the loss of the father who had kindled his appreciation for heroic drama. The description of intense emotion in precise language appealed to the same aspect of Ampère's mentality that later delighted in the expression of laws of nature by means of classification schemes and mathematics. In both cases he felt that a subtle and hidden reality finds expression through a severely structured medium.

The emotional sensitivity responsible for Ampère's extreme response to his father's death re-emerged with even greater strength following his recovery. Intellectually, it found expression in poetry and a general fascination with language; mathematics was temporarily demoted to a position of secondary importance. He entitled one rather ambitious project *L'Americide*, a depiction of Columbus as a tragic figure responsible for the transformation of the American continent into "a vast desert." As Ampère's plot unfolds, Columbus is granted a vision in which he is guided through a tour of both heaven and the physical universe with its multitude of worlds held in law-like order by the creator. He is then given the mission to bring religious enlightenment to the inhabitants of America. The native Americans are said to suffer from their lack of restraint, but they enjoy a resonance with nature unknown to Europeans. Columbus initially overcomes all obstacles with religious zeal, but Ampère's narrative breaks off in the midst of a sub-plot involving an encounter with a mysterious elderly voyager. The incomplete but repeatedly revised first three "songs" of *L'Americide* illustrate Ampère's romantic nostalgia for a world untainted by technological civilization. He shared many of the views of Catholic romantics such as Chateaubriand and Ballanche, men less complex than Ampère in the sense that their conservative reaction against Enlightenment attitudes was not combined with scientific knowledge or ability. *L'Americide* also indicates that Ampère's religious orientation at this point in his life included strong convictions about a divine source for the laws of nature and an idyllic life after death for the righteous. Ampère also wrote politically oriented poetry generally critical of the Terror that had consumed his father. His ability to address this issue indicates his renewed stability and his recovery from

25

the period of shock, but explicit discussion of his father remained a taboo subject; Ampère was not inclined to put the memories he retained into writing.

By 1795 his interest in *L'Americide* had flagged and he gradually dropped it for other concerns. He renewed his correspondence with Couppier, who sometimes visited Poleymieux from his home in Claveisolles, Beaujolais.[2] Ampère's letters were long and detailed during the boring winter of 1795–6. Those that have survived reveal several of his attitudes and preoccupations at that time.[3] He was thoroughly dissatisfied with his limited social circle. Family card games were not very stimulating and Ampère longed for more complex conversation that might include mathematics. He replied to Couppier's letters immediately upon receiving them and agonized over the safe delivery of his own. Formerly his father had sometimes carried out that task for him. The enthusiastic Couppier had made progress in mathematics and physics during the past few years and now he was interested in the principles governing the operation of machinery. Ampère had regained his knowledge of Lagrange's mechanics and he readily attempted to apply it to Couppier's questions. He even proposed that they carry out out a cooperative investigation with Ampère responsible for its theory.

As the winter progressed, the two correspondents became increasingly interested in astronomy. Ampère carried out extensive observations using telescopes he constructed himself. He invented a way to preset his telescope at predetermined angles and thus was able to observe specific stars in broad daylight. His letters included charts of data, comparisons with Lalande's tables, and detailed diagrams concerning the construction of apparatus. Although in later life Ampère was sometimes described as incompetent with experimental apparatus, this is an exaggeration; he could be quite skillful when he found the investigation compelling. To some extent Ampère discussed topics in conformity with Couppier's interests, but he did become genuinely fascinated by astronomy, particularly the comparison of his observations with accepted values. He also carried out some mathematical investigations of the attraction of the moon to the sun and the earth; he became familiar with approximate solutions in terms of infinite series and noted that this was probably all that could be hoped for. There is an almost electric tension in Ampère's letters that indicates how highly charged his emotional and intellectual condition still was, following the impact of his father's death. With the spring of 1796 this energy took a new direction and fueled an enthusiastic courtship that resulted in his marriage to Catherine Carron three years later.

Here too language was essential to Ampère. In a journal entitled *Amorum*, he recorded a detailed account of the first two years of the relationship; thereafter, letters took on an almost sacramental significance. At the outset, Ampère found himself in a state of dissatisfaction and malaise that would recur with considerable regularity throughout his life. In an introductory page to *Amorum*, we find the following fragmentary comments:[4]

> Having arrived at the age when the law rendered me master of myself, my heart deep down was still longing to be so. Free and insensible up to this point, it was bored with its idleness. Raised in almost complete solitude, study and reading, which for so long had been my greatest delights, let me fall into an apathy that I had not experienced, and the cry of nature spread a vague and unbearable anxiety in my soul. One day as I was walking after sunset along a little solitary stream . . .

The journal then begins with the cryptic announcement: "Sunday 10 April. I saw her for the first time." The lady in question was Catherine-Antoinette Carron, always referred to as Julie by both her family and Ampère. The Carron and Ampère families had much in common. Like Jean-Jacques Ampère, Julie's father, Claude Carron, had been a successful silk merchant in Lyon. The Carrons maintained a home in Lyon in the same congested central district frequented by the Ampères and they also kept a country home in Saint-Germain, the village just a few miles north of Poleymieux. As was the case for the Ampères, socializing in Lyon was restricted to the winter months, particularly December and January; during the rest of the year both families preferred village life in the country. Claude Carron, born in 1730, was in the final stages of a paralyzing illness in 1796 and he would die the following year. Julie's mother, Antoinette Boiron, was 13 years younger than her husband and had married him in 1762 when she was only 19. They had seven children; Julie was the youngest of the four who survived childhood. Her brother and one of her two sisters had already married by 1796, leaving only Julie and one sister, Elisabeth, living with Madame Carron. Elisabeth, or Élise as everyone referred to her, was 29 in 1796, six years older than Julie, who had been born on 12 September, 1773. Julie was thus a year and a half older than Ampère, who had just turned 21 in 1796.

Ampère inhabited a distinctly feminine environment during the years of his romance with Julie. Aside from André himself, the Poleymieux household was made up of Madame Ampère, now aged 47, her younger sister Antoinette, and Ampère's 11-year-old sister Joséphine.

27

Although Couppier continued to visit occasionally, male acquaintances at this point were few and relatively insignificant. Ampère gradually became acquainted with Julie's brother, Jean-Etienne (François) who visited from his home in Paris, and more closely with Julie's brother-in-law, Jean-Marie Périsse (Marsil), who had a printing business in Lyon. The Carron family had quite a penchant for using names other than their given ones.

But in April of 1796 Ampère's attention was almost exclusively focused on his first romance. His journal records the formal and rather bland proceedings that took place and the emotional intensity with which he experienced them. His initial strategy was to meet Julie after Sunday Mass and visit the Carron home under the pretext of borrowing books, primarily novels, poetry, and drama. There never seems to have been any doubt in his mind that he and Julie were intended for each other; as early as September of 1796 his journal reports that "I began to open my heart." It is not clear who the recipient of this revelation was. Julie herself was in the final stage of extricating herself from a proposal of marriage from one Charles-Louis Dumas. Dumas had been a physician at the Hôtel-Dieu during the Lyon siege, an activity that left him under some suspician of being a counter-revolutionary. Thereafter he took up quite an illustrious academic career in Montpellier, a locale which he preferred to Lyon, owing to its "not having been stained by innocent blood during the revolution."[5] Julie was glad to be free from Dumas' courtship, and she thoroughly enjoyed the Lyon social whirl during the limited periods in which she was allowed to stay in the Carron home there. She was particularly fond of the theatre and dancing, environments where she was apparently sufficiently attractive to be readily welcomed. Ampère's descriptions of her are hardly objective, but we certainly know that he was completely captivated after one sight of her.

Although Ampère's journal entries are enigmatic and cryptic, we can surmise that Julie did not immediately reciprocate his feeling for her. His general procedure was to visit the Carron household under some pretext and then ensconce himself there, holding conversation with whoever was available. Rarely did he find himself alone with Julie herself. In October of 1796 he reports that "I opened myself entirely to her mother who did not seem to want to deprive me of all hope." During November, "Julie tells me to come less often," and "Élise tells me to spend the winter without speaking further." Indeed, during that winter of 1796–7 Élise was the more receptive of the two sisters. As usual, Julie spent much of the winter season in Lyon. In December Élise and Madame Carron returned to Saint-Germain after leaving

28

Julie in the custody of relatives. Élise's letter to Julie the following day reveals much about the state of affairs at that point. After reporting that she had not as yet caught sight of Ampère lurking anywhere in the neighborhood trying to catch a glimpse of her, Élise gives her own assessment.[6]

> If he arrives and mama leaves the hall, he will take issue with me. I have already prepared a thousand little responses which are always the same. I would like to know who could make him happy without advancing things too far; for he interests me by his sincerity, his sweetness, and above all by his tears which are involuntary. Not the least affectation, no phrases from novels which are the language of so many others. Do as you will, but let me love him a bit before you love him; he is so good!

The following day, having learned that Julie had remained in Lyon, Ampère could not contain himself and exceeded the narrow boundaries of bourgeois propriety by setting off to visit her there. Upon his arrival at ten o'clock in the morning,[7] "she opened the door to me in her nightcap and spoke with me a moment *tête à tête* in the kitchen; I then entered Madame Carron's; Richelieu was discussed; I returned to Poleymieux in the afternoon." Stricken by remorse, he made profuse apologies to Madame Carron in Saint-Germain. She was not particularly disturbed and Élise relayed the conversation to Julie in a letter. Ampère had been on the verge of tears, a state Élise found particularly charming. He had also made some attempt to spruce up his wardrobe, an effort that Élise suspected would be lost on Julie; "I am afraid that your eyes, dazzled as they are by those called *muscadins*, will prevent you from seeing him in the same light as I do."[8]

Ampère confided fully in Élise and she reported to Julie how she defended him from the catty remarks of various visitors. Such incidents apparently were fairly common; in one case during January of 1797, a Madame Lacostat made some condescending remarks concerning some of Ampère's verses and then gave a general assessment:[9] "This young man has a great deal of knowledge and some solid qualities. Yes, yes – solid, but his teeth certainly aren't; he seems like an old man, he is so serious. I still have not seen him laugh." In another letter from January 1797, Élise summed up her reactions to this attitude.[10]

> In short, my good friend, I am a little angry at people who only take note of the exterior and who think that they know everything when they have bowed elegantly and made some smutty remark that one would be better off not understanding . . . The day before yesterday he brought us the first scene of his tragedy and a larkspur shoot I had

asked him for and that we planted together while speaking of nothing other than your beautiful eyes. I certainly am doing all that I can to rouse your *amour-propre*, and if I should persuade you that he no longer thinks about you, maybe you would no longer find him to your taste; but unfortunately for him, I believe the poor man thinks of nothing else and I very sincerely pity him, knowing that apart from Madame Périsse, all those you see are set against him and only take the trouble to wonder if he will make a woman happy or unhappy – which nevertheless is the principal thing . . .

After one has said lightly, "Oh, what a man! How could you resolve to marry him? . . . He has no manners; he is awkward, shy, and presents himself poorly," one seems to have said all and seen all; . . . but if one had to decide for oneself, one would reflect more and quickly leave aside the rest to occupy oneself with character, morals, and even that simplicity which, a moment earlier, seemed to be a lack of breeding.

Quite aware of his rotten teeth, his less than modish wardrobe, and his awkward bearing in society, Ampère took refuge in verse. Considerable effort went into a hymn to Julie, relayed to her by Élise:[11]

> Des cheveux d'or, des yeux d'azur
> Un teint où l'on croit voir des roses
> Nager dans le lait le plus pur;
> Sur les lèvres à demi-closes
> D'une bouche digne des dieux,
> Un sourire naïf et tendre;
> Une voix, pour être amoureux,
> Qu'il suffit une fois d'entendre

Élise's wry comment to Julie was that[12]

I will admit to you that it is the *bouche à demi-close* which surprises me the most, for it seems on the contrary that it is quite radiant; but an ancient poet speaks in these terms and what he says being found to be the thought of Ampère, the latter thought it appropriate to follow his model.

Following Julie's return to Saint-Germain late in January of 1797, Ampère's persistence resulted in a regular routine of visits and conversations. Madame Carron occasionally suggested that he visit less often and Élise had to warn him not to stare at Julie so fixedly in public. Novels and plays were borrowed regularly and sometimes read aloud. Ampère taught Italian to Julie's ten-year-old nephew, Antoine-François Périsse. During April Madame Ampère and Joséphine met the Carron sisters for the first time and thereafter visited or met them quite often at the homes of friends or relatives. It seems that 3 July was a particularly significant day, for Ampère at any rate, since he highlighted that date in his journal. On this day Julie and Élise visited Poleymieux for the first time and took a walk with Ampère and his sister. Ampère's journal entry shows us how precious these events were for him.[13]

> I sat on the grass beside her and ate some cherries that had been at her lips. All four of us were in the large garden when she accepted a lily from my hand. We then went to see the stream; I gave her my hand to jump over the little wall and both hands to climb up again. I was seated beside her on the bank of the stream far from Élise and my sister. At evening we accompanied them to the windmill where I again sat beside her as we four observed the sunset which gilded her clothes with a charming light; she took away a second lily that I gave her in passing to leave for the large garden.

There is no direct evidence concerning Julie's feelings during the period documented so intently by Ampère. She obviously had not shared the initial fascination he experienced; Élise was the only strong advocate in Ampère's favor. Whether Julie pledged herself to him in July of 1797 or not, by the fall of that year his journal implies a general acceptance that they would eventually wed. In September, for example, when a Mr Vial suggested that Ampère cultivate his mathematical skills in Paris, Julie vigorously ushered him out with the reply that they had no need of such advice. She had vague hopes that Ampère would find a niche in the business world, an unlikely prospect considering his complete lack of practical experience.

Intellectually, there is no indication that Julie shared or even vaguely understood Ampère's scientific interests; her own reading was restricted to literature and history. Conversation at the Carron household generally centered on various parlor games, charades, and the construction of clever verse. Ampère held his own there, primarily for the sake of Julie's company. He also felt inspired to write more lengthy verses, many of which have survived. Doubts were appropriately raised about Ampère's aptitude for a business career. By December of 1797 it was decided that he should begin tutoring students in mathematics in Lyon. His experience with friends and relatives indicated that he would be successful and thus might be noticed when openings at provincial *écoles centrales* were filled. Julie's brother-in-law, Jean-Marie Périsse, provided a room for the lessons. Ampère made use of Lacroix's textbooks and gradually developed a following. Meanwhile, his relationship with Julie continued with all due decorum. The surviving entries in his journal terminate in February of 1798. Until then, he dutifully noted daily events, such as his first invitation to dinner at the Carron home in Lyon. Conversations were also mentioned and sometimes paraphrased. Élise was more available than her sister; Julie was caught up in her usual social life during the holiday season.

During March of 1799 Ampère was struck with measles in Lyon; quarantined there, he exchanged his first letters with Julie back in Saint-Germain. Ampère's side of the correspondence gushes with emotion and effusive expressions of his inability to find words to express his happiness. Julie is more restrained: "But you know that my refrain is: *too much haste spoils everything.*"[14] She applied this measured slogan to most decisions in her life; in this respect she acted as a source of moderation for Ampère's boyish enthusiasm. He saved her first letter to him and always referred to it as his *talisman*. It became an important member of a set of "treasures" he would brood over during their times of separation.

The wedding date was set for August of 1799. The Ampère and Carron families were deeply attached to the doctrines and rituals of Catholicism, commitments which the revolutionary government both ridiculed and prohibited. They thus arranged for a clandestine religious ceremony that took place on 6 August. There is no indication that Ampère himself attributed any great importance to the sacramental aspect of this event. Certainly he began to experience aversion to religious ritual shortly thereafter. Civil proceedings took place the following day, 7 August. Witnesses included Julie's brother-in-law, Jean-Marie Périsse, the Lyon printer who would be particularly helpful to the young couple during the following few years. According to Madame Cheuvreux, Ballanche attended the civil ceremony and recited a hymn in the couple's honor.[15]

The marriage lasted less than four years; in July of 1803 Julie died of an ailment vaguely diagnosed as an abdominal tumor. This short period included some of the happiest moments of Ampère's life. It was also filled with anxiety about Julie's failing health and concern for the status of their son, Jean-Jacques, born on 12 August, 1800. Other developments during these years became the basis of Ampère's professional career. He began functioning within the French educational system, a world he had never experienced as a student. He published his first memoirs in mathematics and was noticed by the highly centralized French scientific community. He made close and enduring friendships within a small circle of Lyon intellectuals with literary, philosophical, and religious interests. Ampère responded to this complex environment with an emotional and intellectual exuberance that became his trademark.

The factual details of Ampère's employment are easily summarized. He continued his tutoring in Lyon until February of 1802. At that point he was appointed to a teaching position at the Departmental École Centrale d'Ain in Bourg, about 60 kilometers northeast of Lyon. Leaving the ailing Julie with her family in Lyon and Poleymieux, he spent a year in Bourg teaching elementary mathematics, physics, and chemistry. In April of 1803 his appointment to the Lyon Lycée coincided with the closure of the departmental Écoles Centrales. This long awaited reunion with Julie was cut short by her death the following summer. The ensuing period of mourning and recovery brought Ampère's life in the provinces to a close; in 1804 he departed for Paris.

Due to his separation from Julie during so much of this period, Ampère's correspondence is a major source of information concerning his relationship with her. As had been the case prior to their marriage, Ampère's letters are profuse expressions of his affection and desire to please. He tells Julie where he kisses the pages of his messages to her; her responses become treasured documents to be horded and memorized. The contrasting attitudes of the two toward each other also continued. Ampère worships Julie and typically refers to her as his benefactress; she calls him "my son," with the same maternal attitude she had adopted soon after meeting him.

The birth of a child in August of 1800 was a joyful event for them both. They had expected a girl but expressed no regrets when a boy was born. Jean-Jacques was an obvious choice for a name; it captured both the memory of Ampère's father and his intellectual and emotional debts to Rousseau. Ampère took sincere delight in the early antics of the child. He relished hearing his first spoken words and agonized over childhood illnesses and injuries. There was nothing feigned in his participation in family life. No one was more removed from the stereotype of an unemotional intellect than Ampère. Julie's decision to nurse the child caused some concern for her less than robust health. The birth had weakened her and she never regained her earlier level of activity.

The financial status of the family was another worry. Julie's brother Marsil had arranged for a reasonably priced apartment in Lyon. It was there that Ampère continued to tutor mathematics students as he had for the preceding two years. The marriage contract had specified that Ampère could claim one quarter of the revenue from the Poleymieux estate; this came to only about 300 francs per year. He was also accorded one half of a 10,000 franc inheritance created when his father's estate was reassigned to his mother by the government. Julie's inheritance came primarily in the form of furniture and household items with only 1,200 francs in cash. Ampère did not draw upon the sum due to him immediately; the household was forced to economize carefully with a very limited income.

Ampère's pursuit of an academic career was closely bound up with financial requirements brought about by political transformations. Had the Revolution not destroyed the financial independence of his family, he might very well have led a relatively isolated life at Poleymieux, supported by family investments. Intellectually, he would have been free to follow the fluctuating whims of his own curiosity. Instead, his energy was constrained and channeled by specific educational curricula. His own research and creativity were now directed toward winning acknowledgement from those with influence in dominant research programs and institutions. Initially, this restriction in his freedom did not weigh too heavily on him. He accepted the challenge of an academic career with an enthusiastic sense of accomplishment in providing for his new family. Later, deprived of Julie's companionship, the demands of this career would become a heavy burden.

Ampère developed an amazing ability to carry out creative mathematical research in the midst of severe emotional turmoil. With respect to Julie's health, his concentration may have been aided by his physical separation from her, a condition that allowed him to remain unduly optimistic about the gravity of the situation. Throughout her increasingly incapacitating convalescence, he continued to dream of their future happiness when he would win an academic position in Lyon and reunite his family. Julie was more fatalistic and reminded him that "doctors are not gods."

Indeed, the doctors who were consulted had little beneficial effect, if any. Julie suffered from severe pains in her stomach, and it became badly swollen. A chronic cough, a weak and rapid pulse, and irregular menstruation added to her discomfort. In July of 1801 she was diagnosed as suffering from a stomach tumor attributed to a "milk deposit." She was repeatedly subjected to various dietary regimes, including one restricted to fruit and ice. She was also advised to take the water at various spas, but she was never able to do so, owing to her own weakness and her concern for the well being of her son. Her mother, Élise, and Madame Ampère were her closest companions as her husband left for Bourg to accept his first academic position.

French Post-Revolutionary Educational Reform

Ampère's initial participation in the French educational system came during a period of rapid institutional changes. The chaotic years immediately after the 1789 Revolution were understandably ones in which education was not a first priority. Both primary and secondary education fell into serious disarray. Between 1791 and 1794 there were some proposals to correct the situation, but nothing was accomplished until after the end of the Terror. Major innovations began in 1794 with the creation of the École Centrale des Travaux Publiques, known as the École Polytechnique following the 1795 revisions. Located in Paris, the École immediately became the single most influential center for technical education in France. It featured a faculty that included most of the leading men in French science, particularly those interested in mathematics and the application of new mathematical techniques to physics and astronomy. Ampère's career at the École would begin in 1804, the year of his permanent move to Paris.

Before that transition, however, Ampère's first teaching experiences in an institutional setting took place in one of the Écoles Centrales, the secondary schools provided for each of the 102 departments of the French Republic by the decree of 26 October, 1795. The guiding philosophy of the Écoles curriculum was that of the *idéologues*, a group of politically active philosophers much inspired by Condillac's reduction of the learning process to the analysis of sensations and their accurate expression in clear language. Destutt de Tracy and Cabanis were particularly active in transforming this philosophy into educational practice. The Écoles Centrales were intended to produce French citizens receptive to the Enlightenment attitude that progressive social institutions must be grounded on a materialistic conception of human capacities.[16] Based upon a careful study of language, this attitude was to be fostered by an emphasis on the power of science to dispel superstition and prejudice, particularly those generated by irrational religious belief.

To put this agenda into practice, the Écoles Centrales were directed to offer a highly structured six-year curriculum. The first two years were to be devoted to Latin, technical drawing, and natural history, subjects that had both practical utility and ideological content; students were introduced to accurate systems of classification and precise use of language. Mathematics, physics, and chemistry were scheduled for the second two-year period. The goals here were again both practical and ideological. These subjects were obviously essential to the development and survival of a society that was becoming increasingly dependent upon technology. They also could be presented in a manner that emphasized the replacement of ignorant superstition and traditional belief by scientific arguments based upon logic and the assessment of empirical evidence. In this respect they served the ideological intention to transform France into a secular society in which science was to be the final arbiter over both natural knowledge and the direction of social change.

While this was an impressive agenda in theory, it was only minimally fulfilled in practice. The proposed sequence of courses was almost universally ignored; central direction was not maintained and students chose courses according to their own career interests. Lack of laboratory equipment and the poor mathematical knowledge of the students meant that courses in experimental chemistry and physics were often quite elementary and fell short of the expectations of the *idéologues*. Moreover, relatively few students followed the mathematics courses beyond the elementary stage of computation that was of immediate practical value for a career in business. Ampère thus found himself in an academic environment that provided little direct stimulation to his scientific creativity.

Ampère's Teaching at Bourg

On 18 February, 1802, Ampère was appointed professor of physics at the Bourg École Centrale for the department of Ain. He was selected by the local committee, one of the *juries d'instruction* responsible for selecting the staff for the Écoles Centrales. Although Ampère had no formal education, his reputation as a private tutor was judged to be sufficient preparation. This informal assessment of criteria for teaching would soon come to an end; Ampère was one of the last to rise to the forefront of the French educational system without specialized training of any kind. By 1802 the École experiment had generally been conceded to be a failure. Indeed, by the end of the year Napoleon would dissolve the Écoles Générales and institute a new and much smaller system of Lycées. Uncertainty about the future thus contributed to the generally low morale of the staff at Bourg. Both Ampère and his fellow faculty members had only part of their attention on their teaching and were more concerned about prospects for future employment. Nor were Ampère's colleagues a particularly distinguished group. He associated most closely with Clerc, the professor of mathematics. They collaborated on a paper about infinite series but never brought it into a state appropriate for publication.

Ampère's initial impressions of Bourg society were not favorable. He was taken aback by the level of vulgarity he encountered during his first meal at the home of the vivacious professor of philosophical history, Thomas Renaud. Ampère soon decided to lodge at the École Centrale itself and simply took his meals at the Beauregard residence. There he discovered that Madame Beauregard was a renowned subject of local gossip and was presently encouraging the attention of the fine arts instructor, Mermet. Tongues wagged about the prudish new professor from Poleymieux who had inadvertently situated himself within a major source of scandal. Ampère initially found most of this mildly amusing and wrote detailed descriptions in his letters to Julie. Before long the situation at the Beauregards' turned sour. Madame Beauregard found Ampère too aloof and started complaining about his eating habits. He eventually ceased taking meals with her when she berated him for appearing at table with hands stained by his work in the chemistry laboratory. For several months he ate alone in his own room with the assistance of the École concierge. Ampère soon lost his initial tolerance for Bourg social life and became increasingly disappointed by how petty and mean-spirited it could be.

Part of the challenge presented by Ampère's teaching duties was the novelty of his circumstances. He had never been a student in any formal educational setting. His father had taken him to observe some physics lectures in Lyon, but that was the full extent of his classroom experience. He adjusted remarkably well, but this may have been due to the fact that his students were less capable and demanding than those he had known in Lyon. He devoted his first days in Bourg to putting the chemistry laboratory in order. The school was typical of the Écoles Centrales in that the laboratory was ill equipped. Ampère had to acquire even the most elementary equipment, often at his own expense. He informed Julie that he and Clerc devoted considerable time to their own chemistry experiments, but he provided no details. The content of this research must be inferred from Ampère's few public statements and relevant manuscript material.

In addition to his teaching for the École Centrale, Ampère also taught geometry and arithmetic for a Bourg secondary school in the fall of 1802. The school provided him with reasonable living quarters, much appreciated after his earlier adventures. The extra income allowed him to send regular sums to Julie and took some of the edge off their worries about medical expenses.

Julie's Death

Ampère's hopes that Julie would soon be well enough to join him in Bourg gradually faded; he set his sights on a teaching position in Lyon, where they could establish a common household without a move that would have been beyond her strength. This did not take place until April of 1803; the intervening period of separation was broken only by short visits during Easter and summer vacations.

In addition to the trials associated with Julie's illness, additional tension resulted from Ampère's growing ambivalence toward Catholicism. Most of his correspondence with Julie is dedicated to rather mundane descriptions of his daily routine and affirmations of his love and devotion; however, following his initial public discourse at Bourg he summarized his state of mind in a letter that is far more introspective.[17]

> Ideas of God and eternity dominate among those that hover in my imagination, and after some quite singular thoughts and reflections of which the details would be too lengthy, I determined to request from you the psalter by François de la Harpe which should be at the house, bound, I believe, in green paper, and a book of hours of your choice.

Whatever may have been Ampère's reaction to this reading, he did not straightforwardly clarify his religious status. He did not visit the famous Brou church in Bourg until June of 1802. On the other hand, during that month he made a point of telling Julie that he had prayed at Mass for her recovery. It is typical of him that he felt the need to consult devotional works rather than treatises in doctrinal theology. Religious writing and prayer were usually more important to him as expressions of faith and dedication than as intellectual justification.

By Easter of 1803 the issue came to a head. Julie clearly wished him to receive the sacraments; they had argued about the subject during Ampère's prior visit to Poleymieux. Whether out of consideration for her wishes or from a more sincere return to faith, Ampère did perform his sacramental duties of Confession and Communion shortly after that Easter season of 1803, the last Easter of Julie's life. On 8 April, 1803, two days before Easter, she wrote perhaps her most plaintive letter.[18]

> Although other things may be more sacred, I have an inner feeling, which I cannot rid myself of, that your Julie will always be dear to you, that nothing will be able to make you forget the moments you regarded

as the height of felicity. I mean those when, united by trust, we read each other's heart as well as our own. Yes, my friend, those are the short instants of my happiness. I shared them with you and I perhaps felt them more deliciously. Why do you think they will never return? It is true that discussions of various subjects, the difficulty in persuading you of what I believe, all makes time pass, absorbs the mind, and prevents intimate communication. But, my good friend, we will not always, I hope, be in such an agitated position; your mind will be less so and you will become solidly reasonable with the passage of years and seeing the growth of the son to whom you owe an example and who will demand an account of your opinions. To explain them clearly to him you will need to be persuaded yourself. I see all that in the future; I see myself peacefully between you two whom I regard as my sons. For suffering has aged me and left me the time for reflections that have matured my reason. Thus, although our ages are close, believe, my friend, that your wife has ten years over you! That may be understood in every sense; for freshness, activity, gaiety, the grace of youth, have all disappeared. My heart is the same. It will always love you, and that suffices you, does it not, my good friend? I embrace you with this thought and you respond in kind. I sense that we are in accord.

Four days prior to this letter Napoleon had signed the decree confirming Ampère's long-awaited appointment to the Lyon Lycée. But his reunion with Julie was to be painfully short-lived. Seriously ill, she had sublet their Lyon apartment and was living with her mother in Saint-Germain. As he had done during their courtship, Ampère kept a dated journal of cryptic commentary during the last three months of her life. He received the sacraments and recorded a sincere recommitment to Catholicism. On 13 July Julie died painfully in Lyon, surrounded by doctors and family members helpless to assist her. Ampère recorded the time of her death in his last journal entry and concluded with two Latin verses from the thirty-second Psalm.[19]

> Multa flagella peccatoris, sperantem autem in Domino misericordia circumdabit. [Many sorrows shall be to the wicked: but he that trusteth in the Lord, mercy shall compass him about.]
> D. Firmabo super teoculos meos et instruam te in via hac qua gradieris. [I will instruct thee, and teach thee in the way which thou shalt go; I will guide thee with mine eye.]
> F. Amen.

A combination of guilt and resignation would permanently color Ampère's memories of Julie. He had lost the companion with whom he had recovered from the trauma of his father's death. She had renewed the unique combination of emotional and intellectual enthusiasm that had animated his childhood. Although she had not shared his scientific interests, her companionship had given him an emotional grounding he would never again experience. The sincerity of his devotion to her is evident. Perhaps the fact that they lived together so little during the four years of their marriage helped them preserve the enchanted nature of their relationship in spite of the financial worries that absorbed more of their attention than was ideal.

The fact that Ampère seldom took an active part in the direct management of Julie's household and the rearing of their child also allowed him to devote considerable time to mathematics, chemistry, and physics. He did so with frank aspirations for recognition by those in a position to further his career. From the time of his initial tutoring in Lyon, he quickly came to understand the French system of patronage sufficiently well to be noticed by appropriate men of influence. This is not to say that he was as adept at this procedure as the sophisticates he encountered later in Paris. Nevertheless, he was aware that success in an academic career demanded appropriately publicized monographs and publications in respected journals. The period in which curiosity alone directed his scientific interests had come to an end.

Physics and Chemistry at Bourg

Prelude to Laplacian Physics: Phenomenological Laws, Fundamental Laws, and Coulomb's Electrostatics

No scientific research is carried out in a conceptual vacuum. Ampère's first creative efforts in chemistry and physics were in response to what he perceived to be the central questions in these fields. His manuscripts reveal that, while at Bourg, he was aware that some of the most active French physicists were attempting to standardize the fields of optics, heat, electricity, and magnetism through a study of forces acting between material particles. The possibility that there might be a way to unify these domains immediately fascinated Ampère and continued to do so throughout his scientific career.

Historians of physics have generally adopted Robert Fox's term "Laplacian physics" to refer to the French research program inspired and directed by Pierre Simon Laplace.[20] One of the dominant personalities in the French scientific community during Ampère's career, Laplace was born into a well-established and prosperous family of farmers and merchants in southern Normandy. He came to Paris in 1768 to pursue a career in mathematics; by 1773 he had submitted 13 memoirs to the Académie des Sciences and was elected to the mathematics section at the relatively young age of 24. At a time when there was not yet a field of mathematics devoted to the systematic study of probability, Laplace played a major role in carrying the early development of this topic beyond the rules of thumb of gambling and the preliminary conclusions of earlier mathematicians. We will see that he gave close scrutiny to an early paper by Ampère in this field. Throughout his career he retained his early interest in problems suggested by the mathematical implications of physical laws; aside from a few brief periods, he did not devote himself to experimental investigation of new phenomena.

Inspired by developments during the 1780s and 1790s, Laplace became influential as a physicist only after 1800. The Laplacian research program he directed thereafter commanded attention not only by specific discoveries, but also through advocating methodological norms and conceptual commitments. Because it was the single most important factor in establishing the context in which Ampère made contributions to physics, several questions about the Laplacian school are crucial to understanding Ampère's development. For example, what did these physicists mean when they referred to the "laws" of natural processes? How were their attempts to discover these "laws" related to speculation about the microscopic or unobservable "causes" of directly observable phenomena? How deep was their commitment to the existence of "imponderable" fluids made up of particles that exert short-range forces through empty space, that is, forces that operate even when the particles are "at a distance" from each other? How were these concepts synthesized into a specifically Laplacian mode of scientific explanation?

A coherent response to these questions did not emerge until shortly after 1810. The initial years of Ampère's scientific career thus coincided with the formative period of the Laplacian school, a period in which developments in electricity and magnetism were especially important. In particular, Charles Augustin Coulomb carried out influential experimental and theoretical research in these fields during the 1780s. Ampère knew of this work by the time of his sojourn in Bourg and he responded with objections he never retracted thereafter. A preliminary study of Coulomb thus provides insight into both the essential characteristics of Laplacian physics and Ampère's first serious thinking as a physicist. Unfortunately, Coulomb's methodological terminology was sometimes seriously ambiguous, particularly in his discussions of the "laws" he took to be so central to his research. Some preliminary clarification is called for.

Among French physicists at the turn of the century, physical laws were often stated as mathematical relations between quantitative properties of matter. Within this general category, we can introduce an important distinction between phenomenological laws and fundamental laws. *Phenomenological laws* have their origin in the data of experimental measurements.[21] In many cases, these laws function as a concise way to summarize data trends; in this respect they are descriptive rather than explanatory. Macroscopic phenomenological laws state relations between measurable properties of directly observable phenomena; they might also stipulate how these magnitudes change with time. For example, the law stating that the oscillatory period of a simple pendulum is proportional to the square root of its length is a typical macroscopic phenomenological law. On the other hand, microscopic phenomenological laws include magnitudes that are too small to be observed or experimentally manipulated. For instance, these laws might refer to distances or lengths too small to be measured. When these magnitudes are small enough to justify treating them as mathematical points in comparison to the relevant measurable magnitudes, microscopic phenomenological laws are usually stated in the language of differential or integral calculus. For example, Kepler's second law of planetary motion states that as planets follow their elliptical orbits they sweep out areas at a constant rate, even for time intervals too small to be directly measured. In other words, in its differential form the law makes a descriptive assertion that is more precise than any corresponding macroscopic claim about measurable quantities.

It is possible, but not necessary, for a microscopic phenomenological law to refer explicitly to unobservable entities that are presumed to exist in order for the law to be asserted. The differential statement of Kepler's law is clearly a case that does not include any claim about an unobservable microstructure of matter. But other microscopic phenomenological laws might very well refer to quantitative properties of "atoms" or "molecules." The determination of whether or not an ontological commitment is implied by a microscopic phenomenological law requires a careful study of its terminology and context. This is particularly important for the early nineteenth century when the term "molecule" took on a bewildering variety of meanings. In some cases "molecule" might simply refer to a volume of matter small enough to be represented as an infinitesimal volume element for the purpose of applying the techniques of integral calculus. No assertion about a particulate structure of matter is presumed here; "molecule" is simply a synonym for a volume that is "infinitesimal" in comparison to measurable magnitudes. In other contexts the term "molecule" might refer to a distinct particle with a specific shape, position, or motion. In either case, microscopic phenomenological laws include magnitudes that are not accessible to direct observation. Furthermore, the primary function for a phenomenological law is to describe phenomena for the domain in question. As such, phenomenological laws represent a distinct stage in the assessment of experimental data. Experimental data are always to some extent idiosyncratic with respect to apparatus and procedure. No two experimenters can hope to duplicate each other's data precisely. Nevertheless, data analysis often reveals proportionalities or other relationships between variables that can safely be expressed as phenomenological laws.

The question of whether or not phenomenological laws represent the final stage of sound scientific investigation became a central point of debate during the Laplacian era. From one perspective, these laws seem to cry out for explanation; why do these laws hold rather than other possibilities? In general, an answer to this question would have to be more than another phenomenological law; it should be a principle of sufficient import to be called a fundamental law.

The primary function for a *fundamental law* is thus to provide an explanatory foundation for phenomenological laws within a specific physical domain. The laws of conservation of momentum or conservation of energy often play this role. On the other hand, as is the case for microscopic phenomenological laws, fundamental laws may include references to microscopic entities not directly observable. If the law is truly fundamental, an additional decision has been made to treat these entities as explanatory primitives not subject to further scientific analysis by the theory concerned. This decision is open to revision, but it represents the line of demarcation between scientific explanation and untestable, "metaphysical" speculations about the causes of phenomena. While a microscopic phenomenological law might refer to "molecules," this would not constitute a fundamental law unless these "molecules" were also asserted to have no internal structure or more primitive components relevant to a more fine-grained explanation by the theory in question.

In some cases a fundamental law might only refer to microscopic magnitudes rather than microscopic entities. An important example is the "principle of virtual velocities" that Lagrange took as the foundation for his formulation of statics in the *Mécanique analytique*. Lagrange's principle states that a set of forces will hold a body in equilibrium as long as the sum of the "moments" of the forces is zero. The "moment" of each force is calculated by multiplying the magnitude of the force by an infinitesimally small increment, or differential length, along the line drawn in the direction of the action of the force. Lagrange's principle includes a differential magnitude too small to be observed or experimentally measured; since the principle also acts as an explanatory basis, it is an example of a fundamental law.

Similarly, Newton's theory of gravitation includes the claim that any two volumes of matter, small enough to be considered punctal in comparison to their separation, exert a mutual gravitational attraction which varies with the inverse square of the distance between them. While Newton may have hoped that an ether theory might eventually explain how the gravitational force acts through space, his force law functioned as the fundamental law in his theory of gravitation. He could use this law as a basis for an approximate derivation of the most important relevant phenomenological laws: Kepler's laws for planetary motion and Galileo's law of falling bodies. He could also provide derivations of other macroscopic phenomenological laws, such as the law that the gravitational force between two spherical bodies should vary with the inverse square of the distance between their centers. Lagrange's principle of virtual velocities and Newton's law of gravitation are thus two examples of fundamental laws in which no new microscopic entities are postulated. They draw their power from their scope and their amenability to the techniques of the calculus. As the Laplacian program developed, however, it came to include fundamental laws pertaining to imponderable fluid particles with masses too small to be experimentally detected. This was particularly evident in the domains of electricity, magnetism, heat, and optics.

Coulomb's ambiguous pronouncements on these issues not only left a legacy for the Laplacians to build upon, but also stimulated Ampère's first reactions in 1801 and 1802. Before turning to some specific details of Coulomb's work, the three major phases in the Laplacian program should be outlined.

1 Between 1785 and 1789 Coulomb presented his famous seven memoirs on electricity and magnetism. He repeatedly claimed that the central import of this research was the discovery of the "laws" governing the observed phenomena. In particular, Coulomb discovered that charged bodies such as spheres exert electric forces on each other with the same inverse square distance dependency as is found in the law of gravitational attraction. He attributed these macroscopic phenomenological laws to microscopic forces acting between individual molecules of magnetized

or electrically charged bodies. Although Coulomb believed that these microscopic forces were due to the presence of imponderable electric or magnetic fluids within the relevant molecules, he did not interpret his data as a conclusive demonstration that these fluids are the "causes" of electric and magnetic phenomena.

2 Even before Coulomb's death in 1806, his ambiguous methodological statements were reinterpreted by influential French physicists. Following the invention of the Voltaic pile in 1800, Jean-Baptiste Biot gave an electrostatic explanation of the device based upon Coulomb's concepts. Although Biot and his colleagues remained dubious about the existence of imponderable fluids, they were open to the possibility that calculations based upon hypothetical distributions of these fluids might bring about a "renaissance" of the mathematical mode of scientific explanation championed by Newton.

3 In 1812 Siméon-Denis Poisson carried out a sophisticated mathematical analysis of a two-fluid theory of electricity. The detailed agreement between Poisson's calculations and Coulomb's data inspired Biot to promote the imponderable fluids as the basis for all physical explanations. During the following decade Laplace, Biot, and Poisson clarified the principles of a distinctly Laplacian research program which they applied to to electricity, magnetism, optics, and thermal phenomena.

The first stage in this three-fold process is our concern at present. Coulomb presented his theory of magnetism in a 1777 essay on compasses, the second and seventh of his seven memoirs on electricity and magnetism, and in two subsequent papers on magnetism written during the final years of his life. These memoirs contain the theoretical arguments and experimental data that became the basis for the Laplacian theory of magnetism that flourished between 1800 and 1820. The essential concept, developed most elaborately by Poisson, is that the attractions and repulsions observed between the "poles" of magnets can be explained by a fundamental law for the magnetic forces between particles of imponderable magnetic fluid.

In his 1777 memoir Coulomb's chief concern was to refute the lingering Cartesian explanations of magnetism that relied upon the circulations of fluids in vortices within magnetized bodies.[22] Coulomb encouraged an explanation of magnetic phenomena in terms of central forces acting "at a distance" and modeled on Newton's theory of gravitation. Apart from his rejection of the Cartesian fluids, Coulomb saw little reason to choose between the one-fluid theory of Aepinus and the two-fluid theory of Brugman and Wilcke. Both of these theories made use of repulsive forces between the particles of magnetic fluid and presumed that magnetization results from macroscopic separations of the magnetic fluid or fluids within the magnetized body. Magnetic "poles" were said to be due to localized concentrations or deficits of magnetic fluid. Coulomb found neither version of this account adequate to explain the simple observation that two new magnets are produced when a magnet is broken. Seeking an alternative, Coulomb speculated about the possibility that "each point of a magnet or a magnetized bar can be regarded as the pole of a tiny magnet."[23]

In his 1785 paper Coulomb drew upon his torsion balance data to summarize the observed attractions and repulsions between magnets in terms of a macroscopic phenomenological law in which magnetic force varies in inverse proportion to the square of the mutual separation of magnetic "poles." This law then became the basis for a theory of magnetism that drew much of its mathematical structure from its analogy to Newton's gravitation theory. In particular, there is an obvious analogy between a "density" of imponderable magnetic fluid within a given volume and the more familiar density of atoms within an ordinary material body. Both concepts depend upon the number of particles found within the relevant volume.

Coulomb made a major innovation in his 1789 memoir. He proposed that the magnetic fluid (or fluids) is confined within individual "molecules" where it can become polarized when the material is magnetized.[24] Each molecule thus becomes in effect a tiny magnetic dipole. Furthermore, the poles of adjacent molecules neutralize each other so that only the tips of a magnetized needle are left active. The function of Coulomb's "molecules" is as much mathematical as it is physical. In his application of differential and integral calculus to both electric and magnetic theory, Coulomb calls "molecules" those parts of charged or magnetized bodies that are small enough to be treated as differentials but also have the property of containing various amounts of electric or magnetic fluid. Coulomb's "molecules" are thus infinitesimal volume elements for calculational purposes; nevertheless, they are assigned a fluid "density" due to their confinement of a specific number of fluid particles. This principle of molecular confinement thus acted as a restraint on the motion of magnetic fluid; Coulomb thereby avoided the objection he had raised against those theories that relied on macroscopic separation of magnetic fluids as an explanation for the localization of magnetic action at poles. According to Coulomb, when a magnet is broken, a gap is produced in the previously continuous series of tiny "molecular" magnets; this produces adjacent poles of the two new magnets.

In spite of the appealing simplicity of Coulomb's assumptions, they would have had little significant explanatory value without a reasonable mathematical function for microscopic magnetic forces. The determination of this expression and the analogous electric force was Coulomb's most significant contribution to what became the Laplacian program; it also inspired his methodological comments on the relationship of data to laws. Several regularities stand out amid Coulomb's rather indiscriminant use of the term "law." First, throughout his career Coulomb used the term "law" as a synonym for a "formula" or macroscopic phenomenological law. However, in his first three 1785 electricity and magnetism memoirs he also intended "law" to include microscopic phenomenological laws for the inverse square forces between infinitesimal volume elements or "molecules" within magnetized or charged bodies. After 1788 Coulomb also used the term "law" to refer to more controversial fundamental laws of interaction between individual particles of electric or magnetic fluid.

Coulomb's general strategy was to persuade his readers to accept his data as support for both microscopic and macroscopic phenomenological laws. The transition from macroscopic to microscopic laws was of course facilitated by the mathematics of inverse square forces; Coulomb could simply rely upon Newton's analogous reasoning in gravitation theory. Newton had concluded from Kepler's phenomenological laws for planetary motion that the attraction of the planets to the sun was an inverse square force. He then used one of his most famous corollaries to conclude that a similar force acts between any two particles of those bodies. Coulomb thus could follow an exactly analogous reasoning process for electric fluid particles uniformly distributed over the surface of a charged sphere. Given an inverse square macroscopic phenomenological law for the attraction between two oppositely charged spheres, an inverse square microscopic law could be asserted for infinitesimal surface elements of those spheres.

In the magnetic case Coulomb had to use less straightforward reasoning. He experimentally demonstrated that magnetized needles act upon each other in such a way that, for calculational purposes, the effective magnetic fluids could be imagined to be concentrated at magnetic poles near the ends of the needles. Thus, since the distance between these poles is equal to the distance separating the operative magnetic fluids of the two magnets, the experimental establishment of an inverse square macroscopic phenomenological law allowed Coulomb to also claim support for a microscopic law "according to which the magnetic fluid acts when the magnetic molecules are placed at different distances from each other."[25] After 1788 Coulomb also cautiously cited as "laws" not only macroscopic and microscopic phenomenological laws but fundamental laws for individual fluid particles as well.

Coulomb's reticence about *fundamental laws* was due to his conviction that scientific research should be restricted to the discovery of laws and should not be distracted by ephemeral speculations about the "causes" of the phenomena. He discussed this issue most fully in connection with the debate over the one-fluid and two-fluid versions of electric and magnetic theory. Although Coulomb refused to fully commit himself on an issue that could not be settled by experiment, he admitted that he "preferred" the two-fluid electric theory because it could account for the phenomena without the postulate of a repulsive force between particles of matter, as required by the one-fluid version. After 1799 he also gave tentative support to a two-fluid theory of magnetism. Although he cited the close agreement between his data and his theoretical predictions as verification for an inverse square microscopic phenomenological law for magnetism,[26] he was much more reticent about claiming that this data also confirmed a fundamental force law for individual magnetic fluid particles. In an interesting passage from the last of his seven memoirs, Coulomb gave his clearest statement of how microscopic phenomenological laws contribute to scientific progress.[27]

In order to avoid all discussion, I warn, as I have already done in various earlier memoirs, that every hypothesis of attraction and repulsion according to any law should be regarded only as a formula which expresses a result of experience. If this formula is *deduced from the action of the elementary molecules of a body endowed with certain properties*; if one derives from this primary elementary action all other phenomena; finally, if the results of theoretical calculation are exactly in agreement with measurements that experiments furnish, then one can possibly hope to go further only when one has found a more general law that includes in the same calculation bodies endowed with different properties which, up to the present, have not appeared to us to have any connection between them.

With an emphasis on microscopic phenomenological laws, Coulomb perceived scientific progress unfolding through a cyclic sequence of at least five distinct stages:

1 Accurate quantitative measurements should provide sufficient data to determine macroscopic phenomenological laws for a specific domain.
2 Intelligible and suggestive concepts should be brought to bear to assign mathematically tractable properties to small volume elements or "molecules" of the material bodies manifesting the phenomena.
3 Using these concepts and properties, a mathematically formulated microscopic phenomenological force law should be adopted as a tentative hypothesis.
4 The microscopic force law should be tested by applying it to cases where its implications can be compared to macroscopic measurements or relationships.
5 More general macroscopic and microscopic laws should be sought in an effort to reveal common properties among substances of seemingly unrelated natures and common sources for seemingly unrelated phenomena.

The driving force behind Coulomb's own scientific research thus was the search for microscopic phenomenological force laws. He felt that the task of scientific explanation was to demonstrate that these laws can provide approximate derivations of the macroscopic phenomenological laws of observable phenomena. Only macroscopic laws can be conclusively confirmed by experiment; microscopic laws are constantly subject to revision due to the discovery of new phenomena. The theoretical search for improved microscopic laws may in some cases draw inspiration from the alleged molecular properties that manifest these laws. It was only for this reason that Coulomb was willing to include the presence of imponderable fluids among these molecular properties; in this respect his research in electricity and magnetism was incorporated into his more general methodology.

Coulomb realized that he did not have the mathematical ability to attempt an analysis that would link microscopic phenomenological laws to fundamental laws. This task would be carried out by Poisson in 1812; as we will see, his accomplishment was a major event in the evolution of the Laplacian program. Ampère would be in Paris to witness the developments that intervened between Coulomb's death in 1806 and Poisson's *tour de force* six years later. Prior to that period, however, he already developed a strong and lasting aversion to Coulomb's reliance upon microscopic force laws; he expressed his own position in a long manuscript composed during his Bourg period.

49

Ampère's 1801 Bourg Manuscript

On 12 March, 1802, Ampère gave the inaugural public lecture that was traditional for a new professor at Bourg. The text of the discourse reveals both Ampère's penchant for classification and his conception of the scope and goals of physics. After drawing a preliminary Cartesian distinction between mind and matter, Ampère cited three general types of sciences concerned with material being: mathematics, physics, and natural history. The goal of natural history is primarily classification according to physical properties. Physics, on the other hand, should find among the contingent properties of matter those that "determine the general order of the universe and constitute what are called the laws of nature." Furthermore, the successful search for these laws is "to find in this labyrinth of disconnected facts, lacking mutual dependence, the unique fact of which they are simply consequences and which one should regard as one of the laws of nature." After citing Lavoisier and Newton as two who achieved this feat, Ampère gave an assessment of how physicists should proceed in their footsteps.[28]

> For those of us who gather the fruits of their work without sharing their glory, I believe that above all we should make an effort to reduce to the smallest number possible the principles that should serve as a basis of all explications, and to accustom ourselves to deduce from them the most distant consequences with as much facility as is the case for those that flow from them immediately.

Following this admonition, with its allusion to central principles, Ampère turned to the three divisions of his own course. As would be expected at an École Centrale, he emphasized the practical applications of the discipline. The first section, cosmology, would explore the implications of "the great principle of universal gravitation," including its application to navigation. The second section was entitled mechanical physics; Ampère explained that he would perform simple experiments to illustrate "the general principle of equilibrium and that of motion, at least in so far as it will concern solid bodies, coercible fluids, and light." In addition, however, Ampère commented that[29]

> the slight progress that has been made in the theory of the magnet and of electricity will oblige me, toward the end of this division, to pose two new principles, demonstrated by a multitude of experiments, both with respect to electricity and the magnetic fluid, and which will suffice to explicate all the phenomena presented by these two fluids.

Ampère's confident tone was inspired by a lengthy treatise he had drafted during the preceding year. He had presented a brief sketch of his work at the 24 December 1801 meeting of the Académie de Lyon. A quite distinguished audience that included Volta, the inventor of the electric pile, strained to hear his poorly enunciated reading. According to the Académie's *procès-verbaux*,[30]

> Ampère began the reading of a memoir which he presented only as a weak sketch of a vast system that connects all the parts of physics. The memoir may be divided into two sections. The author announces that the first will contain a new explication of the phenomena of electricity and the magnet, in which they are reduced to the ordinary laws of mechanics. The second is intended for the examination of the influence of electricity on affinities and on the theory of light and colors.

Although Ampère never completed this project, he continued to work on it in Bourg during 1802. Its broad scope included the third section of his physics course, the section he called chemical physics. In his inaugural lecture he said that he would demonstrate for his students that the composition and decomposition of bodies could be traced to three sources:

a   the combination of caloric with other bodies
b   the combination of oxygen with combustible substances
c   the combination of acids with salifiable bases.

Lengthy manuscript drafts of portions of Ampère's treatise have survived and they testify to his independent thinking at this early date. Coulomb, Laplace, and Haüy were some of the influential physicists who had objected to Aepinus' one-fluid theory of electricity, owing to the apparent necessity to postulate both a force between the fluid and ordinary matter and an additional repulsive force between material particles. But Ampère did not simply find this additional force objectionable; he rejected Coulomb's two-fluid thesis as well. In a manuscript fragment labelled *introduction*, Ampère described the motivations for his theory. He was particularly adamant in his rejection of non-gravitational forces as a basis of the theories provided by both Aepinus and Coulomb.[31]

> Three reasons should especially engage physicists to reject both of these systems. The first is this complication of repulsions imagined between molecules of electric and magnetic fluids combined with the attractions of these same fluids for all bodies or for the molecules of another fluid,

both of which act at considerable distances and without any contact between the molecules which are acting on each other.

The second is this very supposition of any action whatever between bodies which do not touch each other. Newtonian attraction has accustomed us to admit in bodies these kinds of properties which do not present to our minds any clear idea. However, I do not believe that this example should authorize the gratuitous suppositions of Aepinus and Coulomb. For if, as a large number of physicists think, the attraction between celestial bodies is caused by a fluid analogous to electricity or magnetism, one would be subject to falling into the most ridiculous vicious circle. If, on the other hand, the attraction that is the cause of all the phenomena we observe is subject to no cause other than the Supreme Will, and if it is the means God used to complete the existence of matter by bringing molecules together, then it should be unique, universal, and constant. Nothing is less fitting than to arbitrarily establish an even stronger attraction between the molecules of a fluid and those of other bodies and a mutual repulsion between the former.

With the single exception of gravitation, Ampère thus rejected scientific explanations that rely upon forces acting at a distance through empty space. His attempt to pose an alternative required a discussion of a material medium for the transmission of effects usually described in terms of forces. Following a discussion of the mechanical paradoxes that result from postulation of a force between imponderable fluid particles and inertial matter, the manuscript quoted above includes two principles, probably the same ones Ampère referred to in the prospectus for his first physics course at Bourg. These are methodological principles intended to act as constraints on any theoretical reliance upon electric or magnetic fluids.[32]

1   One should be allowed to attribute to electric and magnetic fluids only properties observed in other fluids, such as the elasticity common[33] to all gases, and the property of adhering to other bodies, as is noticed in the majority of liquids.
2   One should not suppose any action between these same fluids and the bodies which they do not touch, or between the portions of the homogeneous fluid which are not contiguous, except for that action which is produced by the reaction of the fluid spread in the intermediary space. From this it follows that one should occupy oneself solely with the action of each molecule of matter on those that immediately surround it. Furthermore, since there is no true vacuum near our globe, this action, being propagated by degrees (*de proche en proche*), is sufficient, as will be seen,

52

to explicate all the phenomena pertaining to this theory, and in the future it will be able to make it a branch of mechanics, just as has been done for hydrostatics after the latter science was for so long only an experimental theory.

These principles became lasting convictions for Ampère. Subsequently he would learn to appreciate the value of force laws for pragmatic purposes of calculation; he also would acknowledge that they represent an important stage in scientific understanding. But he would always consider any reliance upon nongravitational forces to be only a preliminary stage prior to a full explanation of the phenomena based on an understanding of the operative medium.

Of course Ampère never achieved his ambitious goal of reducing electricity and magnetism to a "branch of mechanics." His early efforts deserve some attention as a prelude to his more extensive attempt during the 1820s. His 1801 theories of electricity and magnetism were integrated into a comprehensive attribution of optical, thermal, electric, and magnetic phenomena to the action of three elastic, *igniformes* fluids: electric fluid, magnetic fluid, and the caloric fluid responsible for thermal effects. As was the case for the electric and magnetic imponderable fluids, caloric was imagined by theorists such as Black and Lavoisier to be subtle enough to produce effects without itself bearing any detectable mass. Its presence between material molecules counterbalanced their gravitational attraction to each other, and the flow of caloric into a body explained its expansion when heated. Progress within the Laplacian program was driven much more by the analysis of electricity than it was by an attribution of thermal effects to caloric. The commitment of Laplacian physicists to the existence of the caloric fluid was thus highly dependent on debates over the status of the other imponderable fluids in optics, electricity, and magnetism. Laplace himself remained noncommittal in his 1783 collaboration with Lavoisier for their joint publication *Mémoire sur la chaleur*. Nevertheless, discussions of caloric as a material basis for thermal effects were widespread at the turn of the century and Ampère entered into these debates with great vigor.

In a set of notes entitled *Recherches sur les causes de l'attraction et de la répulsion électrique*,[34] he used an analogy to the effects of observable fluids to attribute a force of cohesion between contiguous parts of the electric fluid and an adherence to molecules of matter when in contact with them. The electric fluid thus forms spherical atmospheres around these molecules which can be deformed as the molecules interact. When two molecules approach contact, their electric atmospheres repel each

53

other like two globules of mercury. Most important, each electric atmosphere has a *ressort*, or elasticity, responsible for flow from one body to another. Ampère's attempt to reduce electricity and magnetism to the mechanics of elastic fluids was highly dependent upon his close study of Lagrange's *Mécanique analytique*, originally published in 1788. Lagrange defined a compressible and elastic fluid as one in which the parts "can change in both shape and volume simultaneously, and which always tend to dilate with a known force that ordinarily is supposed to be proportional to a function of the density."[35] In his notes, Ampère added that the elastic effect also depends upon the "state of compression or dilation of the given fluid within neighboring bodies."[36]

According to his own methodological principles, Ampère could have cited the observed elasticity of ordinary gases as sufficient reason to tentatively assign an analogous elasticity directly to the electric or magnetic fluid. Instead, he argued that this property should be assigned to caloric, the thermal fluid which he asserted had electric fluid as one of its components. He initially referred to its other component as the *base du calorique* or *pyrogène* before ultimately concluding that it is the magnetic fluid. Ampère attributed the spring or *ressort* of the caloric fluid to the compression it undergoes due to the gravitational attraction between the molecules surrounding it, a compression that tends to expel the caloric from the body containing it. Lavoisier had used a similar idea to explain the elasticity of gases in terms of a strong attractive force between adjacent caloric particles.[37] Ampère then identified the radiation of caloric from bodies with thermal and optical radiation. The degree of refrangibility of the radiation was attributed to the relative proportion of its electric and magnetic fluid components. No doubt this is the basis for his remarks about colors for the Académie de Lyon. During thermal and optical phenomena, the ratio of electric to magnetic fluid within the emitted caloric falls somewhere between the extremes set by the *chaleur obscur*, which is the most refrangible and contains a relatively small amount of electric fluid, and the *calorique violet*, which is the least refrangible and contains a relatively large amount of electric fluid. Caloric outside the limits set by these two extremes is highly unstable; it quickly returns to acceptable proportions by absorbing or emitting electric fluid.

On the basis of these generalities, Ampère developed his account in two directions; he tried to explain known magnetic and electric effects, and he cited new experimental observations as support for his theory. All of this work was qualitative and thus stands in sharp contrast to Coulomb's application of calculus to derive measurable quantitative implications of his force laws. The acceptable format for explanation in physics was changing rapidly; Ampère would become fully aware of these changes when he moved to Paris in 1804. A brief selection of some of the details of his 1801 treatise will serve to illustrate the speculative thinking that produced a treatise he soon realized was not worth publishing.

On the explanatory front, Ampère imagined each material molecule to be surrounded by a thin atmosphere of caloric with which additional electric or magnetic fluid could be mixed. When a charged or magnetized body is exposed to air, this excess electric or magnetic fluid on each of its surface molecules will have an elasticity that is either greater or smaller than that of the atmospheres surrounding each of the adjacent air molecules. When the first of these two possibilities is the case,[38]

> the expansive force of the super-abundant fluid, overly compressed by the layers of this fluid which cover the molecules of the surrounding air, makes this fluid slide over the surface of these molecules so as to pile up on the side of this surface opposite the body.

The result is an increase in the *ressort* of the air molecule atmosphere at the point of increased fluid surface density; this distorted atmosphere then acts upon the next neighboring air molecule just as the molecule of the charged or magnetized body acted upon it. A series of distorted electric or magnetic atmospheres is thus produced without any electric or magnetic atmospheres actually passing from one molecular atmosphere to another. These distorted atmospheres would have *ressorts* at their point of greatest surface concentration which would equal that of the original atmosphere of the magnetic or electric body were it not for the adherence of the fluids to the air molecule itself. This adherence results in a slightly smaller *ressort* in each successive link of the series; the effect thus gradually dies out and must vanish at an experimentally determined distance from the charged or magnetized body, a distance corresponding to the range of electric and magnetic forces. Ampère of course made no attempt to calculate this distance; his theory gave only a qualitative description of the process that might be responsible for the finite range of the observed attractions or repulsions.

To explain these phenomena in more detail, Ampère relied upon the "artificial" (*factrice*) electricity or magnetism of the distorted fluid atmospheres. The basic principle is that "attraction is always the greatest between bodies for which the electric or magnetic *ressorts* differ the most."[39] For example, Ampère gives the following explanation for the observed repulsion that takes place when similarly charged bodies or similar magnetic poles are brought near each other.[40]

An electrified or magnetized body, never being able to produce in the surrounding air a compression or a rarefaction of fluid equal to that which takes place around its own molecules, can be considered to be pushed toward the air that surrounds it. However, the action being the same on all sides, as long as the fluid of the air is in the same state all around the body, there cannot result any displacement in that body.

However, if one imagines two bodies electrified or magnetized in the same manner, that is, deprived or supercharged with fluid, they will cooperate in their efforts to put the air and the other bodies between them . . . in a state closer to their own than is the state in which each individual body puts the air or the bodies that are on the other side of it.

It follows necessarily that each of the two bodies should tend with more intensity toward the air and other bodies of which I have just spoken, and thus seem to be repelled by the other body, given that attraction is always the greatest between bodies for which the electric or magnetic *ressorts* differ the most.

Similarly, Ampère attributed the attraction between magnets and metals such as iron to the limited degree of inter-molecular mobility of the magnetic fluid within such materials. He relied upon the existence of distorted magnetic fluid atmospheres throughout the interior of a magnet to explain the production of two new magnets when a magnet is broken; thus he did not have to invoke Coulomb's absolute restriction of the magnetic fluid to individual molecules.

Turning to Ampère's assessment of experimental evidence, an important methodological attitude stands out. At this early stage in his career he placed strong emphasis on the successful prediction of new experimental results as a key indication that a hypothetical micro-explanation is correct. For example, in one of the 1801 manuscripts on electricity, he claimed that in proposing a "physical system" as a causal foundation for a given domain of phenomena,[41]

one has all the more means of verifying it as one has used fewer in discovering it. One has only to imagine some new experiments and predict their outcome; it then suffices to try them in order to see if the conjectures one has made are confirmed or not.

Ampère thus recorded long lists of phenomena he considered to be implications of his *igniformes* fluids theory. In many cases he also described experiments that would test these implications. It is noteworthy that none of these proposed experiments required quantitative measurement of the dependency of one variable upon another. Unlike

Coulomb, Ampère was generally interested in whether or not an expected effect took place. Experiment was called upon to provide a simple yes or no answer rather than a table of quantitative data. For example, Ampère proposed as the ninth in one list of possible experiments:[42]

If it is also *pyrogène* [magnetic fluid] that is combined with lime deprived of carbonic gas, then communicating some electric sparks to it should produce some caloric . . . The strongest proof of this system is that an electric spark tends to decompose all combinations of two gases such as water and ammoniaq [*sic*] either into hydrogen, oxygen, or nitrogen developing at the same time while combining with some free caloric which is always supercharged with *pyrogène*. But the same spark tends to further combine two other mixed gases such as oxygen and nitrogen when they are both held in this state by some *pyrogène* because it releases this *pyrogène* from them. It seems that in this production of nitrous gas it is only oxygen which looses it because it holds it there less strongly . . . It remains to find out if the decomposition of water and ammoniaq produces some cooling. That of the galvanic pile is quite appreciable and naturally should be more so since the caloric there is absolutely decomposed and in the other case is only deprived of its excess *pyrogène*.

In letters to Julie, Ampère sometimes reported success in these experiments, particularly those performed with his colleague Clerc. He gave no specific details; one can only speculate about which ones were performed. As the above examples illustrate, the linkage between Ampère's elaborate theory and its experimental implications was tenuous at best. Surviving manuscripts suggest that he gave close study to the discharge of Leyden jars. But his disregard for quantitative predictions and measurements made it possible for Ampère to concoct an appropriate explanation for the outcome of virtually any of the questions he proposed as experimental tests.

On 25 July, 1802, after learning of the prize of 60,000 francs that Napoleon had offered for a discovery in electricity comparable to Volta's invention of the pile, Ampère wrote to Julie with his typical enthusiasm:[43] "This is precisely the subject that I was treating in the work on physics that I started to print; but it must be completed and my theory must be confirmed by new experiments." Ampère's overly optimistic assessment of his own research is a clear indication of how out of touch he was with recent developments in physics. Coulomb, Biot, Poisson, and Laplace were creating the mathematical physics of the nineteenth century by bringing new mathematical techniques to bear upon the imprecise empirical disciplines of electricity, magnetism, heat, and optics. Perhaps it was just as well that during the spring of 1802 Ampère's attention shifted to mathematics.

Mathematics at Bourg and Lyon

Mathematics was the only field in which Ampère actually published during his time in Lyon and Bourg. The grandiose treatise that he initially hoped would provide new foundations for chemistry and physics lacked focus and was abandoned by the end of 1802; mathematics provided a more careful route that eventually brought him full entry into the French scientific community.

Ampère's decision to pursue mathematics was motivated by his career. He enjoyed teaching chemistry and remarked to Julie that he preferred it to other branches of physics. His chemistry course at Bourg during the summer of 1802 went particularly well; Lalande observed some of his lectures and was favorably impressed in spite of an accident that splashed scalding water into one of Ampère's eyes. But Ampère also found that teaching physics and chemistry consumed far too much of his time and energy; by teaching mathematics he would not be required to prepare experimental demonstrations and maintain equipment. His tutoring was successful in both Bourg and Lyon, but the major drawback was the repetitive and uninspiring routine of teaching elementary mathematics. Aside from these pedagogical factors, Ampère also recognized that he had a trustworthy intuition about what specific mathematical problems he could solve; his knowledge of the state of research in other fields was less reliable. Understanding the context for Ampere's first mathematics publications requires a brief summary of the state of the subject in 1800.

Generally speaking, the eighteenth century was a period in which the insights of the previous century were developed and expanded but were not supplemented by ideas of comparable impact. The most influential seventeenth-century contribution was the creation of the calculus by Newton and Leibniz. Both differential and integral calculus quickly became central to the research of most mathematicians; Ampère first encountered these ideas as a child while reading *Encyclopédie* articles written by d'Alembert. Most of the major currents in eighteenth-century mathematics sprang from techniques or problems raised by the calculus and its physical applications, a trend which continued into the nineteenth century. Although Ampère was no exception to this generalization, he did choose a quite unique set of specific research topics. As we will see, he demonstrated his individuality as much in the problems he avoided as in those he adopted. A major reason for the immediate and widespread interest in the calculus was the fact that it could be applied to so many important physical problems. Newton had invented these new mathematical techniques in direct response to problems encountered in his development of mechanics and gravitation theory. Interest in the physical problems at stake often contributed to mathematicians' initial willingness to tolerate less than rigorous statements of relevant principles.

The central idea in differential calculus is of course the derivative of a function. Early in the eighteenth century, the "function" concept was allowed a liberal and ambiguous interpretation; sometimes a dependent variable was said to be a function of an independent variable when its values depended on those of the independent variable in any manner whatsoever. Distinctions between continuous and discontinuous functions developed gradually with the recognition that many claims about functions were not true unless qualified to apply only to continuous functions. But an accurate definition of continuity was not achieved by Augustin-Louis Cauchy until well into the nineteenth century. This initial lack of precision was also particularly evident in the central notion of the derivative. The physical motivation for the idea came from mechanics, where the velocity of a moving body is studied. The average velocity over a finite period of time is defined as the ratio of the change in position to the change in time. Newton depicted this relationship graphically and argued that the slope of a line drawn tangent to the graph of position with respect to time should indicate the instantaneous value of the velocity at a particular point in time, that is, the value of the tangent angle should correspond to the velocity value achieved by letting the time period in question become arbitrarily small when dividing the change in position by the change in time. If the position is assumed to be a function of time, the velocity is said to be the derivative of the position with respect to time. Because velocity is the rate at which position changes with respect to time, the notion of an instantaneous velocity was initially considered by many mathematicians to be counter-intuitive. A precise statement of the relevant limiting process in a rigorous definition of the derivative was not given by Cauchy until the 1820s. Ampère would be a close colleague of Cauchy's at the École Polytechnique during the period in which this work was accomplished. Although he may very well have had the ability to help provide the rigorous foundations of calculus, Ampère's interests never focused there and his contributions to this topic were minimal.

In addition to the derivative and the associated differential calculus, Newton and his contemporaries studied the inverse procedure of taking the "integral" of a function. As was the case with the derivative, the integral concept initially had a geometric interpretation. For example, when the velocity of a moving body is plotted as a function of time, the integration of velocity with respect to time is given by the area under the graph and provides the distance travelled by the body during a specific time interval. Integrating the velocity function thus reverses the differentiation process in which the velocity function is generated by taking the derivative of position as a function of time. When velocity is a constant, the area under the graph of velocity with respect to time is, of course, just the simple area of a rectangle. In more complicated cases, the area can be approximated by dividing it up into very small areas that can be approximated by rectangles. As was the case with the derivative, there is a limiting process involved in taking an integral; as the number of small rectangular areas used becomes increasingly large, taking their sum results in an area that becomes increasingly close to the actual area or the actual integral of the function involved. The difficulty of carrying out integration and differentiation thus depends upon how complicated the function is.

Newton determined simple algorithms for differentiating power functions such as x^n. More complicated cases were addressed by the application of another of Newton's major contributions, the binomial theorem. According to this theorem, a function such as $(x + 1)^n$ can be expanded as an infinite series of the form:

$$ax + bx^2 + cx^3 + \ldots$$

Integrating or differentiating the initial function was then carried out by taking the derivative or integral of the terms in the series and summing these results. In cases where the terms of the series become small after the first few, the integral or derivative could be conveniently approximated by ignoring all subsequent terms in the series. Throughout the eighteenth century these procedures were not justified with rigorous proofs, but they were accepted for their interesting and successful applications to physical problems. For example, Newton often showed that an algorithm worked for several specific values of a parameter and then concluded "by induction" that the procedure was generally true for all values of the parameter.

During the eighteenth century, the problems generated by the initial applications of the calculus gradually expanded into distinct but related fields of mathematics. Integration and differentiation techniques were studied in great detail and contributed to the interest in infinite series and the summation of these series. Similarly, infinite series were used to study and solve differential equations; these are equations which include the derivative of a function or higher derivatives given by differentiating the derivative function. In cases where a function depends upon more than one variable, "partial" derivatives can be calculated with respect to any of the independent variables; partial differential equations then are equations in which these partial derivatives appear. They too could sometimes be solved by expanding the relevant functions using the binomial theorem and then carrying out the relevant operations on the terms of the series. Infinite series thus became increasingly important and were studied in their own right. In particular, mathematicians such as d'Alembert and Lagrange paid some attention to the distinction between converging and diverging series. The motivation for the distinction is the need to justify neglecting all but the first few terms in a series expansion of a function. When this approximation is sufficiently accurate, it offers an expedient way to derive approximate results about what would otherwise be unmanageable functions. However, the technique depends upon the assumption that the later terms in the series make no significant contribution even though they are infinite in number.

In one of the *Encyclopédie* articles Ampère read as a child, d'Alembert makes the following remark about converging series, a remark that indicates how the concept was stated without the use of rigorous definitions of limit.[44]

> When the progression or series approaches some finite quantity more and more, and consequently the terms of the series, or quantities of which it is composed, go on diminishing, one calls it a convergent series, and if one continues to infinity, it will finally become equal to this quantity.

This description is too vague to capture the distinction sought, and Cauchy did not accomplish this until the 1820s. But as long as the functions involved were in fact convergent, mathematicians continued to get valuable results. For example, in spite of his hazy notion of convergence, d'Alembert was also responsible for the famous "ratio test" for the convergence of a series. In 1768 he argued that if the ratio of a general term in the series to the preceding term is equal to a fixed value less than unity, the series converges.[45] For series of this type, it is permissible to drop terms after the initial few for the sake of approximations. Ampère was well aware of this theorem at the time he began working on his first significant mathematics paper in 1801.

61

Owing in part to the emphasis on power series, there was a general eighteenth-century tendency to consider the calculus as simply an extension of algebra. This was particularly the case until the notion of limit became the unifying concept for both calculus and the study of infinite series. Lagrange was the single most important figure for this algebraic tradition. He was one of the first lecturers at the new École Polytechnique during 1795–6, and the following year he published an edition of his lectures as the *Théorie des Fonctions analytiques*. One of the central themes in his approach was to avoid what he called "metaphysical" assumptions about the infinitesimal quantities associated with the idea of a derivative. To do so, he defined the derivative and subsequent higher derivatives of a function in terms of the coefficients in the series expansion of the function. This attitude was shared by Ampère, and in 1806 he took this procedure as a starting point in one of his most influential mathematics papers. Generally speaking, Lagrange was by far the most important influence on Ampère's early efforts as a mathematician. Although in 1802 Ampère was not as yet concerned with the conceptual basis of the calculus as such, in his first memoir he conclusively demonstrated his thorough familiarity with power series and his skill in the manipulation of their coefficients.

Ampère's 1802 Considérations sur la Théorie mathématique du Jeu

Ampère's first mathematics essay was a consideration of the probabilities associated with a gambler's attempt to avoid elimination from a game of chance by losing all of a finite initial fortune. The problem may first have occurred to Ampère during parlor games in Poleymieux; he commented in a letter to Couppier that he sometimes lost his entire allotment of tokens in the course of an evening's games. The composition of the memoir spanned most of 1802, and Ampère refers to it frequently in letters to Julie beginning in April of that year. At that point he asked her to send him all his mathematics reference books and notes, material he had left behind in Lyon and had not consulted seriously for five years. His renewed interest was sparked by a desire to impress the examining committee for the forthcoming mathematics position at the Lyon Lycée. His initial insight was the discovery of a "direct solution" to a problem he had posed for himself seven years earlier.[46]

Ampère began by considering a gambler who is assumed to wager

the same specific fraction of his original fortune in every game. That
is, his initial resources are divided into m equal parts and he wagers
one of these m parts in every game. What is the probability that the
gambler will be eliminated from further play after any specific number
of games? In any particular sequence of games the gambler will be
eliminated if he wins p games and looses (m + p) games, where p can
be any possible number of victorious games. Ampère let q represent
the ratio between the probability to win a game and the probability to
lose. These two probabilities thus can be written as q/(1 + q) and
1/(1 + q) respectively. The first and primary goal of Ampère's essay
is to calculate the probability of the gambler being eliminated as a
function of q.

Ampère's strategy is to write the general joint probability for p wins
and (m + p) losses and then consider all sequences in which this combination can take place; a complication arises, owing to the fact that
some sequences of wins and losses are ruled out if they include a
subsequence in which the gambler has been eliminated. For instance,
if the first m games are lost, there is no need to consider any longer
sequences of p wins and (p + m) losses. Ignoring these complications
initially, and using the probabilities for winning and losing, the probability for p wins and (m + p) losses is:

$$\left(\frac{q}{1+q}\right)^p (1 + q)^{-(m+p)} = \frac{q^p}{(1 + q)^{-(m+2p)}}$$

Ampère next calculated the number of sequences that result in elimination after (m + 2p) games, that is, after p wins and (m + p) losses.
He represented this number by $A^{(m+2p)}$. The probability of elimination
after (m + 2p) games thus is given by:

$$A^{(m+2p)} q^p (1 + q)^{-(m+p)}$$

The well-known combinatoric formula gives for the number of ways
to have (m + 2p) games distributed into p wins and (m + p) losses:

$$\frac{(m+2p)!}{(m+p)!\,p!}$$

But to calculate $A^{(m+2p)}$ requires subtracting from the above quantity
all the various cases where elimination takes place after (m + 2r) plays
for any value of r less than p and greater than or equal to m. Such
cases are those where an additional (p − r) wins and (p − r) losses after

63

elimination were imagined to be a possible sequence in the initial calculation. Ampère thus subtracted the number of sequences corresponding to all possible values of r. This correction summation results in a value for the total number of possible sequences given by:

$$A^{(m+2p)} = \frac{(m+2p)!}{(m+p)!\,p!} - \sum A^{(m+2r)}\frac{(2p-2r)!}{(p-r)!\,(p-r)!}$$

This equation is not of any practical value due to the fact that values for $A^{(m+2p)}$ appear within the summation. To rectify this, Ampère discovered another way of writing $A^{(m+2p)}$ by carrying out an analysis similar to the preceding one for $(m + 2p - 1)$ games. The resulting equation is then subtracted from the equation given above to give the following result.

$$A^{(m+2p)} = \frac{m\,(m+2p-1)!}{p!\,(m+p)!}$$

The probability of elimination after p wins and $(m + p)$ losses then is given by:

$$A^{(m+2p)}\,q^p\,(1+q)^{-(m+2p)}$$

Ampère then calculated the general probability of elimination by considering the above probability for all values of p. His goal was to determine the probability of elimination by calculating the value of the sum of the resulting infinite series. Before doing so he used the ratio test to show that the series converges when 6p is larger than $m^2 - 3m + 2$. Ampère does not mention that this condition will only be fulfilled when the number of wins is significantly larger than m, where $1/m$ is the fraction of the gambler's resources wagered in each game.

Ampère expanded each term involving $(1 + q)$ using the binomial theorem. Initially he did not notice that the expansion must actually be done for two different cases, depending upon whether q is larger or smaller than unity. Early in 1803, when Ampère finally submitted his memoir to the First Class of the Institut, Laplace noticed this oversight and mentioned it in a letter to him. The flaw caused Ampère considerable consternation; in a letter to Julie he remarked that he initially read Laplace's comment as the *éloge* of his memoir. But he soon recovered when he noticed that only a few pages of the paper needed to be altered and learned that his brother-in-law Marsil was willing to quickly print up a corrected insert. In the altered version Ampère considered the two possible series and used an interesting recursion formula to show that in both cases the coefficient of each power of q vanishes when the probability sums are written out for all values of p. This leaves for the value of the probability only the first term in each series. For q less than 1 this value is unity and for larger values of q the probability is q^{-m}. In other words, if the odds of winning the game are less than the odds of losing ($q < 1$), then the gambler will inevitably be ruined eventually. Ampère went on to consider variations in the initial conditions of his problem, but these results are of secondary interest.

Ampère's isolation in Bourg was one reason for his multiple revisions of the memoir and his hesitation to consider it complete. Couppier was too polite to make objections, if indeed he had any. Ampère suspected that Clerc did not fully understand the paper and thus gave untrustworthy praise. These suspicions grew when Lalande read an early version of the paper following Ampère's presentation to the Société d'Émulation de l'Ain on 26 July, 1802. The Société was one of the few centers for intellectual life in Bourg and Ampère was pleased that his memoir resulted in his election as an associate member. Lalande made a guest appearance at the meeting, studied the memoir with Clerc, and gave a positive assessment that won Ampère membership. Ampère's enthusiasm waned when Lalande suggested that he include some numerical illustrations of the solution so as to make the results more accessible. Ampère had good reason to believe that Lalande simply did not understand the derivation and wanted to see some simple examples. Ampère was slightly offended and felt that this would give his memoir the style of a schoolboy primer.

This meeting between Ampère and Lalande symbolizes another stage in Ampère's encounter with the French scientific community. Lalande at this point was 70 years old and had enjoyed an active and controversial career.[47] Born in Bourg in 1732, he blazed a reputation as a scrupulous astronomical observer. He became a professor at the Collège de France in 1760 and published his *Traité d'Astronomie* in 1764. It became a standard reference work and was updated in 1771 and 1792. Lalande emphasized accurate data and techniques of calculation that avoided theoretical controversy. Ampère had carefully studied the *Traité* during 1795 and 1796 and this may have provided an initial source of collegiality between the two men. In addition, however, Lalande had been very active in the Masonic Loge des Neuf Soeurs, a pre-Revolutionary lodge that not only had a significant scientific membership but also served as a podium for Lalande's outspoken atheism. Lalande asserted that all true philosophers were atheists, and he saw scientific progress as inextricably linked to the triumph of clear thinking over religious superstition. In July of 1802 Lalande was not appreciative of recent political events. One year earlier Napoleon had signed the Concordat that once again officially linked France to Catholicism. In April of 1802 he published the full legal details of the incorporation of Catholicism as a state religion. By 1805 Napoleon had to muzzle the prolific Lalande by forbidding him to publish.

Aside from Lalande's suggestions for revision of the probability essay, Ampère did not leave any records of their conversations. Nevertheless, for a man struggling with religious doubt, it must have been an interesting encounter. Lalande represented an atheistic solution to Ampère's struggle to reconcile his scientific rationalism with his emotional attachment to Catholicism. Ampère did not follow Lalande's lead at this point; he remained highly ambivalent, but he yielded to Julie's entreaties that he receive the sacraments during Easter week of 1803. He did not pursue the agnostic option until his arrival in Paris in 1804.

Several aspects of Ampère's first significant publication are worth noting. First, it is clear that at this point in his career he had a thorough knowledge of the technique of power series expansions based on the binomial theorem. He could manipulate these series at will and was clever at using recursion relations to transform them into interesting results. Second, he was aware of the pitfalls of divergent series. Laplace's critique drove this point home conclusively. Third, Ampère demonstrated a preference for what he called "direct" proof rather than proof "by induction," that is, rather than show that a result held for various values of his parameter p and then assume that it held for all values, he went out of his way to derive the general result directly. This was a preference that was characteristic of him, and it would recur throughout his career.

Finally, we should note that Ampère's display of mathematical sophistication far outweighs the practical significance of his results. For a physicist there is a sterility about this type of calculation that he would later find unfulfilling. As early as May of 1802 he reported to Julie that he was bogged down in "arid calculations." Following Julie's death it is not surprising that Ampère sought out spiritual as well as intellectual companionship.

Mechanics and the Calculus of Variations

Writing to Julie in February of 1803, Ampère mentioned that he had started a *petit mémoire* on the calculus of variations prior to leaving

Lyon the previous year. In contrast to his repeatedly amended probability memoir, Ampère kept the scope of his second paper within manageable bounds. As a result, he could submit versions of it to the Société d'Émulation de l'Ain on 12 March, 1803, and to the *Institut* on 16 May. The memoir was not published until 1806, and by that point Ampère had lost his initial enthusiasm for the topic.[48] Nevertheless, the paper played an important role in his early campaign for recognition as a mathematician. During 1803 he was characteristically jubilant about the magnitude of his accomplishment and he was confident that it would have a stabilizing effect on his relationship with Delambre. In its published form of 1806, Ampère's memoir consists of preliminary generalities followed by a critique of Lagrange's application of the calculus of variations to the principle of *vitesses virtuelles*, the fundamental principle of Lagrange's formulation of statics. Ampère's motivation came from his careful study of Lagrange's 1788 *Méchanique analitique* and he frequently refers to that treatise in the course of his paper.

The calculus of variations was one of Lagrange's primary and permanent interests; he was intrigued both by the mathematical formalism of the subject and by its relevance to mechanics. In 1762, with Euler's encouragement, Lagrange published two important papers in which he presented his new approach to the subject and applied it to mechanics.[49] He kept variational principles in the forefront of his research in mechanics during the subsequent two decades. By 1782 he had all but completed his *Méchanique analitique*, making variational principles the foundations of his formulation of both statics and dynamics. Lagrange moved to Paris just prior to the publication of his treatise in 1788; thereafter, his teaching responsibilities and his interest in foundational rigor resulted in the innovations included in his 1797 *Théorie des Fonctions analytiques* and the algebraic definition of variation in the 1806 edition of his *Leçons sur le Calcul des fonctions*. It is typical of Ampère that he did not sustain his initial interest in calculus of variations and did not participate in Lagrange's reformulation of the foundations of the subject. Nor did Lagrange cite Ampère's work in his own publications; nevertheless, the issues raised in Ampère's paper were given detailed treatment in the 1806 edition of the *Leçons sur le Calcul des fonctions* and the 1811 publication of the first volume of a second edition of the *Mécanique analytique*.

Although the 1788 *Méchanique analitique* served as Ampère's primary introduction to the calculus of variations, he was well aware of other sources, particularly Euler. As is generally the case for topics in eighteenth-century mathematics, origins are to be found in problems generated by Newton's applications of the calculus. For example, Newton was interested in determining the curve y(x) with fixed endpoints that would produce the least possible surface area when rotated about the x-axis. The problem requires the determination of the function y(x) such that a definite integral taken with respect to a function of both y(x) and its derivative becomes a minimum. Similarly, Euler and Jakob and Johann Bernoulli found that they needed to minimize other integrals to solve the brachistochrone problem, isoperimetric problems, and the determination of geodesics on a surface. In general, these problems can be characterized as a search for the function y(x) that makes the value of a definite integral of the following type as small as possible.[50]

$$J = \int_{x_1}^{x_2} f(x, y, y')\, dx \tag{7}$$

Lagrange generally referred to problems of this type as isoperimetric problems;[51] he reformulated them by introducing the notation $\delta y(x)$ to represent a variation in the entire curve y(x) over the range between the endpoints (x_1, y_1) and (x_2, y_2). The various possible curves linking the endpoints thus could be written as $y(x) + \delta y(x)$. More generally, the variation operator δ could be applied to any function, including integrals of the form cited above. In an analogy to the process of determining the minimum or maximum of a given function using differentials, another way to pose variational problems involving these integrals was to say that the variation of the integral must vanish for the desired minimizing function. Typically the variation operation was interchanged with the integral symbol and applied to the integrand prior to integration. Although in 1806 Lagrange would make an effort to give the variation operation an algebraic formulation independent of geometric figures, diagrams similar to those used by Euler were often used to explain elementary applications of the concept. For example, Lacroix depicted δ as an operation analogous to the differential operator d in the sense that δ produces changes in all values of a function such as y(x) by replacing y(x) with another slightly different curve; on the other hand, the differential operator d produces a change through "the passage from one point to another on the same curve."[52] In the context of mechanics, Lagrange relied heavily on the fact that these curves could be thought of as the result of a body in motion.

The 1788 edition of the *Méchanique analitique* was the product of Lagrange's conviction that mechanics could be formulated in terms of a small number of central principles. As such, it became the primary text for the Lagrangian tradition in mechanics. In this respect it stands in contrast to the approaches that the historian Ivor Grattan-Guinness labels "Newtonian" and "energy mechanics."[53] Lagrange emphasized neither Newton's laws of motion nor the conservation and transformation of *forces vives*, the eighteenth-century term for twice the kinetic energy of a moving mass. Instead, Lagrange chose the variational principles of *vitesses virtuelles* and least action as the foundations for statics and dynamics respectively. The *Méchanique analitique* had a symmetric formality and elegance that was deeply appreciated by Ampère.

Ampère was strongly affected by his initial reading of Lagrange's treatise. His first sustained study of the text was disrupted by the report of his father's death. For an adolescent of Ampère's sensitivity, the nightmarish intrusion of the guillotine into the formal cadence of Lagrange's abstractions had a shattering impact. After a period of severe withdrawal, his return to the text made it a central component of his thoughts about the relationship of mathematics to physical science. This is not to say that Ampère dogmatically accepted Lagrange's point of view; he was too independent a thinker to become an uncritical disciple. Nevertheless, he was permanently affected by his early exposure to Lagrange's goals and the style with which he pursued them.

For example, Lagrange's introductory *Avertissement* included vigorous statements of his ambitious proposal to reduce mechanics to the consequences of a few mathematically stated principles.[54]

> There already are several *Traités de Méchanique*, but the plan of this one is entirely new. I propose to reduce the theory of this science, and the art of resolving the problems that relate to it, to general formulas, the simple development of which gives all the equations necessary for the solution of each problem . . . No figures will be found in this work. The methods that I present require neither constructions nor geometrical or mechanical reasoning, but only algebraic operations subjected to a regular and uniform procedure. Those who love analysis will see with pleasure Mechanics become a new branch of it and will be grateful to me for having extended its domain in this manner.

Lagrange divided his agenda into the two domains of statics and dynamics; statics would provide the context for Ampère's reaction in 1803. Lagrange presented that topic as an increasingly complicated exploration of the consequences of his central principle, the principle of *vitesses virtuelles*. Given forces P, Q, R, . . . acting on a point mass along directions represented by dp, dq, dr, . . . , and with L = 0,

M = 0, N = 0, . . . representing additional constraints on the system, Lagrange wrote the general equation of equilibrium as:

$$Pdp + Qdq + Rdr + \ldots + \lambda dL + \mu dM + \nu dN + \ldots = 0$$

where λ, μ, and ν are Lagrange multipliers and the forces and constraints can be functions of position coordinates and their differentials.[55] Lagrange in effect used this equation to treat the constraints formally as if additional forces act on an unconstrained body. His reliance on the multiplier technique is in analogy to the procedure he followed in the determination of the minimum or maximum value of a function when constraints are imposed. For example, when the stationary points of a function f(x) are sought subject to the constraint g(x) = 0, Lagrange considered the expression f(x) + λg(x) to be dependent on the two variables x and λ and then set derivatives with respect to these variables equal to zero to determine the appropriate value of x. Similarly, he included the constraint conditions in his equilibrium equation by adding in their differentials multiplied by a multiplicative factor which would later be allowed to vary.

To explore applications of his general equation, Lagrange introduced variational notation by noting the analogy between the mathematical variation in a curve, $\delta y(x)$, and the change in the position of a material body as considered in mechanics.[56] If a single dimensional body is described by a curve such as y(x), for example, then the motion of the body can be thought of as transferring this curve to a new location, a process Lagrange represented by $\delta y(x)$. After commenting on the commutivity relation for the d and δ operations, Lagrange rewrote the general equation of equilibrium by integrating over all the differential mass elements of the body:[57]

$$\int (P\delta p + Q\delta q + R\delta r + \ldots + \lambda \delta L + \mu \delta M + \nu \delta N + \ldots)dm = 0$$

To eliminate integrations with respect to variations, Lagrange relied upon integration by parts. The simplest such transformation, for example, is a replacement of the following type:

$$\int X d\delta x = X\delta x - \int \delta x dX$$

Analogous expressions for higher order differentials quickly become more complicated. It is typical of Lagrange's algebraic procedure that these relationships are manipulated as algorithms without any attempt to attribute physical significance to them. Differential expressions such as dx, dδ, and dδx are multiplied and divided like any other algebraic quantity. In general, Lagrange's procedure is to carry out these integrations by parts, rewrite the results as a sum in δx, δy, and δz, and then set the coefficients of each of these independent variations equal to zero. Later mathematicians would refer to the resulting relations as the Euler-Lagrange equations. Euler had arrived at them in 1744 and Lagrange had given his own derivation in his 1762 papers. For example, if the integrand of the integral to be minimized is $f(x, y, y')$, where y' is dy/dx, and the integration is taken with respect to x, the Euler-Lagrange equation is:

$$\frac{\partial f}{\partial y} - \frac{d\left(\frac{\partial f}{\partial y'}\right)}{dx} = 0$$

Additional terms result when f is a function of higher derivatives of y. When constraint conditions are imposed, elimination of the Lagrange multipliers from a set of simultaneous equations results in a relationship for x, y, and z that represents the state of equilibrium. The integration by parts procedure also generates terms that must be evaluated at the endpoints of the integration. These terms disappear when the endpoints are fixed, but in the more general case they contribute to the full solution of the problem. Lagrange argued that they could be analyzed through a procedure analogous to that carried out on the integral terms from which they are independent.

One of Lagrange's illustrative examples in the *Méchanique analitique* caught Ampère's attention. The problem is to determine the shape of a flexible wire of fixed length when suspended from both ends and acted on by a gravitational force at each point of its homogeneous mass.[58] Since the wire is assumed not to be susceptible to stretching or compression, the major constraint on the system is that variations in the position of the wire do not affect the value of a differential length element ds, where $ds^2 = dx^2 + dy^2 + dz^2$. In Lagrange's notation, this means that $\delta ds = 0$. The general equation of equilibrium then becomes:

$$\int (X\delta x + Y\delta y + Z\delta z + \lambda \delta ds) = 0$$

To determine the function $y(x)$ that satisfies this condition, Lagrange first rewrote δds as:

$$(dx\delta dx + dy\delta dy + dz\delta dz)/ds.$$

71

He then reversed the order of d and δ in the resulting three new constraint integrals and used integration by parts to rewrite these integrals. For example,

$$\int \lambda dx/ds \ d\delta x = \lambda dx/ds \ \delta x - \int \delta x \ d(\lambda dx/ds)$$

where the first of the two new terms must be evaluated at the two endpoints of the integration. Lagrange then incorporated these expressions into the general equation of equilibrium and set the coefficients of δx, δy, and δz equal to zero. After eliminating λ, he arrived at equations for the *chainette* in the form:

$$dy/dx = \frac{B + \int Y dm}{A + \int X dm}$$

It is typical of Lagrange's emphasis on formal generality that he did not introduce the simplifying assumption that the gravitational force is vertical and conforms to the y axis. This assumption would make Y proportional to y while X and Z vanish; the solution then quickly emerges in the form:

$$Y = a\sqrt{1 + (dy/dx)^2}$$

At some point in his study of these derivations, Ampère became dissatisfied and felt that significant revisions were in order. The resulting memoir was rather disjointed and did not demonstrate or illustrate all the generalities he asserted in his introduction. He began by pointing out that in studying the *Méchanique analitique* he initially thought that he had discovered an easy way to uniformly avoid the integration by parts procedure relied upon by Lagrange. Further reflection convinced him that more careful attention had to be focused on the various contexts in which variational methods are employed. In particular, Ampère distinguished between "purely geometric" problems and problems in mechanics where specific additional restraints arise, due to the structure of the material body concerned. In the second section of the memoir he gave a new procedure for analyzing the general equation of equilibrium and then provided a lengthy discussion of the *chainette* example, the determination of the shape of a suspended flexible wire.

Some of Ampère's preliminary comments indicate the state of his mathematical aspirations in 1803. His language stands in sharp contrast to claims by d'Alembert and Lagrange that they intend to rid mathematics of "metaphysical assumptions". According to Ampère,[59]

72

What is called a fact of analysis should always be traced back, if one wants to have a proper idea of it, to the metaphysical principles of this science. It is evident, in fact, that since the use of algebraic characters can add nothing to the ideas that they represent, one ought always to find in the attentive examination of the conditions of each question the reason for all the results one is led to by calculation.

To clarify these metaphysical principles, Ampère drew three distinctions to delineate what he considered to be distinct procedural domains. Unfortunately, he did not stipulate very carefully how these three dichotomies are related. He began by noting that in geometric problems the Lagrange multipliers are held constant while in mechanics they are allowed to vary.[60] Ampère did not specify the type of geometric problem he had in mind although throughout the memoir he cited maxima and minima problems as typical geometric ones. In an effort to uncover the "metaphysical principles" at the root of this difference, Ampère noted that in geometrical problems it is possible to stipulate arbitrary relations or conditions for the variations δx, δy, and δz; on the other hand, mechanics deals with material bodies for which this is not possible. In other words, the constraint equation of the form

$$\delta \int ds = 0$$

has different implications for material bodies than it does for purely geometric curves. In the case of an inextensible wire, for example, no additional relationships can be imposed between the variations of the three components over and beyond the equation of restraint itself. In the realm of abstract geometry, on the other hand, additional relationships can be imposed. In mechanics problems, the coefficients of δx, δy, and δz thus can and must be set equal to zero when they appear in expressions that vanish as a whole; since no additional relationships exist among the variations, a linear combination of them cannot produce a vanishing quantity unless each of their coefficients vanishes. But Ampère then claimed that in mechanics problems:[61]

> the necessity one is under to separately set equal to zero the coefficients of all the variations leads necessarily to as many equations as there are coordinates; and as there are in each problem a determined number of these variables that must remain independent, and that consequently cannot be employed to satisfy these equations, the undetermined factors by which the condition equations have been multiplied present the sole means of supplying them, and it is for this reason that one is obliged, in this case, to consider them variable.

73

It is not clear why Ampère said that independent variables cannot provide the basis for a solution to a mechanics problem; it would seem that they do precisely that. Perhaps what he intended to say was that when an equation of constraint reduces the number of truly independent variables, the Lagrange multiplier for that equation must be allowed to vary, and thus to function as an independent variable when the constraint equation is included in the analysis of the general equation of equilibrium.

As a third way of drawing a distinction between mechanics and geometry, Ampère restated his point by saying that the correct procedure in mechanics attributes variations to all the independent variables rather than only to dependent variables, as in geometric problems.[62] Ampère apparently had in mind cases such as the determination of the maximum surface area produced by a rotation of a curve $y(x)$ with endpoints fixed by the independent variable x. In such a case the variation of x is taken to be zero and the various curves $y(x)$ are produced by the variation $\delta y(x)$. Ampère never clarified how his insistence that δx be included in the analysis of mechanics problems is related to his initial distinction when he said that Lagrange multipliers must be allowed to vary in mechanics problems. Ampère chided Lagrange for considering only problems in which dependent variables are given variations in his 1797 *Théorie des Fonctions analytiques*; he then offered the following overly optimistic projection about his own future contributions.[63]

> The general theory of the calculus of variations thus is not yet established on absolutely rigorous principles; there remains in this regard in the ensemble of mathematical truths a lacuna that the subject of this memoir does not allow to be examined here, but with which I intend to occupy myself at another time. I will then be able to enlarge more upon the difficulties that remain in this last branch of transcendental analysis, and to propose the means that appear to me to be the most appropriate to make them disappear, means of which it seems to me to be impossible to give an idea without going back to the first principles of the differential calculus, and without applying in a manner more simple than has been done up to now the notation of Leibniz to the theory of Lagrange.

We will see that Ampère did publish one paper in 1806 on the foundations of the calculus, but he never linked that work to the calculus of variations. On the other hand, in that same year whatever thunder Ampère might have been saving was abruptly stolen when Lagrange provided an algebraic foundation for the calculus of variations in his *Leçons sur le Calcul des fonctions*. Furthermore, in the *Méchanique analitique*, Lagrange himself took differentials of his multipliers in the process of treating examples such as the suspended wire; Ampère's complaint that Lagrange did not include mechanical applications in his 1797 mathematics text was considerably overblown.

Although Ampère thought he had revealed an essential metaphysical foundation for the differences between the way the calculus of variations is applied to geometry and mechanics, his own presentation of an alternative to Lagrange's reliance on integration by parts did little to illustrate his point. The heart of Ampère's revision falls within the first five pages of his second section.[64] He claimed that by allowing the condition equation coefficients to vary prior to applying the integration by parts procedure, he would in effect eliminate Lagrange's dependence on that process.[65] Ampère adopted a condition of constraint equation of the form $\delta(Ldx) = 0$, an equation that contributes a term of the following form to the general equation of equilibrium: $\int \lambda \delta(Ldx)$.

Ampère expanded dL into a summation of differentials multiplied by the appropriate partial derivatives and then substituted the δ operator for d:

$$\delta L = M\delta x + N\delta y + P\delta p + Q\delta q + \ldots + N'\delta z + P'\delta p' + \ldots$$

where $p = dy/dx$, $q = dp/dx$, etc. This is a procedure identical to that of Lagrange in his 1762 papers except that Ampère follows Euler's procedure and expands his function in terms of differential quotients such as p, where $dy = pdx$, while Lagrange used differentials such as dx, dy, d^2x, etc.[66] Second, Ampère set $\delta y = p\delta x + \omega$ and $\delta z = p'\delta x + \omega'$ and inserted these substitutions into the expansion of δL. He then rewrote the constraint integral as:

$$\int \lambda \delta(Ldx) = \int \lambda(Ld\delta x + \delta Ldx)$$

and used integration by parts, ironically, to show that:

$$\int \lambda Ld\delta x + \lambda \int dLd\delta x = \lambda L\delta x - \int Ld\lambda \delta x$$

He then could rewrite the constraint integral as:

$$\int \lambda \delta(Ldx) = \lambda L\delta x - \int Ld\lambda \delta x + \int N\lambda \omega dx + \int P\lambda d\omega + \ldots$$

Ampère noted that the integration with respect to λ is what distinguishes his approach, an approach that should reduce to equations appropriate for geometrical problems when λ is taken to be constant. Furthermore, by setting the coefficients of δx, δy, and δz in the general equilibrium equation equal to zero, his procedure results in equations equivalent to Lagrange's.

As an example, Ampère considered the suspended wire where the equation of restraint is δ(Ldx) = 0 with a length element given by Ldx = ds. After considerable translation of notation he did indeed arrive at Lagrange's conclusion that the wire takes the shape of the *chainette*. Ampère admitted that in this particular case Lagrange's integrations for each component are so simple that there is no great advantage to be had by changing to the new procedure. Ironically, he also remarked that "the limits of this memoir" did not allow him to address more complicated cases where the superiority of his methods would become apparent. Instead, he concluded by deriving various properties of the *chainette* function: the proportionality of subsumed area to arc length, the proportionality of wire tension to ordinate value, and the minimum or maximum properties of surface area after rotation about an axis.

The somewhat mundane details of these final derivations lend emphasis to the fact that his early mathematics papers served as a rite of passage for Ampère. Lacking the credentials that might have been provided by formal education, he needed to qualify as someone familiar with recent developments in mathematics. Under scrutiny as a possible professor of mathematics, he had to demonstrate that he could present ideas convincingly in an expository mode. He succeeded in this task. His overly optimistic assessment of the originality of his contributions could be pardoned as the posturing of an isolated provincial. What was important was the fact that he appeared to be someone who could be entrusted with the daily chores of mathematical education.

This is not to imply that the contents of his paper went unnoticed. In 1806, when Lagrange published a revised edition of his *Leçons sur le Calcul des fonctions*, he added two chapters in which he addressed the calculus of variations anew.[67] Although he did not cite Ampère as the motivation for his remarks, Lagrange did devote considerable attention to the procedural habit of not considering variations in the independent variable x.[68] Lagrange gave two derivations of how the consideration of variation in x adds nothing to the general equations produced when that variation is ignored; only the results of endpoint considerations are affected. Lagrange did not belabor the point; he was more interested in placing the general equation for the determination of maxima and minima on a basis of Taylor's series expansions and in pointing out the relationship between the solution to iso-perimetric problems and the specification of the conditions under which a function is integrable. He dealt with the issue raised by Ampère in passing and without embarrassing him.

Second, in 1811, when Lagrange published the first volume of a second edition of the *Méchanique analitique*, with an altered title of *Mécanique analytique*, he included a new section addressed to the demarcation issues Ampère had raised.[69] In sharp contrast to the position taken by Ampère, Lagrange argued that the proper way to classify mechanics problems was simply as a type of minima problem typical of the calculus of variations. He began by reviewing the procedure for determining the function that maximizes or minimizes an integral. For example, when the integrand U is a function of x, y and derivatives of y, with the differential dx assumed to be constant, Lagrange expanded δU as a sum of differentials and collected the resulting factors of δx, δy, and δz. As usual, the minimizing relation is provided by the Euler-Lagrange equations derived by setting these factors equal to zero. After noting that the δx equation adds nothing to what is provided by the other two variables, Lagrange concluded that there was no need to consider δx, but that:[70]

> nevertheless, it may be useful to consider all the variations simultane-ously, with respect to the limits of the integral, because there may result from each of them particular conditions for the points that correspond to those limits, as we have shown in the last lesson of the *Calcul des fonctions*.

This passage, written just a few years before Lagrange's death in 1813, harkens back to foundational issues Lagrange had debated over a half century earlier. In their approximations of curves by polygons, Leibniz and Euler had treated a differential such as dx as a fixed element or "constant" determined by the projections of sides of these polygons on the x-axis. Other choices were possible, such as considering dy or the arc element ds as constant. The early algorithms for the calculus were devised in such a way as to be independent of these choices.[71] In 1756 Lagrange had explained to Euler that in variational problems with variable endpoints the consideration of both δx and δy was useful.[72]

Turning to the corresponding problems in mechanics, Lagrange wrote the summation of forces multiplied by virtual velocities as the varia-tion of a function π:

$$P\delta p + Q\delta q + R\delta r + \ldots = \delta\pi$$

He then carried out an analysis of the restraint $\delta L = \delta dm = 0$ using the multiplier λ just as Ampère had done in his 1806 paper. Once again, three equations result from setting the factors of δx, δy, and δz equal to zero, but now expressed using the new function π. The fact that three equations result seems to indicate a difference from the corresponding geometric problems. But Lagrange's conclusion was quite different from Ampère's:[73]

> But I observe, first of all, that because of the indeterminate λ, the three equations reduce to two through the elimination of this indeterminate; and although in general the equations of condition always replace those which disappear through the elimination of indeterminates, the condition introduced here, $\delta dm = 0$, that is, dm constant, cannot furnish a particular equation for the solution of the problem, because, according to the spirit of the differential calculus, it is always permitted to take an arbitrary element as constant, since, strictly speaking, only the ratios of differentials to each other, and not the differentials themselves, enter into the calculations. Thus, the three equations will be reduced to two, and will only serve to determine the nature of the curve, as in the problems *de maximis et minimis*.

Lagrange's 1811 comments in effect acted as a rebuttal to Ampère's claim that the constraints imposed in mechanical problems were different in type from those of pure geometry. For Lagrange, the constraint $\delta dm = 0$ generated no more serious issues than had Ampère's worries about the variation δx.

Ampère apparently accepted Lagrange's last publications as sufficient to fill what he had called the serious "lacuna" in the foundations of the calculus of variations. As we will see, this may have been due in part to the fact that after 1804 Ampère lost interest in creative contributions to the subject and submitted to the monotony of teaching it at the École Polytechnique. He eventually published a set of lecture notes for educational purposes in 1825; they indicate that, in spite of his 1806 criticisms, Ampère found the general procedure of Lagrange's 1762 presentation the most useful for teaching purposes.[74] In his introductory remarks Ampère did mention in passing that in mechanics problems, where the entire body is in motion, the endpoints also change their position so that δx must be considered.[75] He made only brief mention of Taylor's series techniques and relied upon Euler's 1744 notation to set up the variation problem $\delta \int V dx = 0$ where $V(x, y, p, q, \dots)$ is a function of x, y, and $p = dy/dx$, $q = dp/dx$ and higher derivatives. In contrast to Lagrange, who wrote V as a function of differentials, Ampère maintained his earlier preference for Euler's differential coefficients p and q. Ampère expanded V into the form

$$\delta V = M\delta x + N\delta y + P\delta p + Q\delta q + \ldots$$

After carrying out Lagrange's usual integration by parts, he set the result equal to zero to derive the Euler-Lagrange equation in the form[76]

$$N - dP/dx + d^2Q/d^2x - d^3R/d^3x + \ldots = 0$$

Ampère also discussed the endpoint relations in a manner that drew praise in a footnote to Ampère's paper by the editor of the *Annales de Mathématiques pures et appliquées*, J. D. Gergonne. According to Gergonne, Ampère was the first to give a clear justification for simultaneously setting equal to zero both the integral terms and the non-integral terms that result from the integration by parts stage in the analysis.[77] It was not until the second half of the nineteenth century that mathematicians justified this step through a proof and application of what has since come to be known as the fundamental lemma of the calculus of variations.[78] In modern terminology, this lemma states that for any function $g(x)$ with continuous second derivative on the interval $[a,b]$, where $g(a) = g(b) = 0$, where $f(x)$ is continuous, and where the following relation holds,

$$\int_a^b f(x)\, g(x)\, dx = 0$$

then $f(x) = 0$ for all x in the interval $[a,b]$.

Ampère did not attempt a proof of this lemma and there is no indication that he differed from his contemporaries by feeling a need for it. Nevertheless, aside from the step that requires the lemma, Ampère's derivation resembles that followed by subsequent mathematicians, that is, if the Euler-Lagrange equation is accepted for invariant endpoints, a step that would later be justified by the fundamental lemma, then the general endpoint relations follow by recognizing that the Euler-Lagrange equation will continue to hold for variable endpoints.

Although we will see that Ampère was not enchanted by his teaching responsibilities in mathematics, we should not conclude that he did not stay abreast of his subject. Although he did not live up to his 1803 hopes that he would make creative contributions to the development of the calculus of variations, his competence in that field was well established.

Delambre and Ampère's Appointment to the Lycée de Lyon

Ampère's presentation of his calculus of variations memoir to the *Société d'Émulation de l'Ain* on 12 March, 1803, was followed the next day by good news. Delambre informed him that he would be recommended for the mathematics position at the Lyon Lycée. Delambre was confident that his choice would not be overruled; "the government has never yet changed anything in all that I have done, and surely it will not begin to do so with you."[79]

Delambre proved to be Ampère's most influential patron. Born in 1749 in Amiens, Delambre had the typical generation of seniority over his protégé. With very little financial support, he had managed to acquire a Parisian education in literature and languages before becoming interested in astronomy. Lalande, Ampère's other examiner at Bourg, had adopted Delambre as his own assistant and protégé during the 1780s. Delambre became well known for both his observational and computational skills, particularly his analysis of the motion of Jupiter, Saturn, and the moons of Jupiter. In 1792 he was elected to the Académie des Sciences as a *membre associé* in the *sciences mathématiques* section. Following the Revolution, he achieved considerable notoriety by participating in the creation of the metric system and the accompanying new trigonometric and logarithmic tables. He was entering into the prime years of his career when he met Ampère in 1803.

Most important, in January of 1803, Delambre was elected the first permanent secretary for the mathematical sciences of Napoleon's Institut. Postponing a full discussion of Parisian scientific institutions to chapter 3, it suffices at present to realize that as a *secretaire perpétuel* of the Institut, Delambre was one of the most powerful scientists in France. He was not exaggerating when he said that his approval of Ampère's appointment was tantamount to a signed contract. Although it is not entirely clear why Delambre was so favorably impressed by Ampère, he surely recognized Ampère's mathematical ability, ability Delambre was quite capable of assessing accurately. In this respect his judgment represented the first thorough test of competence Ampère had undergone since Laplace's critique of his probability memoir. Upon Delambre's return to Paris, Ampère's fawning letters were accepted as appropriate to their relationship. He had acquired the patron who would bring him to Paris the following year.

Meanwhile, one additional mathematics memoir was completed as a prelude. In his calculus of variations memoir Ampère had worried in passing about defects in the foundations of the calculus and had made vague allusions to his plans to fill this "lacuna". But in the essay he began immediately thereafter these foundational issues were not addressed. On the contrary, Ampère took the opportunity to revel in the power of the calculus to tackle problems in analytic geometry. Leaving aside the precise meaning and definition of the "derived function," Ampère was content to apply the accepted rules of differentiation in extensive calculations, work that kept his intellect partially occupied during the last few months before Julie's painful death in July of 1803. The memoir was completed by the following fall. Delambre presented it to the Institut on 28 November, 1803, and Ampère left a copy with the Bourg Société d'Émulation de l'Ain one month later. It was eventually published in 1808.[80]

Ampère stated as his general goal the investigation of the most appropriate way to express what is "truly characteristic" of a two-dimensional curve. He was dissatisfied with representations that include terms dependent on the arbitrary orientation of the coordinate system axes. His own preference called for a rather complicated analysis of any given point on the curve. He felt that the curve could best be characterized by specifying, for any point on the curve, the axes and parameter of the osculatory parabola at that point. The study of osculatory curves was a common topic in early nineteenth-century analytic geometry; in general, given any curve, an osculatory curve is one that intersects the given curve at a single point. Ampère pointed out that the circle is not sufficiently complicated to reveal the "essential characteristics" of some curves. For example, curves involving "contact of the third order" require equality with the osculatory curve for the first, second, and third derivatives, as well as equality between the values of the curves themselves. This requirement results in four equations to be satisfied at any given point; these equations can be used to stipulate the position, orientation, and dimensional parameter of the osculatory curve at that point. Since only three equations are required to stipulate the position and radius of an osculatory circle, the fourth equation would, in effect, provide no further information about the initial curve. On the other hand, an osculatory parabola is not symmetric with respect to rotation; thus the fourth equation would be required to fully determine the parabola's orientation.

With this insight in mind, Ampère calculated general formulas that would stipulate the position, orientation, and scaling parameter of the osculatory parabola that makes "contact of the third order" at any point of a given curve. After some discussion of the ellipse and cycloid as examples, Ampère discussed other properties of the osculating

81

parabola and concluded by showing how to determine whether the osculatory parabola will fall "inside" or "outside" a given curve.

None of this was ground-breaking innovation. Nonetheless, Ampère thoroughly demonstrated his mastery of the calculational techniques of the calculus, a mastery that was expected from anyone planning an academic career. We should also note his recurring interest in classification schemes, a topic that would be central to his subsequent work in both mathematics and chemistry. If we compare Ampère's early mathematics publications to his aborted treatise on electricity and magnetism, the contrast is striking. Although he was aware of Coulomb's work, Ampère had not adjusted to the new standards of mathematical physics. No scientific journal would have accepted his grandiose speculations for publication. Nevertheless, his early aversion to action-at-a-distance forces would become a permanent fixture of his methodological expectations for physical explanation. On the other hand, in mathematics Ampère demonstrated that he was capable of original contributions. His work drew the attention of Laplace, Lalande, and Delambre, men with high standards. Lalande and Delambre observed him in the classroom and lecture hall and were favorably impressed.

Had not Julie's death intervened, Ampère's appointment to the Lyon Lyceé would have initiated a period of long delayed happiness. His subsequent depression contributed to his decision to take the earliest opportunity to leave Lyon for new surroundings in Paris. Later he would regret this decision. The Lyon friends who attempted to fill the emotional void left by Julie's death were missed painfully. Although Ampère gradually adjusted to the priority disputes and infighting of the Parisian scientific community, he always longed for a return to the intellectual life he experienced in Lyon.

Lyon Mystical and Medical Traditions

For a man in Ampère's distraught condition, Lyon's intricate labyrinth of mystical and esoteric societies offered a complex spiritual environment to explore. Mesmerism, or the study of "animal magnetism," was just one of many cult practices that had flourished during the 1780s. Although spiritualist networks were disrupted by the Revolution, devotees continued to practice in Lyon during the years Ampère spent there; he could have taken part in them if he had chosen to do so. Some of these organizations deserve a brief discussion; although they failed to capture Ampère's serious attention, they did form part of the context in which his own spiritual life developed.

During the 1780s, two important centers for spiritualist activity in Lyon were the Lyon mesmerist society, La Concorde, and the Loge Élue et Chérie, a spiritualist society founded by Jean-Baptiste Willermoz (1730–1824). The interactions and numerous joint memberships between these and other groups were the result of both doctrinal flexibility and shifting interests and loyalties. La Concorde attracted a wide variety of unorthodox individuals. While it included Rosicrucians, Swedenborgians, and theosophists, the majority of whom were intent on communication with the spirit world, other members were more interested in innovative medical treatments and the experimental study of mental faculties.

La Concorde drew its general orientation from the practices introduced by Franz-Anton Mesmer, author of *Mémoire sur la Découverte du magnétisme animal* in 1779. Headquartered in Paris, Mesmer claimed that by applying appropriate massaging and stroking procedures to the human body's "polarity," a skillful practitioner could direct a subtle all-pervasive fluid in such a manner as to induce a state of crisis that would beneficially affect human nerves and thus rectify nervous disorders. He considered the similarity between the human body's responses and those of a magnet to be sufficient to justify the label "animal magnetism" for these phenomena. Mesmer soon became a notorious public figure and he elicited reactions ranging over the full gamut from adulation to scorn. The palpitations and convulsions of swooning female patients became a favorite subject of satire among the skeptics.

Mesmer's rhetoric could sound impressively technical to people without scientific experience. For example, one of the 27 propositions of his theory seemed to promise the synthesis so much in demand among physicists and chemists of Lavoisier's generation:[81] "This system will furnish new enlightenment concerning the nature of fire and light, as well as in the theory of attraction, flux and reflux, the magnet and electricity."

Mesmer's language in this excerpt closely resembles that of Ampère and numerous other theorists at the close of the century, but the similarity ends when the empirical basis of such claims is considered. Mesmer's fortunes suffered a severe blow in 1784 when a royal commission that included Lavoisier and other members of the Académie des Sciences observed a set of alleged mesmeric phenomena and found no empirical evidence for the existence of Mesmer's fluid. Nor did they find any connection between mesmeric procedures and the ecstatic behavior of the patients.[82] Ironically, mesmerist activity in Lyon accelerated just at the time of Mesmer's fall from Parisian esteem. In Lyon, the mesmerist society La Concorde relied upon the sensitivity of the mesmerizer to diagnose disease while the patient was in a mesmerized state. There was no physical contact with the patient and no reliance upon an alleged movement of a magnetic fluid. During the summer of 1784 La Concorde carried out what it considered to be successful treatments of mesmerized horses at the École Vétérinaire, directed by Louis Bredin. Following private administration of mesmerist techniques, a public operation and a confirmatory autopsy were performed.[83] Skeptics were not convinced and argued that there was no need to have recourse to secret procedures and subtle fluids to predict that an old horse was suffering from intestinal worms.

In Paris the mesmerist Puységur alleged that the beneficial effects of the mesmerizing procedure included clairvoyance when the crisis state became one of sleep. Mesmeric sleep became a primary research topic in Lyon several months later. Rather than depending upon a material fluid, the mystic state was said to be due to a divine influx that temporarily returned the individual to the glorious state enjoyed by Adam.[84]

The Masonic, mesmerist, and spiritualist networks were seriously disrupted by the Revolution. Willermoz was accused of royalism during the seige of Lyon and was forced to go into hiding. He returned after the Terror and continued to function within the less structured post-revolutionary spiritualist community in Lyon. He still initiated seekers into the mysteries of his discipline, but they lacked the enthusiasm of days past. There is no indication that any of Ampère's circle sought him out.

Two characteristics particularly set Ampère apart from spiritualist and mesmerist groups. In contrast to the Masonic emphasis on ritual, he was not attracted to secrecy, initiation rites, and the notion of esoteric knowledge revealed to a chosen few. He preferred to debate publicly accessible texts; he was quite conservative in the texts he considered of importance for revealed theology, limiting himself primarily to Biblical sources. Nor did he feel drawn to the practices engaged in by the Lyon mesmerists and assorted mystics searching for direct illumination or contact with a spiritual world.

It would not have been difficult for Ampère to explore those options. Julie's brother-in-law, Jean-Marie Périsse Dulac, was a close member of Willermoz's circle and remained so until his death in 1800.[85] Ampère's close friend, Claude-Julien Bredin, was the son of Louis Bredin, the director of the veterinary school who had supervised and notarized the 1784 mesmerist experiments with animals. With these connections, it might be expected that Ampère and Julie would have recourse to mesmerist techniques at some point during her long illness. However, there is no evidence that they did so. On the contrary, the physician they put most faith in was the celebrated Dr Jacques Henri Désiré Petetin (1744–1808), a prolific author who argued that mesmeric phenomena were simply due to one of many possible types of catalepsy he could explain on the basis of a purely materialistic electrical theory.

According to Madame Cheuvreux, Petetin had been Julie's physician since her childhood.[86] If this is true, Petetin must have made himself available to the Carron family shortly after his arrival in Lyon in 1774. By the time of Julie's illness, he had become a successful and prominent figure in the Lyon medical establishment. Among other positions, he was *secrétaire perpétuel* of the Société de Médecine de Lyon and co-editor of the *Journal des Maladies régnantes* with two colleagues, Gilibert and Vitet. Ampère made a concerted effort to engage Petetin in Julie's case during July of 1801. At that point Petetin was nearly 60 years of age and in poor health; he was not willing to see Julie unless she came to Lyon. A consultation did not take place then as she was unwilling to travel. She did consult him during March of 1802 after her move to Lyon. By April he planned to begin treatment, but in May he still had not begun, much to Ampère's impatience. In June he prescribed herbal baths, pondered the possible use of leeches, and then severely restricted her diet, allowing plenty of ice and cool drinks. Julie found it all boring, ineffective, and unnecessarily expensive. By July Petetin admitted that he had misconstrued Julie's condition; he now conformed to the consensus diagnosis of a *dépôt de lait*. Julie was disgusted and by August she was hoping he would simply disappear. By November she calculated that he had made over 65 visits; fortunately he did not charge his usual full rates. He was in attendance during the last six days of her life in Lyon.

Inept as Petetin was in Julie's case, he and Ampère may nevertheless have had some interesting conversations about mesmerism, electricity, and related topics. Petetin had achieved considerable renown as a specialist in catalepsy and related nervous ailments involving paralysis and convulsions. In 1787 he felt compelled to publish a long *mémoire* in which he summarized the pertinent phenomena, presented an explanatory theory, and proposed treatment procedures.[87] The structure and content of Petetin's *mémoire* are worth noting as part of the context for Ampère's early unsuccessful attempt to organize a treatise on electricity in 1801. The reduction of allegedly magnetic phenomena to the action of electric fluid is a particularly significant theme.

We have seen that in the voluminous notes he compiled during 1801, although Ampère made numerous references to "experiments," he never clarified how specific principles of his complex theory were supported by his observations. Nor did these experiments require quantitative measurements. Similarly, Petetin cited twenty *expériences* in support of his assessment of the primary phenomena associated with *somnambulisme hystérique*. According to Petetin, the cataleptic state is accompanied by a transfer of the location of all sensory input to the stomach. His patients allegedly could only see, hear, and taste when objects were brought near the stomach region. Petetin mentions that he only discovered these phenomena by chance in the process of treating a patient.[88] To bring these remarkable events "within the natural order,"[89] Petetin adopted a set of 17 "suppositions" related to his primary one that in cases of hysterical somnabulism the patient's stomach fills with the electric fluid that ordinarily flows from the brain to the sensory organs. As the electric fluid radiation from external objects collides with the radiations from the excessively electrified stomach, the stomach is transformed into an extremely delicate sensory apparatus, permitting sight through ordinarily opaque objects and in the dark.

In support of these suppositions, Petetin offered four new *expériences*. For example, a chain of people linking hands are reported to be able to transmit a message from the hand of the speaker to the stomach of the patient; if a substance that does not conduct electricity is introduced into the chain, the message is not received. Petetin concluded incidentally that the usual view that air is the medium for sound transmission must be rejected in favor of an all-pervasive electric fluid medium. Furthermore, Petetin interpreted mesmeric procedures as processes that simply stimulate the initial contraction of the brain on the electric fluid, processes that do not involve any magnetic phenomena.[90] In the second part of his *mémoire*, Petetin lists 20 new principles concerning the electric fluid motions that generate sensory input. By way of remedial treatment, Petetin recommends icy baths, applications of leeches, cold drinks, and ample consumption of milk as a reliable regime to return the electric fluid to its normal distribution.

As he reported in a subsequent 1808 publication, Petetin was undismayed by the poor reception his first book received. He continued to investigate catalepsy and he made a striking discovery during the seige of Lyon in 1793. Confronted by a particularly difficult case, he achieved success when he administered electric sparks and shocks.[91] By this point he had also discovered that his patients often had extreme sensory sensitivity in the tips of their fingers and toes as well in the stomach. He cited as evidence experiments with chains of people similar to those he had used for transmissions to the stomach. In response to allegations from followers of Mesmer, Petetin was adamant in insisting that he was not "magnetizing" his patients; his experimental demonstrations, diagnoses, and treatments were all interpreted in purely electrical terms.[92]

Inspired by his own investigations and the announcement of Volta's discovery of the pile in 1800, Petetin became intrigued by the properties of electric fluid to such an extent that he began publishing on the subject during 1802 and 1803. These compositions may well have delayed his acceptance of Julie's case. By 1803 he had quite a reputation for his ice bath and electric shock treatments of severe paralysis and convulsions. The fact that Julie's ailment was vaguely located in her stomach may be no more than a bizarre coincidence; Petetin's prescribed regime had little effect and apparently he did not think electric shocks were appropriate. Unfortunately, Ampère left no records of any discussions he may have had with Petetin. The most detailed diagnosis of Julie's "tumour" was written by Petetin's colleague, Vitet, an outspoken critic of mesmerism.

One conclusion that can be drawn from all of this is that Ampère may very well have found some scientific interest in Petetin's materialistic electric-fluid analysis of mesmerism, somnambulism, and induced catalepsy. He surely noted that electrical phenomena were thought to play an important role in biological processes. On the other hand, he was not inclined to explore mesmerist techniques as a mode of spiritualist encounter. Nor did Ampère record any attempts to communicate with Julie after her death. More tentatively, we might suspect that Ampère took serious note of how Petetin interpreted the phenomena attributed to the alleged "magnetic" fluid of the mesmerists as processes to be understood entirely in terms of the motions of electric-fluid particles. Ampère carried out an analogous reduction of magnetic phenomena to electric currents during the 1820s. He never cited Petetin as an inspiration, but unacknowledged memories may have lingered.

It is not known how carefully Ampère studied Petetin's publications. We have already noted the unfortunate multiplication of suppositions, principles, and "experiments" that thoroughly mingled the stipulation of the relevant phenomena with the presentation of an explanatory theory. Ampère's 1801 treatise was equally disorganized. Although he gradually learned to be more methodical, he would always have difficulty expressing his scientific ideas in writing.

Meanwhile, unmoved by the insights offered by mesmerist and mystical societies, Ampère sought autonomy within a small group with a different agenda. The nucleus was formed by Ampère and his two closest friends, Bredin and Ballanche. Brief sketches of Ampère's associates are appropriate at this point; these men remained friends for life and would be Ampère's regular correspondents during his subsequent years in Paris.

Ampère's Lyon Friendships

Claude-Julien Bredin was born on 25 April, 1776, in the village of Alfort. His father, Louis Bredin, moved to Lyon four years later and became well known as the director of the Lyon École de Médecine Vétérinaire. Young Bredin was a brilliant but extremely melancholy child. In later life he recalled an incident at about age four or five in which he was so overcome by the sad nature of life that his attendants could not comfort him. He devoured romantic and spiritual literature but also excelled in technical subjects, achieving a veterinary diploma at age 15. Far more active than Ampère and Ballanche at that age, he managed to join the French cavalry in the Alps. At home on leave during the Lyon insurrection, he rejected his father's objections and joined the rebels in the unsuccessful defense of the city. Following the suppression, he managed to win an appeal against execution and was returned to his veterinary position in the Alps.

He returned to Lyon in August of 1795 and accepted a position teaching anatomy at the veterinary school directed by his father. Joseph Buche cites an 1830 letter from Bredin to Ampère in which Bredin recalls that it was in 1795 that he met Ampère in Lyon. It is more probable that they did not meet until 1803 when Ampère moved to Lyon from Bourg. At that point they became very close and remained so throughout their lives. Shortly thereafter Bredin became infatuated with Mela, a young orange peddler, and he married her amid considerable dismay on the part of his bourgeois family. Mela soon displayed a violent and unstable temperament. The couple shared living quarters at the École with Bredin's parents and this resulted in notorious quarrels. Bredin succeeded his father as director of the École Vétérinaire in 1813 and held this position until his retirement in 1835. He and Mela parented nine children who amplified the domestic discord. By 1816 Mela was declaring herself to be illumined by divine inspirations; in 1824 the couple separated amid consternation that their public altercations were damaging the reputation of the École.

Through all this turmoil Bredin preserved a patience that prompted Ballanche to refer to him as the Job of Lyon. He shared with Ballanche the attitude that suffering was a sign of divine attention and served as a means of initiation into eventual salvation. His Christianity was eclectic and drew heavily from the mystical traditions that flourished in Lyon.

Pierre-Simon Ballanche had childhood experiences strikingly similar to those of Ampère. Born in Lyon on 4 August, 1776, Ballanche was just over a year younger than Ampère. He was not a healthy child and his parents allowed him to spend about eight months of each year at their country estate in Grigny, on the Rhône river south of Lyon. His experiences at Grigny resembled Ampère's idyllic years at Poleymieux. Both men were enchanted by the beauty of their surroundings. Direct exploration of nature and ready access to his father's library constituted the bulk of Ballanche's education. Like Ampère, he was drawn to classical poetry, drama, and Rousseau. Furthermore, just as Poleymieux served as Ampère's sanctuary during the violence of the Lyon insurrection, Ballanche was sheltered at Grigny. Although removed from direct exposure to the brutality Lyon experienced, both men were deeply affected. Ballanche's father, Hugues-Jean Ballanche, was an active associate of the Delaroche family, founders of the printing house Halles de la Grenette. Following the suppression of the revolt, he and a friend, Charles-François Millanois, purchased the Delaroche share of the house and planned to continue the business under Ballanche's leadership. Both men were accused of royalist loyalties; Millanois was executed and Hugues-Jean Ballanche was only spared by the testimony of his workers concerning the compassionate relationship he had maintained with them. The printing business was temporarily confiscated by the government but was returned to Ballanche in 1795.

Upon his return to Lyon from Grigny, Pierre-Simon Ballanche was shocked by the destruction the city had undergone. His health deteriorated and he was subjected to a trepanning operation for caries of the jaw; the operation was crude and the left side of his face was permanently scarred. He withdrew into almost total seclusion for three years and became so lost in reading and introspection as to experience visions. His physical deformity amplified his sensitivity and his character became thoroughly retiring and diffident.

His literary interests gradually became known and his sphere of friendships grew to include Camille Jordan, Lenoir, and Dugas-

Montbel, the famous translator of Homer. By 1801 he had thought at length about the relationships linking religion, history, and artistic expression. With the support of his literary circle he felt confident enough to have his father publish his *Du Sentiment considéré dans ses rapports avec la littérature et les arts*. The book was not well received and Ballanche was loath to publish for many years thereafter. In 1802 he became a partner in his father's printing house and ironically became the long-suffering publisher of René Chateaubriand, the author of *Génie du Christianisme*, the widely acclaimed 1801 manifesto that bore striking thematic similarities to Ballanche's *Du Sentiment*.

After several morose years of reading, introspection, and sporadic writing, Ballanche met Madame Récamier during her stay in Lyon between June 1812 and January 1813. He became utterly devoted to her and followed her to Paris in 1817. In 1820 he introduced her to Ampère's son Jean-Jacques, who promptly became part of her retinue. Ballanche paid regular homage at her salon and fell naturally into a subservient role in the shadow of more glamorous rivals such as Chateaubriand. Madame Récamier was a major inspiration for Ballanche's literary career; his output included *Antigone, Le Vieillard et le Jeune Homme, Essais de Palingénésie sociale, La Vision d'Hébal*, and *La Ville des Expiations*. These works were all duly noted but none of them became widely acclaimed. As a whole, they were recognized as support for the Catholic revival of the Bourbon Restoration. Ballanche repeatedly explored the nuances of Christian images and doctrines, particularly those addressing the origin and purpose of evil. His writing was to a large extent a continued effort to become reconciled to the injustice and suffering he had experienced in his youth. It is not clear when Ballanche and Ampère first became acquainted. According to Madame Cheuvreux, Ballanche was present at Ampère's marriage ceremony at Poleymieux and composed a celebratory hymn.

In addition to Bredin and Ballanche, two other friends had a lasting impact on Ampère's personal life. Jaques Roux-Bordier was born in Geneva on 21 August, 1773. His father, Etienne Roux, was a botanist, and Jaques Roux developed an intense interest in that subject at an early age. His parents made sure that he spent considerable time in the Swiss countryside; like Ampère and Ballanche, Roux experienced a deep instinctive response to the beauty of complex natural relationships, particularly among plants. His family was wealthy enough to allow him the leisure to pursue his interest in botany without a professional position. By 1791 and 1792 he was taking increasingly far-ranging botanizing trips outside Geneva. He never published professional articles and thus had minimal impact on his discipline.

Nevertheless, he systematically compiled extensive notes and amassed an impressive collection of plants. His expeditions continued through 1814 and included strenuous mountaineering. He slackened his pace thereafter and his moods of depression increased. His suicide in 1822 was not entirely unexpected.

Roux became part of Ampère's circle in Lyon during one of his prolonged residences there. Aside from his status as a botanist and his general interest in literature, he possessed a rare knowledge of German philosophy, particularly Kant, whom he could read in the original German. He became a sharp critic of Ampère's efforts as a philosopher.

Finally, we should note that another of Ampère's Lyon friendships was with Joseph-Marie Degérando (1772–1842). Born in Lyon in 1772, Degérando shared the fervent concern for religious issues that was characteristic of Ampère's circle. He came close to death at the age of 16 and vowed to dedicate his life to good works if he survived. He entered the Saint Irenée seminary but was forced to leave when it was suppressed during the Revolution. By 1793 he was 21, and it was during the seige of Lyon that he became a comrade and permanent friend of Camille Jordan. As was the case with so many of Ampère's close associates, Degérando and Jordan fought to defend Lyon from the government troops of the Convention. Degérando was wounded, but he survived to eventually join Jordan in exile in Switzerland. After three years abroad, he returned to Lyon in 1796 and joined Jordan in Paris the following year. Jordan was a prominent member of the Council of Five Hundred and the *coup d'état* of 1797 sent the two friends into exile once again, this time to Germany. Degérando married, returned to France in 1798, and enlisted in the army.

In spite of his chaotic living conditions, Degérando managed to preserve an active intellectual life. In 1799 he won the Institut prize competition with an essay on the influence of signs on the formation of ideas. His effort caught the attention of literary circles in Paris, and Degérando was released from the army to assume a far more convenient position with the Bureau Consultatif des Arts et Manufactures. He published a revised version of his essay in 1800 and in 1804 published his *Histoire comparée des Systèmes de Philosophie*,[93] an expansion of another prize-winning essay for the Berlin Academy. Ampère relied heavily on this three-volume history for his introduction to philosophical debate. Its publication coincided with Ampère's move to Paris in 1804, a time when metaphysics had become his primary passion.

Degérando also placed his influence at the service of Ampère's career. In January of 1803 he wrote a glowing letter supporting Ampère for the mathematics position at the Lyon Lycée.[94] Aside from the fact that the letter gave Ampère's first name as "Antoine," he could hardly object to being cited as "un véritable phénomène sur l'horizon de la science." Subsequently Degérando continued to smooth Ampère's career path in Paris; unfortunately, he also introduced Ampère to the woman with whom he formed a disastrous second marriage.

The Société Chrétienne

The eclectic nature of Ampère's circle provided an environment he sorely missed after leaving Lyon. Although their wide-ranging interests included science, philosophy, literature, art, and history, they focused on Christianity during the religious revival of the early Napoleonic years. They did so in a manner that prompted Ballanche's biographer, Gaston Frainnet, to refer to a specific "Lyonnais character."[95]

> To go to the heart of the matter, the dominant character of these men was above all a profound spirituality, that is, a type of dogmatism which held them all closely attached to their religion despite doubts that sometimes tortured their intellect. In addition to this there was an inalienable foundation of honesty and rectitude, and a liberal royalism, a friend of progress and an enemy of violence.

Although short-lived, the most vigorous collective expression of this introspective character was the Société Chrétienne which met for the first time on 24 February, 1804. Our main source of information about the Société is Claude Alphonse Valson's biography of Ampère, originally published in 1886.[96] Valson made extensive use of the minutes of Société meetings and a long manuscript by Ampère, documents that were stored in the archives of the Lyon École de Médecine Vétérinaire until destroyed by a fire in 1884.[97] Valson lists seven original members of the Société: Ampère, Bredin, Ballanche, Chatelain, Deroche, Grognier, and Barret; ten associate members were soon admitted. Ampère acted as president and Bredin was secretary. The group adopted a rather scholastic procedure. Each of the original members was assigned a proposition or question to investigate; then he was to develop a statement of his own position with respect to the issue at stake and defend this position against criticism at a meeting. For example, Bredin took as his problem "the importance of knowledge of the destination of Man."

Ampère's task was to present an "exposé of the historic proofs of revelation." As Ampère interpreted it, this meant that he was to marshal the available evidence showing that Christianity was divinely inspired. Valson's biography includes a transcription of the manuscript Ampère produced. Unfortunately, it includes only a few of Ampère's numerous quotations. The document is divided into three sections: proofs drawn from the Old Testament, proofs drawn from enemies of Christianity, and proofs drawn from the writing of Christians. Ampère considered "proofs" found in the writings of unbelievers or enemies of the Church to be particularly telling because of the negative bias of the writers. He wondered why pagan authors reported the virtuous behavior of the early Christians if they in fact did not behave in that manner. In response to doubts put forward by one of the Société's more sceptical members, Ampère retorted: "And yet what is this morality in comparison with that of the God-Man? Such is the question that M. C. should examine and which should put the seal on the proofs of the divinity of Christianity."[98]

Ampère's arguments from Christian authors drew heavily upon accounts of miracles attributed to Jesus and his disciples, proofs he classified as "factual proofs" (*preuves de fait*), that is, those that can be appreciated by all types of people, including those who are not intellectuals.[99] Ampère held that the surviving texts would not agree on the essentials of these miraculous events if they were fabricated. In this respect Ampère's discussion dimly foreshadows that of David Strauss several decades later. In his 1835 *Life of Jesus* Strauss carried out a lengthy analysis of the miracle question and drew quite different conclusions.

Perhaps the argument that best expressed Ampère's own perspective on Christianity is the one he associated with Pascal. The problem of evil and the origins of the human proclivity to err or sin were issues Ampère took very seriously. His inconstant allegiance to Christianity was to a large degree a function of how well he felt that it provided adequate answers to these questions. His state of mind early in 1804 is best summarized by quoting at length from the relevant portion of the Société manuscript.[100]

All modes of proof combine in favor of Christianity. To see the truth, the metaphysician needs only to examine the manner in which this divine religion simultaneously explains the grandeur and the baseness of man, and the idea it gives us of the relations of God with his creatures and of the intentions of Providence.

Why can man mentally embrace all centuries if he was limited to an existence of a few years? Why, if he was born for the highest destinies, do his penchants bend him, almost everywhere and at all epochs, under the shameful yoke of the most vile passions? Can one fail to recognize in this baneful depravity, which a philosopher cannot contemplate without blushing to some extent to be a man, the profoundly ulcerated wounds inflicted upon the liberty and conscience of this being, created with such sublime faculties, by the unfortunate fall concerning which revelation has revealed the mystery to us? Such was the demonstration that pleased Pascal. He felt that, without that which Christianity teaches us about what man can be, and of the event which has degraded him, one could not conceive how the same intelligence, the same will, which animates Newton and Vincent de Paul, is perverted to the point of giving birth to those monsters of cruelty or depravity who, a thousand times below the most vile animals, have been the horror of their peers or the disgrace of humanity.

By way of context, we should note that Ampère's reliance upon the revealed doctrine of original sin is quite different from a contemporary response to the problem of evil by Thomas Malthus. In his *Essay on Population*, first published in 1798, Malthus argued that the existence of evil provides the opportunity for Christians to demonstrate their divine potential for charity and compassion. Ampère was more distressed by cases in which these potentials are not made actual and he developed a permanent distaste for Malthus.

To conclude this discussion of Ampère's Lyon circle, we should mention the contribution of its most prolific writer, Pierre-Simon Ballanche. During February of 1804 Ballanche had two research assignments for the Société: "Should there, could there be revelation?" and "the influence of Christianity on the human race." Ballanche had published *Du Sentiment* three years earlier; the relationship of Christianity to culture was a major theme in that work and thus was not a new topic for Ballanche. Among Ballanche's most fundamental convictions, several stand out as particularly influential. He was convinced of the existence of a God who had created humanity in accordance with a divine plan, a condition Ballanche referred to as Providence. Second, religious traditions all originate from divine inspiration; they simply express essential truths in myriad different languages and cultures. The key to the recognition of these truths, these *faits primitifs*,[101] is to notice what all or most traditions hold in common. Ballanche readily admitted that this process calls for "divination," a sympathetic intuition into the essence of alien traditions.

Ballanche thus shared with Bredin a predilection for esoteric religion. Ampère was less inclined in that direction and struggled with his religious doubts within a more conservative context. He quickly drifted into agnosticism following his departure for Paris in the fall of 1804. He maintained close contact with his Lyon circle, and his

94

correspondence with Bredin is a particularly rich source of information about his personal life in Paris. The two friends carried out an energetic dialectic as they attempted to reach an intellectual and spiritual equilibrium. We will return to this theme in chapter 4; at present we should take stock of the powerful institutions and individuals Ampère encountered in the scientific capital of the world.

Part II

Paris
(1804–1820)

3

Laplacian Physics

The Académie des Sciences and the Institut National

Ampère made his initial contacts with the Parisian scientific community during the completion of his 1801 probability memoir in Bourg. By that time, French science had been structured by government institutions for over a century. Although the Revolution temporarily suppressed the Old Regime's Académie Royale des Sciences and then replaced it in 1795 with the "First Class" of the new Institut National des Sciences et des Arts, there was a significant amount of continuity between the two institutions, a continuity that was preserved when the Académie was reinstated following the Restoration in 1816. The Napoleonic Institut and the Restoration Académie were the two principle public forums for Ampère's activities as a mathematician, chemist, and physicist.[1]

Following its inception in 1699, the Académie Royale des Sciences quickly became an important component of the network of chartered Académies under Louis XIV. As a state institution, it was assigned consultation responsibilities for patents, instrumentation, and applied technology. It was also intended to act as a forum for the advancement of the sciences quite independent from utilitarian goals. The small membership and the internal organization of the Académie reflected the hierarchical elitism of Old Regime society. There were approximately 56 fully active members. The *secrétaire perpétuel* or "perpetual secretary" was the most influential single individual; assignment of Académie duties, preparation of annual *histoires* of Académie accomplishments, and *éloges* of deceased Academicians received close attention. Although seniority and rank were major concerns for the internal politics of the Académie, it was also divided along disciplinary boundaries that indicate what French scientists recognized as legitimate categories of scientific research. These categories did evolve slightly over time; the most convenient way to summarize the changes that took place between 1699 and 1803 is to refer to a table based upon one constructed by the historian Maurice Crosland.[2]

As indicated in Table 3.1, initially the Académie's disciplines were divided into two categories that were preserved through subsequent revisions: mathematical sciences and physical sciences. The classification of mathematics (*géométrie*), mechanics, and astronomy as "mathematical sciences" indicates the survival of a taxonomic tradition with Aristotelian roots. The investigation of mathematical relationships was extended to a treatment of the motion of terrestrial and celestial bodies with the understanding that causal explanation of these motions was not part of the agenda. In spite of the mathematical turn taken during the Scientific Revolution, during the eighteenth century in France "physics" still retained its broad Aristotelian connotation of seeking the causes of both organic and inorganic phenomena in the essential properties or structure of matter.[3] Botany and anatomy were grouped with chemistry in the "physical sciences".

On the other hand, the century also witnessed the expansion of Newton's narrow and precise "natural philosophy" into a wide range of "experimental" sciences such as optics, electricity, magnetism, acoustics, and heat. In contrast to mechanics and astronomy, these topics did not yet receive sophisticated mathematical treatment. The emphasis was on empirical investigation and carefully staged demonstrations. A second sense of the term "physics" gradually came to apply to these topics. As indicated in the *table*, the reforms promoted by Lavoisier in 1785 introduced the new category of "general physics" to represent a new combination of empirical investigation and quantitative description.[4]

The Revolution subjected the elitist attitude of the Académie to destructive scrutiny. Although the utilitarian function of the Académie set it apart from the academic institutions of the Old Regime, resentment by rejected applicants and the general climate of revolutionary egalitarianism resulted in the closing of the Académie in 1793.[5] France remained without a state-supported scientific institution for over two

Table 3.1 French academic disciplines

Académie Royale des Sciences		First Class of the Institut	
1699	1785	1795	1803
Mathematical Sciences			
1 Geometry 2 Mechanics 3 Astronomy	1 Geometry 2 Mechanics 3 Astronomy 4 General physics	1 Mathematics 2 Mechanical arts 3 Astronomy 4 Experimental physics	1 Geometry 2 Mechanics 3 Astronomy 4 Geography and navigation 5 General physics
Physical Sciences			
4 Chemistry 5 Botany 6 Anatomy	5 Chemistry and metallurgy 6 Mineralogy and natural history 7 Botany and agriculture 8 Anatomy	5 Chemistry 6 Natural history and mineralogy 7 Botany and vegetable physics 8 Anatomy and zoology 9 Medicine and surgery 10 Rural economy and veterinary medicine	6 Chemistry 7 Mineralogy 8 Botany 9 Rural economy and veterinary medicine 10 Anatomy and zoology 11 Medicine and surgery

years. During this chaotic revolutionary period, transformations took place that produced a gravely altered scientific community. The disruption of the old academic system of loyalties and procedures was accompanied by a general reassessment of the role of intellectuals in French society. During the worst period of the Terror, the time in which Ampère's father perished, the fragmentation of the scientific community left individual scientists to fend for themselves without a collective identity or a unified political voice. Lavoisier's arrest in 1793 and his subsequent execution gave a clear indication that scientific expertise offered no protection from violent assault on vestiges of the Old Regime. Government agencies hired scientific consultants on an individual basis, with political loyalty as an important prerequisite. Scientists correspondingly adopted a more specialized and bureaucratic identity.

Some attempts were made to found or expand independently supported associations of scientists. The most notable and influential of these was the Société Philomathique, founded in 1788. It struggled under the multiple burdens of precarious social conditions, severe inflation and printing costs, and the general suspicion of aristocratic sympathies that plagued most private organizations. It did survive, however, and it became a forum for some vigorous debates between Biot and Ampère.

With the elimination of Robespierre in 1794, the Revolution entered the reactionary phase that culminated in Napoleon's *coup d'état* in 1799. By 1795 there was a general consensus that the creation of a new state-supported cultural institution was appropriate. The National Convention ratified the Institut National des Sciences et des Arts on 22 August, 1795. The new Institut was structured into three classes. Science made up the "First Class" with 60 members; the "Second Class" of moral and political sciences had 36 members, and the "Third Class" of literature and fine arts was allotted 48 positions. Science was thus the largest class with six members in each of ten sections. By 1803 Napoleon had become thoroughly dissatisfied with the political debates provoked in the Second Class by the strong *idéologue* contingent there; nor was Napoleon willing to tolerate a public display of atheism and materialism by the *idéologues* during his rehabilitation of French Catholicism. He was powerful enough to order the suppression of the Second Class in 1803; its literary and historical sections were transferred to the Third Class and a three-member geography section was added to the First Class. Ampère's arrival in Paris came shortly after this disruption of what had been an ambitious effort by the *idéologues* to make their philosophy a foundation for both scientific methodology and political practice. As we will see, Ampère initially found their company more interesting than that of mathematicians and physicists.

The fact that Napoleon could intervene in Institut procedures in such a drastic fashion is indicative of what the historian Roger Hahn has called "the paradox of the Institut."[6] The security and prestige achieved by election to the Institut was largely a ceremonial recognition of prior accomplishments and a trustworthy political attitude. Innovative research and creative debate was far more likely to occur at other locations than the generally sedate meetings of the Institut. The new facilities of the École Polytechnique, for example, had begun to attract ambitious mathematicians and physicists during the very period in which the scientific community had been dispersed by the suppression of the Académie. Although membership in the First Class was a highly prized symbol of arrival in the French scientific community and opened doors to other positions, it did not provide an ideal context for men of Ampère's loquacious and argumentative temperament.

Napoleon's 1803 revisions of the Institut's First Class preserved the basic distinction between the two divisions of mathematical and physical sciences; each division appointed its own *secrétaire perpétuel*. Delambre was the initial appointee for the mathematical sciences and he remained in that position until his death in 1822. Georges Cuvier was his counterpart for the physical sciences. The *physique expérimentale* section was reassigned its former Académie title of *physique générale*; its location within the division of mathematical sciences reflected the continued expectation that physics would include mathematical analysis.

Educational Institutions: The École Polytechnique

We have followed Ampère's first encounters with formal education while teaching in the short-lived Bourg École secondaire and the Napoleonic Lycée in Lyon. In spite of his lack of formal training, he had been fortunate enough to be available at the appropriate time. By 1804 his mathematics memoirs and his teaching record had made a favorable impression in Paris; in that year he was appointed *répétiteur* of analysis at the École Polytechnique, becoming one of the last to acquire a teaching position there before standardized educational degree requirements were introduced. By 1804 the École had been functioning for a decade and Ampère confronted a complex and closely scrutinized environment.

The school had opened in 1794 as the École Centrale des Trauvaux Publics. Its primary purpose was to provide preliminary training for military and civil engineers; a relatively small number of graduates were expected to follow a teaching career. After the large admissions of the first few years, between 100 and 200 students were admitted in each annual *promotion*; they were ranked according to their performance in an entrance examination that emphasized mathematics. Upon acceptance, students were required to stipulate whether they planned a military or a civil career. In 1799 the original three years of training were reduced to two. Graduates were expected to receive more training in one of the specialized engineering schools known as *écoles d'application*. The best positions were reserved for those who emerged near the top in the final examinations.

The École atmosphere was both competitive and highly structured. Beginning in 1804, students were organized into *brigades*, lived in barracks, and marched to classes in military formation. Thereafter, Napoleon repeatedly intervened to alter the curriculum for the sake of training that had direct military applications.[7] Advanced courses in mathematics and chemistry were gradually shortened or made optional to allow more training in fortifications, technical drawing, and cartography. Most of the students probably shared Napoleon's perspective. Many of them had little taste for abstract mathematics; some attended the school only briefly to take accelerated practical courses prior to artillery school.[8] It would be difficult to imagine a learning environment more of a contrast to the one Ampère had enjoyed at Poleymieux.

Although the École curriculum relied upon some of the same terminology used to stipulate the disciplinary divisions of the First Class of the Institut, there were significant departures in content. For example, although a distinction was drawn between mathematical and physical sciences, physics was placed entirely within the latter category. The physical sciences were divided into *physique générale* and *physique particulière*. The content of specific courses varied as the role of mathematics was debated from both methodological and pedagogical perspectives. According to an early discussion published in the *Journal de l'École Polytechnique* by the *répétiteur* of physics, Étienne Barruel, general physics should be restricted to the study of properties common to all material bodies: extension, impenetrability, mobility, inertia, and gravity.[9] According to Barruel, *physique particulière* "strictly speaking" is the study of matter under specific conditions: electricity, magnetism, optics, heat, elasticity, porosity, and affinities. These topics thus corresponded to the Institut division of *physique générale*. Barruel also stipulated two other areas of *physique particulière*: "analytic physics" or chemistry, and the fields of "observational physics", including anatomy, meteorology, acoustics, geology, acoustics, and pneumatics. These topics roughly correspond to the division of physical sciences within the First Class of the Institut.

The major distinction within the mathematical sciences was between analysis, together with its applications to geometry and mechanics, and the descriptive subjects of technical drawing, architecture, fortification, and descriptive geometry. Teaching positions were distributed over the categories of analysis and mechanics, descriptive geometry, *travaux graphiques*, *travaux civils*, fortifications, architecture, physics, and chemistry.[10] The distribution of classes required of students was initially weighted heavily in favor of chemistry and descriptive geometry, particularly due to Monge's influence. This emphasis changed significantly during the École's first decade. By 1806, almost half of the students' time in class was taken up by analysis and mechanics.[11]

Ampère's 1804 appointment as *répétiteur* of analysis placed him in a controversial position. From the time of the school's inception, the appropriate level of required mathematical training was a constant topic of debate. Because the École was intended to provide preliminary training for all the *écoles d'application*, course requirements had to be justified under this agenda. Laplace and Lagrange preferred all French engineers to share a sophisticated knowledge of mathematics, including calculus and differential equations. On the other hand, there were objections that much of this higher mathematics was of little practical value. Chaptal, who taught chemistry at the École during the 1790s, felt that the emphasis on mathematics as a necessary condition for acceptance and success at the École was distorting the entire French scientific community. Aptitude for mathematics had in effect become a prerequisite for a career in all branches of science, even those such as industrial chemistry where experimental ingenuity was far more important.[12] Generally speaking, those advocating a high mathematical content were victorious, but a vocal contingent remained unconvinced.

With these conditions in mind, we should not expect that the students in Ampère's mathematics classes provided a particularly receptive audience. They realized the importance of mathematics for their academic careers, but they were unlikely to inspire creativity or contribute to a critique of the foundations of the techniques they were required to master. Ampère enjoyed the security of steady employment but with the burden of an essentially captive audience. Perhaps no single individual was more influential in supporting this state of affairs than Laplace.

Laplace and Berthollet: Patronage and the Société d'Arcueil

We have already encountered Laplace as a critic of Ampère's 1801 probability memoir and as a colleague of Coulomb. Prior to the Revolution, he had been elected to the mechanics section of the Académie des Sciences. In 1784 he published a very important treatise entitled *Théorie du Movement et de la figure elliptique des planètes*. It included a general consideration of gravitational attraction based upon what would later be referred to as the "potential" function, and extensive use of series expansions in terms of Legendre polynomials. These techniques became characteristic of the Laplacian style of physics; they were introduced to a broad audience through Laplace's monumental *Traité de Mécanique céleste*, published in five volumes between 1799 and 1825. The third volume, published in 1802, contained the mathematical foundation for Poisson's important 1812 analysis of electrostatic forces.

In 1795 Laplace quickly took a leading role in the new Institut. He was elected vice-president at the organizational meeting and president in 1796. Within the First Class he was very active in the mathematics section and on several committees, including those responsible for the choice of prize contest topics in mathematics and physics. Several of these contests became thinly disguised efforts to foster research in the direction Laplace desired. With Napoleon's coming to power in 1799, Laplace rose to new positions of influence. His relationship to Bonaparte was complex and uneven, as is to be expected for two proud men with conflicting priorities and agendas. Their attitude toward each other was more condescending than warm. Although Napoleon quickly found Laplace unsatisfactory as a Minister, a less politically demanding and more ceremonial lifetime position in the Senate was reserved for him. He became Chancellor of the Senate in 1803 and his annual income rose to over 100,000 francs. With Napoleon's support, Laplace directed the passage of the 1799 law that reorganized the École Polytechnique. Laplace and his scientific colleagues were willing to tolerate the militarization of the École Polytechnique in exchange for continued support for science at the expense of the humanistic facets of French culture.

Laplace has frequently been characterized as a man who allowed himself to be used politically in exchange for personal profit. While there may be some truth to this charge, it is also true that he was thoroughly dedicated to the advancement of science in the direction

106

he thought most valuable. He became a central figure in the complex system of patronage that played an unofficial but vital function in the French scientific community. Both he and his close friend, Claude-Louis Berthollet, channelled a significant portion of their fortunes into scientific projects.

In 1803 Berthollet purchased an estate five miles south of Paris in Arcueil. He set aside rooms for physical and chemical laboratories and made them available to young researchers such as Gay-Lussac and Thenard. Laplace purchased the adjoining property in 1806; under these comfortable and informal conditions, the two men held meetings for a group that during the summer of 1807 became known as the Société d'Arcueil. The full list of membership is worth recording. The original members were: Claude Louis Berthollet, Laplace, Alexander von Humboldt, Jean-Baptiste Biot, Louis Jacques Thenard, Joseph Louis Gay-Lussac, Augustin de Candolle, Hippolyte Descotils, and Claude Berthollet's son, Amédée Barthélemy Berthollet. Subsequent additions were: Étienne Louis Malus, Dominique François Jean Arago, Jacques Bérard, Jean Antoine Chaptal, Pierre Louis Dulong, and Siméon-Denis Poisson. The group was deliberately kept small and of high quality. Aside from the senior German scientist Humboldt, all the original members were born within five years of Ampère's birthdate in 1775. Unlike the Société Philomathique, which primarily provided a forum for individual edification and career advancement, members of the Arcueil group shared an agenda they promoted within the larger scientific community.

That agenda was set by Berthollet and Laplace, who acted as patrons for the younger men. We have already summarized the general stages of the Laplacian research program and the motivational impact of Coulomb's experimental work on electricity. Berthollet gave the Société a balance between physics and chemistry. His own interests centered on the relationship between cohesion and chemical forces and the measurement of affinities by means of combining weights. He shared with Laplace the conviction that chemical forces or affinities must ultimately be traced to short-range forces. Far less mathematically inclined than Laplace, Berthollet encouraged research that combined chemistry and the expanding domain of *physique particulière*. Investigations of the effects of pressure, temperature, and light on chemical reactions called for careful quantitative measurements that Berthollet's equipment made possible. Furthermore, by seeing that their protégés were given authoritative positions in the Institut and the educational system, Laplace and Berthollet guaranteed that their agenda was promulgated throughout the French scientific community.

Some general features of the research carried out at Arcueil are worth noting. Chemical analysis was a central concern, as exemplified by Dulong's work on barium sulphate and the 1809 memoir on chlorine by Gay-Lussac and Thenard. Gay-Lussac's famous discovery of the law of combining volumes for gases was also a result of research at Arcueil. As we will see, Ampère's scientific interests prior to 1810 overlapped with those of the Arcueil group more in chemistry than in any other area. On the other hand, the experimental search for quantitative data to exemplify phenomenological laws was an Arcueil characteristic that never appealed to Ampère. Biot and Humboldt's study of terrestrial magnetism, Biot's measurements of the velocity of sound, Biot and Arago's measurements of optical refraction by gases, capillarity measurements, Malus' study of polarized light, and Biot and Gay-Lussac's spectacular measurements of variations in magnetism and atmospheric oxygen content with altitude during their 1804 balloon ascent are all examples of the Arcueil style.

The members of the Société d'Arcueil were a carefully chosen elite group. As such, they offer a collective foil against which to gauge Ampère's status following his arrival in Paris in 1804. He gravitated toward quite a different group of associates and there is no indication that he was ever considered for Arcueil membership. He lacked both the requisite experimental skill and the sedate decorum expected in the homes of Senators of the Empire. It is not surprising that the period of his rise to eminence in the French scientific community followed the decline in the Société's influence after 1813.

As has already been pointed out, prior to 1812 the Laplacian program benefited both from careful experimental work and the clarification of methodological standards. It is appropriate at this point to turn to the second stage of this development with an emphasis on its most articulate spokesman, Jean-Baptiste Biot.

Haüy, Biot, and the renaissance de physique véritable: *1787–1812*

During the 1780s, Coulomb had argued that physics should be a search for laws rather than causes. Speculations about the electric and magnetic fluids were far less important to him than the discovery of phenomenological laws. Even microscopic phenomenological laws could be stated in "molecular" language that was noncommittal about the existence of imponderable fluids. In general, this attitude prevailed in Paris until Poisson's 1812 memoir on electricity. Nevertheless, subtle

108

departures from Coulomb's norms prepared the way for Poisson's accomplishment. René-Just Haüy and Jean-Baptiste Biot were two influential advocates of these changes. Although Haüy was not a member of the Arcueil group, he shared many of its expectations; he and Biot were convinced that a "renaissance of true physics" should be accomplished by purging scientific research of the *esprit de système* that had diverted it from the course so brilliantly charted by Newton.[13] Their methodological directives were inspired by a desire to distinguish legitimate physical "theories" from speculative *systèmes* devoid of explanatory value.

René-Just Haüy took an early interest in botany before specializing in mineralogy. His important publications in that field during the 1780s won him membership in the botany section of the Académie in 1783. In 1795 he entered the Institut in the section of natural history and mineralogy. Haüy's eclectic interests provided a basis for his discussion of methodology. In 1787 he published a more organized treatment of Aepinus' one-fluid theories of electricity and magnetism than was to be found in the original memoirs. His introductory remarks included an attempt to stipulate the criteria that characterize methodologically sound scientific theories. According to Haüy, legitimate theories are based upon a few fundamental "facts" that point to a unifying structural basis for what might otherwise seem to be a random collection of data.[14] In his *Traité de Physique* Haüy tried to contrast the organizational power of a theory with the imaginative speculation represented by a *système* which goes beyond the facts of observation. Unfortunately, his tidy distinction between theories and *systèmes* was not so easy to apply to actual controversies. For example, Haüy claimed that the central explanatory concept of the theories of gravitation, electricity, and magnetism was the "fact, proved undeniably by observation" that the operative forces have inverse square dependencies on distance. Furthermore, Haüy claimed that when a physicist uses the word "force," this "merely expresses a fact for which one makes no attempt to find a cause."[15] But Haüy's "facts" are hardly the result of direct observation. On the contrary, in the case of gravitation, for example, a long series of inferences led from observations of planetary positions to Kepler's laws and then to Newton's law of gravitation. Similarly, Coulomb carried out his laborious measurements of electrical and magnetic attractions and repulsions before inferring macroscopic and microscopic phenomenological laws. On the other hand, Newton and Coulomb did not make dogmatic assertions about the causes of gravitational or electric forces. It is this agnostic attitude about causes that Haüy praised as the advantage of theories.

109

Nevertheless, Haüy did admit that in some cases "hypotheses" could play a productive role in scientific reasoning. For example, he tentatively adopted two-fluid hypotheses for both electricity and magnetism; in this respect he exemplifies a gradual acceptance of the possibility that the imponderable fluids might become incorporated into legitimate scientific theories. By 1790 influential French physicists were following Laplace's lead in the expectation that future scientific explanation might be anchored upon a stipulation of material particles which they were not afraid to call the "causes" of both microscopic and macroscopic phenomenological laws of attraction and repulsion. For these theorists, fundamental forces between particles of the imponderable fluids might be incorporated into legitimate scientific theories as long as they were quantified and had measurable empirical consequences. The pejorative term *système* was reserved for speculations about the mode of transmission for these fundamental forces. Nevertheless, among the relatively conservative members of the French scientific elite, there was not yet a widespread belief in the reality of imponderable fluids prior to 1800. After all, the Académie had just recently rendered its negative verdict on the "magnetic fluid" postulated by Mesmer and his associates. Berthollet himself had attended mesmerist sessions and declared them fraudulent. Speculations about etherial media were particularly taboo due to the resemblance to Cartesian vortices, everyone's favorite example of a bogus *système*. Furthermore, application of Coulomb's microscopic phenomenological force laws did not require commitment to a more basic ontology of fluid particles.

Jean-Baptiste Biot clearly articulated the ambivalent state of affairs during the first decade of the new century, the period in which Ampère entered the Parisian scientific community. During the 1820s he and Ampère advanced rival theories of electromagnetism; prior to that time, Biot's carefully planned career offers an interesting contrast to Ampère's less conventional efforts.[16] Born in Paris in 1774, Biot was only one year older than Ampère. His parents were reasonably well-to-do members of the bourgeoisie; Biot's father invested his resources wisely and intended to prepare his son for a business career by giving him a literary education at the Collège Louis-le-Grand. Nevertheless, following his graduation in 1789, Biot remained for an additional two years of study and became enchanted with mathematics. Following a strenuous year in the Army he returned to Paris and was admitted to the new École Centrale des Trauvaux Publiques, the original name for the École Polytechnique.

Unlike Ampère, Biot found the atmosphere at the École exhilarating;

he completed the course of study in a single year, graduating in 1795. After some additional study at the École des Ponts et Chaussées, he was selected to teach mathematics at the École Centrale de l'Oise in Beauvais beginning in February of 1797. He found teaching at an École Centrale no more satisfying than Ampère did in Bourg during 1801. Before leaving Paris, he had become acquainted with the mathematician and prolific textbook writer, Sylvestre-François Lacroix. Lacroix was Biot's first patron and in 1799 he provided an unusually prompt and very favorable report on a memoir Biot read to the Institut on difference equations. The other two members of the review committee were Laplace and Napoleon; Biot had attracted attention in a rather elite circle. In May of 1800 Biot was elected to the First Class of the Institut as an *associé non resident* in the mathematics section. Between 1801 and 1803 he wisely followed Laplace's lead and shifted his attention from mathematics to topics in *physique générale*. For example, Volta's invention of an electric "pile" in 1800 became the subject of Biot's first extensive physical investigation. In 1802 he presented a report to the Institut in which he interpreted the operation of the pile as a sequence of static distributions of electric fluid periodically interrupted by discharges in such rapid succession as to result in effects that produce the illusion of a steady "current" of fluid flow.[17] Biot's theory was so widely accepted in France that very little additional exploratory research on the operation of the pile was performed there prior to 1820. Studious cultivation of Laplace's patronage brought election to the Institut in 1803.

We have already noted some of the experimental research that made Biot a central member of the Société d'Arcueil by 1807. During 1805 and 1806, Laplace gave explanations of atmospheric refraction and capillary action in terms of short-range forces between "molecules." This convinced him that microscopic forces offered the key to understanding all physical phenomena. Although he believed that these microscopic force laws were actually the result of gravitational forces acting between tiny "molecular" volume elements, Laplace eventually accepted these "molecular" phenomenological laws as the fundamental laws of capillary theory, citing as justification the impossibility of determining the shapes of the relevant "molecules." In this context the only property that distinguishes the "molecules" of simple substances from larger quantities of matter is that these "molecules" have a uniform density uninterrupted by the pores of empty space that separate them in larger aggregate bodies. Although Laplace and Berthollet were stubborn opponents of "chemical atomism," they did believe in the particulate structure of matter. What they objected to

111

was the use of assumptions about the properties of these particles as a basis for a theory of chemical reactions. Furthermore, at the turn of the century Laplace did not feel that the time was ripe for a wholesale shift in the foundation of scientific explanation from the "molecular" level to the level of individual imponderable fluid particles.

Biot was very articulate on this complex issue. He used the term "cause" in two different senses, only one of which was pejorative. In general, changes in the state of motion of inert matter are to be attributed to "causes," that is, to forces expressed either as phenomenological or as fundamental force laws. On the other hand, in his famous 1809 polemic, *Sur l'Esprit de système*, Biot wrote in a vein that brings to mind earlier admonitions by Coulomb.[18]

> The true object of the physical sciences is not the search for first causes, but the search for the laws according to which the phenomena are produced ... Universal gravitation, thus established and verified, itself becomes a fact. The cause of it alone is occult, and mathematicians are not logically compelled to specify it, since they have no need to know it in order to discover and assign the particular laws of the phenomena which alone have any interest for us.

The "law" of universal gravitation Biot refers to is the inverse square attractive force between volumes of matter small enough to be considered "infinitesimal" in comparison to the distance between them; speculation about the cause or mode of transmission of this force is rejected as non-scientific metaphysics. For Biot, Newton's demonstration that planetary and projectile motion are calculable consequences of the fundamental law of gravitation became the procedure for all physicists to emulate. An analogous guiding principle had not yet been provided by imponderable fluid hypotheses; no calculation had as yet linked observable thermal, electric, and magnetic effects to hypothetical inter-particle forces. If electric and magnetic fluids were ever to serve as a foundation for scientific explanation, specific distributions of particles would have to be subjected to force calculations. Instead, the insidious *esprit de système* still dominated electricity and magnetism. As a result, physicists "have imagined certain elastic fluids bestowed with attractive or repulsive properties" which they actually consider "merely as convenient hypotheses to which they take care not to attach any ideas of reality, and which they are ready to modify or abandon completely as soon as the facts are shown contrary to them."[19]

During the early years of the new century, Biot, Laplace, and Haüy thus shared a skepticism about imponderable fluids that was more forcefully expressed than Coulomb's had been. Unfortunately for Ampère, their depiction of the flaws of the *esprit de système* applied all too accurately to his own efforts to create a unifying theory of electricity, magnetism, optics, and heat. His imaginative invocation of elastic molecular atmospheres to transmit forces was not quantitative and he made no effort to calculate measurable predictions. The sobering effect produced by his new awareness of Parisian standards was partially responsible for Ampère's lack of activity in physics during his first decade in Paris. Poisson's 1812 memoir on electricity made the new standards painfully clear.

Poisson's 1812 Electricity Memoir

Siméon-Denis Poisson was born in Pithiviers, about 50 miles south of Paris, on 21 June, 1781; he was six years younger than Ampère. He came from an undistinguished but very supportive family; his father spent time as a common soldier and then administered a judicial position similar to that of Ampère's father. Poisson's mathematical talents were recognized during his attendance at the École Centrale in Fontainebleau. His performance in the entrance examination for the École Polytechnique placed him first in the *promotion* for 1798. He immediately became renowned for his mathematical skill and insight; in 1800 he was allowed to skip final examinations and was appointed to a teaching position at the École as *répétiteur-adjoint*. Poisson did not follow the usual career route to an *école d'application*; nor did he have to leave the Parisian environment for service in the provinces. Lagrange and Laplace had been particularly impressed; two more powerful patrons could hardly be imagined.

Poisson took to publishing with alacrity. Eventually he would publish over 300 memoirs and two editions of his *Traité de Mécanique*. His interests and aptitudes were perfectly suited to the mathematical physics advocated by Laplace. Poisson's early work in mechanics, mathematical astronomy, and capillarity was appropriately rewarded. By 1806 he was teaching mechanics at the École Polytechnique in place of his own teacher, Fourier. He received a position of *géomètre-adjoint* at the Bureau des Longitudes in 1808 and teaching positions at the Collège de France and the Paris Faculté des Sciences in 1809. As a teacher, Poisson was renowned for his clarity and dedication; in 1811 he published the first edition of his *Traité de Mécanique*, based upon lecture notes.

113

Poisson and Biot complemented each other within the ranks of Laplace's disciples. Biot was a skillful experimenter, and he was especially astute at extracting phenomenological laws from complex data. He had a forceful literary style and he applied it to an energetic defense of Laplacian methodology. Unfortunately, he was not very sensitive to priority claims from other scientists; his heavy-handed habit of claiming credit for important discoveries resulted in some bitter personal feuds. This was especially the case with Arago in optics, Fourier in the theory of heat, and, to a lesser degree, with Ampère in electrodynamics.

Poisson, on the other hand, was content to remain totally unfamiliar with the vicissitudes of experimental research. It is quite unlikely that he ever attempted an experimental measurement; nor did he try his hand at drafting experimental designs. He was judged incompetent as a draftsman quite early in his studies at the École Polytechnique; he was only excused from the school's drawing requirements because his projected career was in pure science rather than public service.[20] Shunning the laboratory, Poisson devoted himself with religious fervor to the solution of mathematical problems arising from the application of Laplacian concepts, particularly short-range forces. In a eulogy of Poisson written in 1840, Libri offered some insight into the mentality that pursued this research.[21]

> Poisson never wished to occupy himself with two things at the same time; when, in the course of his labors, a research project crossed his mind that did not form any immediate connection with what he was doing at the time, he contented himself with writing a few words in his little wallet. The persons to whom he used to communicate his scientific ideas know that as soon as he had finished one memoir, he passed without interruption to another subject, and that he customarily selected from his wallet the questions with which he should occupy himself. To foresee beforehand in this manner the problems that offer some chance of success, and to be able to wait before applying oneself to them, is to show proof of a mind both penetrating and methodical.

Perhaps no sharper contrast to Ampère's spontaneous and passionate style can be imagined. Poisson's single-mindedness also stands out in contrast to Ampère's eclecticism. While Ampère preferred the company of philosophers, Poisson was known to advocate the slogan that "Life is good for only two things: doing mathematics and teaching it."[22]

Election to the First Class of the Institut was on Poisson's agenda as early as 1806. He had strong backing from five of the six members of the mathematics section: Laplace, Lagrange, Lacroix, Legendre, and Biot. Nevertheless, in spite of the advanced age of Bossut, the sixth member of the section, there was no immediate prospect of an opening there. In 1811 the discovery that Malus had consumption presented the possibility that Poisson might be elected in the physics section. To encourage that event, the mathematics section proposed a topic for the next year's prize competition that was prearranged for Poisson's benefit. The topic statement is worth quoting in detail since it indicates both the general mathematical bent of the Laplacian school and their specific agenda in electricity.[23]

> To determine by calculation and to confirm by experiment the manner in which electricity is distributed at the surface of electrical bodies considered either in isolation or in the presence of each other – for example, at the surface of two electrified spheres in the presence of each other. In order to simplify the problem, the Class asks only for an examination of cases where the electricity spread on each surface remains always of the same kind.

Poisson was already well along in a response to this problem when Malus' death was announced at the 24 February, 1812, meeting of the First Class. On 9 March he presented a first installment of his results under the title *"Distribution de l'électricité."* Poisson began by explicitly adopting the two-fluid theory of electricity. In contrast to Coulomb's cautious emphasis on phenomenological laws, the explanatory foundation of Poisson's theory was the fundamental law for the attractive and repulsive forces between individual fluid particles. From his point of view, a successful match between his calculations and Coulomb's data would provide positive evidence for the existence of the imponderable fluids of electricity. To facilitate calculation, Poisson constructed an idealized model for a static distribution of the electric fluids in equilibrium on the surface of a conductor. Conductors are assumed to offer no resistance to the motion of electric fluid particles; any excess quantity of one of the fluids over the other is thus assumed to be driven to the surface of the conductor by its inter-particle repulsive forces. Poisson used the term "free fluid" to refer to this excess quantity of either of the two fluids. The free fluid will have a small thickness that varies from point to point on the surface and corresponds to Coulomb's measurements of surface charge density; the internal surface of the layer will take whatever shape is required to produce equilibrium for each particle with respect to all the repulsive forces acting on it.[24] At equilibrium there is no net electric force on any fluid particle within the conductor, including those on the internal boundary of the surface layer.

To apply his fundamental law to this model, Poisson invoked the procedures Laplace and Legendre had followed in their general study of forces, particularly attractive gravitational forces. He calculated force components as partial derivatives of "a certain function" which he left unnamed, but which, beginning in 1828, Green would refer to as a "potential" function for a particle distribution. For inverse square forces, the potential function V is evaluated at any given point by summing the inverses of the distances from the point to all fluid particles. In practice, Poisson calculated the electric potential by imagining the surface fluid layer to be a mathematical surface of infinitesimal thickness with a surface density equal to the small thickness of the actual layer. He then disregarded the particulate nature of the fluid and imagined it to be continuous even within infinitesimal surface areas. This allowed him to write a general expression for the potential by an application of integral calculus to the surface of the conductor. He used the following notation:

(r',θ',ω'):   the spherical coordinates of a point on the surface of the conductor

(x,θ,ω):   the spherical coordinates of the point O where the potential is to be evaluated

y':   the electric fluid layer thickness (or surface fluid density) at the point (r',θ',ω')

X: the reciprocal of the distance between the point O and a point on the surface of the conductor with coordinates (r',θ',ω').

Using this terminology, the potential V is determined by performing a double integration over the surface of the conductor. The result is[25]

$$V = \iint X y' r'^2 \sin\theta' \, d\theta' \, d\omega'$$

Poisson now addressed two general preliminary problems. First, he showed how to use the vanishing of the potential within the conductor to calculate the distribution of electric fluid particles over the surface, that is, the thickness of the surface fluid layer as a function of position on the surface. Here Poisson drew upon his knowledge of series expansion techniques to rewrite the potential using an expansion of the function X as a series in Legendre polynomials. He then invoked the requirement that the forces at any point within the conductor must

116

vanish; since these forces are determined by differentiating V, the coefficient of every power of x in the expansion of V must also vanish. Another Legendre series expansion transformed this condition into the problem of determining the fluid layer thickness as a function of μ' and ω' for a given conductor surface, a problem he solved in detail for the relatively easy special case of a "spheroid" differing very little from a sphere. Second, using an expression for the potential outside the conductor, he showed that the external force exerted radially from a point on the surface is proportional to the fluid layer thickness at that point.

With these two general problems solved, Poisson turned to the specific cases mentioned in the Institut's prize announcement, cases for which Coulomb had taken measurements. He did so in exhausting detail by calculating fluid surface densities for several different combinations of spheres either in contact or isolated after an initial contact.[26] Poisson's modest conclusion was that the agreement of his conclusions with Coulomb's data "furnishes an important confirmation of the theory of two fluids."[27]

Poisson's accomplishment was the deciding factor in his election to the First Class of the Institut in the general physics section. Historians have noted that Poisson's election was an indication that the previous emphasis on empirical research in the physics section was giving way to the mathematical analysis favored by Laplace.[28] Poisson was notoriously incapable of experimental research; with his election, the connotation of "physics" was officially recognized to include purely mathematical investigations. Poisson's own work conveyed little appreciation for electricity at the phenomenological level. He operated entirely within the mathematical context of an idealized model; for Poisson, Coulomb's "data" were simply numbers. Malus' sudden death actually caused the Institut election to take place sooner than expected. Poisson thereby became a member of the Institut before the prize could be awarded for the electricity topic; he thus became ineligible to receive the prize and it was decided not to award one.

Only one other entry had been delivered for consideration. François Joseph Gardini, an Italian physician and physics teacher, submitted an almost entirely qualitative essay that indicated a serious lack of understanding of the Institut's expectations.[29] Gardini's essay offers an interesting point of comparison to Ampère's 1801 notes on electricity. Like Ampère, Gardini speculated at great length about alleged analogies between the imponderable fluids of electricity, magnetism, light, and fire. Although he imagined a means of transmission for electric phenomena through a disturbance of the ordinary state of atmospheric

117

equilibrium, he attempted no specific computations of forces. Ampère at least was slightly better informed of the state of electrical science in 1801 than was Gardini. The Italian scholar made no references to Aepinus, Coulomb, or Haüy. Ampère was aware of their work and wrote in response to what he considered to be an illegitimate reliance on unexplained forces by Coulomb. Nevertheless, Poisson's memoir made it clear to Ampère that causal speculations without quantitative analysis would not be well received by the scientific community. This is not to say that Ampère lacked the requisite mathematical knowledge. His 1802 essay on probability showed his clear grasp of series expansions and transformation techniques. But between 1804 and 1812 Ampère had made no progress toward anything comparable to Poisson's achievement.

Biot and the Principles of Laplacian Physics

Responses to Poisson's electricity memoirs took place on several different levels. He was admitted into the exclusive Société d'Arcueil, a context in which experimental research had predominated hitherto. In retrospect, the optimism that prevailed at Arcueil at that time makes 1812 stand out as the zenith of the Laplacian research program. Biot was their most energetic spokesman and he reached a wide audience with publications that were not restricted to technical journals. We have seen how during the first decade of the century he and Haüy promoted mathematical articulation of hypotheses, but were highly dubious of imponderable fluids as a foundation for these investigations. Biot's position changed dramatically after 1812. During the next three summers, he relied upon imponderable fluids as an explanatory basis as he composed his *Traité de Physique expérimentale et mathématique*.

In the introduction to the *Traité*, Biot cited Poisson's results as "the strongest induction, indeed, the only one we may have, for believing in the real existence of the two fluids, invisible and imponderable." More generally, Biot argued that physics had become a discipline in its own right and should be distinguished from the "general and experimental physics" that guides the instrumentation and experimental techniques of all the sciences. For Biot, physics *per se* should now be limited to "studying the phenomena produced by the actions of invisible, intangible, and imponderable principles such as electricity, magnetism, caloric, and light."[30]

118

Beginning in 1812, Biot thus adopted a conceptual framework that gave a new orientation to the "renaissance of true physics" he advocated. Conformity to the mathematical standards of the new physics was now inseparably linked to a reduction of phenomena to the action of fundamental forces exerted between particles of the imponderable fluids. In earlier writings, Biot had reserved the term *principes* for force laws; in the *Traité*, *principes* took on an additional meaning and referred to the imponderable fluids themselves. Through this dual usage, Biot sought to convince French physicists that the "principles" of a physical theory must include a stipulation of the material basis for fundamental forces. The combination of forces and particles now became a necessary condition for the distinction between a viable theory and a speculative *système*. For Biot, the mathematical study of fundamental force laws applied to specific particle distributions would return physics to its Newtonian task of "calculating" rather than "explicating."[31]

In light of the revision that took place in Laplacian objectives after 1812, it would be misleading to attribute a Comtean positivism to Laplace and his disciples. Auguste Comte himself was still warming the benches of the École Polytechnique in 1816; he did not give his first public lecture on *philosophie positive* until 1826. In his seminal paper on Laplacian physics, the historian Robert Fox was careful to point out that positivism did not become a general characteristic of French physics until after the rejection of the Laplacian program during the 1820s.[32] Ampère would of course contribute to that revolution; only thereafter would Biot make defensive positivistic statements in an effort to obscure his earlier commitments. For example, in an 1821 publication during his unsuccessful debate with Fresnel over the wave theory of light, Biot claimed rather disingenuously that "I have never claimed in my research to establish anything other than experimental laws."[33] His colleagues were not deceived.

The following schema offers a general characterization of a physical theory from the point of view of Laplacians such as Biot, Poisson, and Laplace himself.

1 The theory must be based upon a careful preliminary establishment of the macroscopic phenomenological laws of the domain in question; or, if no law-like relationships are discovered, at least quantitative data pertaining to the phenomena must be gathered.

2 The theory must explicitly state a set of "principles" of three main types:

119

  (a) a description of the primitive material units that are taken to
       be causally responsible for the observed phenomena;
  (b) a stipulation of the fundamental laws that apply to these
       units, including fundamental force laws;
  (c) a stipulation of any additional fundamental laws due to re-
       straints imposed by boundary or initial conditions.
3 A mathematical formulation and application of the "principles"
   of the theory must provide derivations of all relevant macroscopic phenomenological laws or the magnitudes of all relevant
   empirical measurements.

   In some cases, these derivations may be premised upon
   microscopic phenomenological laws pertaining to "molecules"
   of the substances concerned. However, any microscopic phenomenological laws used in this manner must be compatible with
   the principles of the theory; in fact these laws should be readily
   conceivable as due to properties bestowed upon material "molecules" by the elementary units of the theory in conformity with
   fundamental laws.

In addition to these general requirements, Laplace, Biot, and Poisson
held that physics could be partitioned into four distinct domains of
heat, light, electricity, and magnetism, corresponding to the four types
of imponderable fluid. The fundamental forces to be cited in the principles of a theory would either be forces between particles of a given
fluid or between those particles and "molecules" of ponderable matter. Biot admitted that this taxonomy of independent domains was a
provisional one that might eventually have to be revised, due to the
discovery of significant "analogies" between seemingly disparate fields
of phenomena.

Second, owing to their reductionist inclinations, Laplacians were
prone to locate scientific rigor and precision in fundamental laws rather
than in macroscopic phenomenological laws. Biot even argued that to
establish the certainty of a phenomenological law:[34]

> one must reduce it to the general laws of mechanics; that is, one must
> derive it from the general conditions of motion and equilibrium re-
> quired by these laws. For, when such a reduction can be completely
> effected, it necessarily places in evidence the character of the FORCES
> by which the phenomena are produced, which is the ultimate limit that
> human science can attain.

For Biot, the reduction of a phenomenological law to the action of
fundamental forces ensures the accuracy of the law beyond the

limitations imposed by the order of magnitude of experimental accuracy. Similarly, in 1808 Laplace drew the following implication from his alleged derivation of Huygens' phenomenological law for double refraction from fundamental laws of mechanics.[35] "Until now, this law was only the result of observation, approaching truth within the limits of the errors that even the most precise experiments are subject to; now it can be considered a rigorous law."

Ironically, although Poisson's 1812 analysis of electricity set the standard for Laplacian theories during the subsequent decade, it was never duplicated with comparable precision in any other domain. Poisson did not attempt a corresponding treatment of magnetism until 1824. During the 1820s, one of the major controversies within the new domain of electrodynamics would concern the relationship between unspecified fundamental laws and the "molecular" phenomenological laws that were actually used to calculate magnitudes of observable phenomena.

Another Laplacian *desideratum* was that the explanation of experimental discoveries in a given domain should not repeatedly require the adoption of additional hypotheses to modify theoretical principles. Biot insisted that detailed familiarity with an experimental domain should allow a scientist to discern conceptual relationships within the more or less arbitrary sequence of discoveries produced by empirical research. It then becomes possible to organize a more perspicacious presentation of the facts so as to highlight those that reveal salient features of the operative fundamental forces. Later Ampère would refer to this reconstruction as a "methodical exposition" (*exposé méthodique*).[36] Ideally, an accurate *exposé méthodique* should call attention to the most significant phenomena in a domain; further study should aim at the discovery of the *fait primitif*, the "primitive fact" ultimately responsible for all the observed phenomena. We have already noted how Haüy and Biot praised Newton's astute recognition of the *fait primitif* of universal gravitation stated as a fundamental law from which phenomenological laws could be derived. In general, it is only after a perceptive *exposé méthodique* has been outlined and the *fait primitif* of the domain has been established that theoretical principles should be postulated. During the height of the 1821 debates over optics and electrodynamics, Biot gave the following résumé of this process.[37]

> Observation and sometimes chance reveal the phenomena; the experimental method develops them and determines their physical laws; however, the ultimate mystery of the *elementary forces* that produce them can only be revealed by the power of thought.

Furthermore, in an oft-cited passage, Biot explained that this effort "in some manner of divination is the goal of almost all physical research."[38] It is in this "divination" of fundamental forces that analogy and simplicity play an important, albeit an ill-defined role. To cite the most obvious example, Coulomb's theories of electricity and magnetism were based upon inverse square fundamental forces strikingly analogous to Newton's gravitational force. A fundamental force also should be "analogous" to the phenomenological laws highlighted by an *exposé méthodique* of the domain. It is no small decision to declare that empirical research has reached a stage where a definitive *exposé* is possible. This decision represents an attempt to resolve a central tension between two *desiderata* of Laplacian methodology. On one hand, the divination of theoretical principles should be based upon accurate knowledge of all known phenomena so as to make use of "analogy." On the other hand, theoretical principles should not repeatedly require revision in the light of empirical discoveries. These two demands allow for significant disagreement about whether the discovery of new phenomena provides evidence for or against a theory.

For example, beginning in 1815, disagreement over this issue generated much of the controversy between the corpuscular and wave theories of light. As we will see in chapter 6, Ampère observed this debate closely; he would discover that quite often arguments about whether theoretical modifications in Laplacian theories had become excessively *ad hoc* could not be resolved simply by a comparison of data and calculations. The decision was also influenced by prior commitments and aesthetic judgment of the theory's mathematical elegance and conceptual simplicity. But in 1804 these issues received little of Ampère's attention. During his first decade in Paris his personal life and the study of philosophy were his two top priorities.

4

Ampère in Paris

With his move to Paris in 1804, Ampère began the second half of his relatively short life of 61 years. At that point he was 29; he lived in Paris, more or less discontented, for the rest of his life. Unlike many of his new colleagues, he did not come to Paris as a young and impressionable student. His character had already been molded by his unique personal and intellectual experiences in Lyon and Poleymieux; he never made a happy adjustment to his new surroundings.

The period between 1804 and Ampère's death in 1836 was a time of complex political and social change for all Frenchmen. As a Parisian, Ampère witnessed the consequences of Napoleon's coronation as Emperor, his military successes and subsequent defeat, the initial Restoration of the Bourbon Louis XVIII in 1814, Napoleon's return for the Hundred Days, the second Restoration of 1815, and the Revolution of 1830. While Ampère's introspective mentality rendered him indifferent to many of these events, political developments inevitably impinged upon his career through the institutions that employed him. Dependent upon the French state for his livelihood as a teacher and *savant*, Ampère knew that his survival was contingent upon political propriety. His political orientation was understandably ambivalent. The circumstances of his father's death and the experiences of his friends in Lyon contributed to his aversion to violence and militarism. Bonaparte could not possibly win his respect. On the other hand, in

123

1814 Ampère adjusted to the role of royalist in support of Louis XVIII. The thought of a paternal and enlightened sovereign came easily to him. Furthermore, Ampère's return to the religious practices of his youth allowed him to participate in the Catholic revival of the Restoration period.

The Paris that Ampère experienced was a city of astounding disparities. Without drastically expanding in area, its population increased from approximately 547,000 in 1801 to 866,000 in 1836.[1] The vast majority of these people lived in conditions of either extreme poverty or minimal subsistence. Physical conditions of daily life were in many respects still those of a medieval city. Sanitation was primitive, old cemeteries were still within the city limits, and violent crimes were common. It is not surprising that Ampère always lived near his primary place of employment at the École Polytechnique. His own economic status placed him within the small minority of Parisians who lived quite comfortably; nevertheless, to maintain this status he had to pursue vigorously a multitude of teaching and administrative posts.

His initial appointment at the École was as *répétiteur* of analysis. A *répétiteur* was essentially a tutor to students who were lectured to by the *professeur* of the subject. Ampère taught at the École continuously until 1828, with promotion to the rank of *professeur* in 1815. Although as a *répétiteur* his salary was only 1,500 francs, it eventually climbed to between 5,000 and 6,000 francs. Nevertheless, to place these figures in context, the historian Maurice Crosland cites an 1816 report on the cost of living in Paris; an annual income of 15,000 francs was necessary to support a bachelor comfortably and 40,000 for "the support of a plain family establishment."[2] Ampère had to patch together a reasonable income by acquiring several simultaneous positions, a procedure the French referred to as *le cumul*. With support from his old Lyon friend Degérando he became secretary of the Bureau de Consultation des Arts et Manufactures in 1806. In 1808 he also was appointed *inspecteur* for the Université Impériale and soon was promoted to *inspecteur général*. His summer months were occupied by long inspection tours of provincial *lycées*. He also gave occasional lectures or courses at the Athénée des Arts and the École Normale. In 1814 he was elected to the Institut as a mathematician; we will consider the research that made this possible in chapter 6. In 1824 he became *professeur* of physics at the Collège de France. But Ampère's tumultuous personal life calls for attention before we consider his professional career in more detail.

Adrift in Paris: Religion and Remarriage

Ampère's early letters to the friends and relatives he had left behind indicate that he quickly become bored with his teaching duties and sorely missed his old environment. Surprisingly, in view of the isolated provincial life he had led up to that point, his first encounter with Paris did not induce any particular awe. In a droll letter to Élise on 2 December he mentioned that he had seen the Pope and Bonaparte pass by in their carriage on the way to Notre Dame for the coronation, a ceremony to which Ampère was thoroughly indifferent. That evening the newly crowned Emperor would make a triumphant tour of the city and would pass below the windows of Élise's brother's residence; Ampère admitted that he really should pay a visit and observe "all this pomp" first-hand. It is unlikely that he did so. He took little interest in public display and distrusted Bonaparte.

The fact that he made few friends among the faculty at the École was due more to his eclectic interests than to a lack of social skills. He quickly made contacts with philosophers and was welcomed into the group of *idéologues* associated with Destutt de Tracy. The epistemological basis of religion, the subject that had so fascinated him prior to his departure from Lyon, was now supplanted by abstract metaphysics and materialistic psychology. His "thirst for certitude" was not quenched by mathematics and he needed to fill the void created by his new indifference to religion. In common with many ambivalent Catholic intellectuals of his generation, he sought a secular basis for the idealism he had initially adopted from a religious perspective. As we will see in chapter 5, Ampère was drawn to the company of Maine de Biran. Psychologically, this shift in Ampère's interests was accompanied by a less humble attitude toward the intellectual life. The thought that human reason might be able to successfully understand its own mode of operation struck him as an intoxicating possibility. The change in his demeanor was noted by his old friends. Bredin made a journal entry following one of Ampère's visits to Poleymieux.[3]

> He is more changed than I thought. Last year he was a Christian; today he is only a man of genius, a great man! What can have disturbed his reason? . . . What have become of the sublime sentiments that filled his soul? He sees only glory; he is an idolater of glory! He takes pride in sounding the mysterious depths of human intelligence.

Postponing a detailed discussion of Ampère's philosophy for chapter 5, we should note that it had several facets. He was interested in the argumentative structure of scientific reasoning, a topic he pursued through a comparative study of the methods of analysis and synthesis and a critique of "explicative hypotheses." He also indulged his penchant for classification by creating elaborate classification schemes for the sciences and the corresponding mental faculties. His capacity for introspection drew him into even more complex taxonomies for the psychological processes that accompany perception and cognition.

Bredin was concerned that this "metaphysics" was detrimental to his friend's spiritual and psychological health. Ampère's investigation of perception struck Bredin as dangerously deterministic.[4]

> Your redoubtable enemy is metaphysics ... I regard certain ideas you have on the power of some faculties as very contrary to your happiness ... I thought I should point out that you particularly tend to regard man as being invincibly compelled to such and such a determination by his organization, his temperament, etc. It seems to me that you take for the original nature of man what is only a baneful habit.

Bredin was disheartened by Ampère's rejection of Catholicism, a development Bredin attributed to secular philosophy. Letters between the two friends abound with religious admonitions on Bredin's part and Ampère's unhappy replies that he is helplessly in the grip of his intellect and is unable to return to the carefree faith of his youth. This condition persisted for over a decade, accompanied by much wringing of hands and emotional fanfare. For Ampère, the doctrine of the eternal punishment of the damned was a particularly difficult impediment to a return to Christianity. He included the following passage in a letter to Bredin in 1807.[5]

> I see nothing contradictory in religious truths. God can do all. But I no longer see reasons that lead me to believe that the Catholic religion is inspired by him. The vague objections that I refuted in a better time present themselves to my eyes. I tell myself that, if this religion was the work of God, it would have propagated better without making a dogma of eternal suffering. How this idea disgusts me, for if the human race must always be unhappy, he would not have created it.

Not that Ampère enjoyed any sense of happy liberation from his old beliefs. Writing to Bredin in 1808, his mood was one of despair rather than victory.[6]

I feel annihilated. It is only with the most painful effort that I can connect my mathematical ideas; my soul is frozen. Do you recall the time when I so enjoyed the pleasure of doing good, or what I thought to be so? This memory will always be dear to me. Others have abandoned God for the delights of this world; I have only renounced him to procure for myself the most bitter sorrows.

This passage was written shortly after the termination of Ampère's disastrous second marriage, a period when he would have welcomed the comfort of religious belief. The turmoil of his family life was a constant factor during the entire first decade after his move to Paris. His creative work in both mathematics and chemistry during this period attests to his ability to concentrate in the midst of domestic chaos.

It was Degérando who introduced Ampère to the woman who became his second wife. Degérando had just completed his *Histoire des Systèmes de philosophie* and he seems to have taken Ampère under his wing upon his arrival in Paris. Their friendship dated from the days of philosophical discussion in Lyon and Ampère trusted him. Unfortunately, Degérando's judgment about an appropriate mate for his friend went disastrously awry. Jean-Baptiste Potot was a member of the Lyon Académie and now resided in Paris with his wife and their 26-year old daughter Jeanne-Françoise, who went by the name Jenny. Degérando introduced Ampère to the household and by the fall of 1805 he had become thoroughly infatuated and intent upon marriage. In the light of Ampère's emotional nature and his lonely condition, it is not surprising that he sought female companionship. It is far more difficult to understand his more specific attraction to Jenny Potot. The surviving descriptions of her are uniformly disparaging. Admittedly, the record is one-sided in that it is provided by Ampère and his friends in reaction to the sad sequence of events that transpired. Nevertheless, that record does paint a remarkably bleak portrait. Ampère's biographer de Launay bluntly characterized the entire Potot household as an intellectual void and a nightmarish web of vanity and petty intrigue.[7]

It was a household as bourgeois as possible, in the pejorative sense in which artists understand that word: narrow ideas, prejudices, pretensions, living only for money and vanity, not having the least ideas of the sciences, exactly the opposite of what it should have been for a big child as simple and modest as Ampère.

Degérando probably hoped that Jenny Potot would fill the role of companion and housekeeper and would provide the domestic stability in which Ampère's creativity could flourish. Ampère, on the other hand, proceeded with the same passion that had animated his relationship with Julie. While Julie herself had been taken aback, she did develop a genuine affection for him, albeit a somewhat maternal one. Jenny Potot and her parents reacted with a far more calculating mentality. Our most detailed account of the affair is provided by an 1808 summation provided by Ampère for the court proceedings that were under way at that point.

According to Ampère, he was accepted as a suitor in spite of misgivings about his financial status. By 26 April, 1806, it was agreed that a wedding should take place following a preliminary transfer of 7,200 francs to Jenny's father and Ampère's agreement to a complex marriage contract. The contract included clauses intolerable to both Ampère and his mother, and when he expressed his objections, an elaborate scene was staged with dramatic details worthy of a Balzac novel. Ampère was summoned to the Potot residence and was informed that Jenny was so distraught over his delay that she had not eaten for eight days. She allegedly wished to bid him her final farewell but her father claimed that this might endanger her health. Nevertheless, Ampère was eventually allowed to see her and gave up his resistance to the marriage at the sight of her "bathed in tears." He signed the contract on 31 July in spite of Degérando's objections.

The wedding took place on 1 August with witnesses of a stature appropriate to the vanity of everyone concerned. In addition to Lagrange and Delambre, they included the Interior Minister, Champagny, and the Governor of the École Polytechnique, General Lacuée. These were not men who made public appearances casually; clearly Ampère had attracted attention in high places. Unfortunately, the marriage degenerated almost immediately after the new couple began living together in the Potot household. According to Ampère's 1808 statement, Jenny promptly informed him that she had no intention of bearing children. The emphasis on finances that preceded the wedding suggests the possibility that she felt no physical attraction to Ampère and may have objected to sexual relations with him altogether. He persisted however, and she became pregnant within about two months. Outraged, she withdrew to her mother's bedroom, consigned Ampère to a small office, and allowed him into her presence only at meals. His mail was opened and servants were instructed to reject his visitors; she ceased speaking to him altogether during May of 1807. On 26 June he conformed to Jenny's demand that he leave the Potot residence and took a room provided for him by Champagny; it was while living there that he learned from a porter of the birth of his daughter, Anne-Joséphine-Albine on 6 July.

Given that we have only Ampère's side of the story, Jenny Potot's behavior elicits little sympathy. Ampère cited as her motivation only an insistence that[8]

> he renounce his opinions in morality; that is, the first principles of all honesty and all virtue. For Mr. Ampère can render this testimony that what he had manifested of his opinions in this regard were anything but exaggerated; furthermore, since he had never required that his wife adopt his opinions, how had she the right to prescribe what he should think?

Whether her rejection of Ampère was based on physical, financial, or "moral" disappointment, there is no denying the thorough manner in which she distanced herself from him. Her aversion to him was combined with a complete indifference to their daughter who remained in foster care. Meanwhile, Ampère persistently tried to convince her to take their marriage seriously. These efforts continued as late as 1813 and were accompanied by illusory hopes of success. The surviving letters tell us virtually nothing about why Ampère was attracted to her. His Catholic heritage may have compelled him to some extent to try to save the marriage out of a sense of duty. On the other hand, his veiled references to the limited power of human will may indicate that he felt incapable of overcoming a purely physical attraction. Bredin recorded his dismay that Ampère found "moral beauty exclusively in the sensibility, provided that it is exalted and without rule. Your sentiment of the beautiful is only the enthusiasm of passion."[9] Ampère's mother was sympathetic to her son's plight, but she also straightforwardly attributed the blame to his own lack of discrimination.[10] "Do not let yourself become depressed. Be a man. If you had not been so weak, you would not have been led like a plaything."

Even though he continued informally to try to reconcile his wife to living with him, Ampère did also take legal recourse during 1808. In April of that year he obtained a court order requiring her to take up residence with him. She appealed, and it was in response to this development that Ampère wrote the history of their relationship that is our primary source of information. The upshot of the proceedings was that in July he was granted a separation decree and received custody of Albine. During the fall of 1807 Ampère had convinced his mother and his sister Joséphine to come to Paris to take care of his household, bringing with them his son Jean-Jacques. Madame Ampère had been doubtful about doing so and Ampère's friends in Lyon felt he was acting too selfishly. She survived in Paris for less than two years, dying on 4 May, 1809. Julie's sister Élise had died of tuberculosis in January of 1808, and these deaths contributed to Ampère's depression. Although he found temporary solace in mathematics, chemistry, and metaphysics, the relief was always fleeting. Bredin thought about his friend a great deal; he gave the following analysis in a letter to him early in 1811.[11]

> Don't I know all too well that this miserable existence, so painful for all those condemned to it, is a thousand times more sorrowful still for you than for the greatest number of others? If your turn of mind and the vivacity and the prodigious activity of your imagination give you numerous pleasures, these are only incomplete enjoyments, false and passing, which leave your heart empty of happiness . . . My poor friend, the more I think about your manner of feeling, the more I feel that you are the most unfortunate of the men I know.

Although he continued to write fawning conciliatory letters to Jenny Potot, Ampère also made other female acquaintances. During 1809 he began seeing a woman he referred to in his letters as *la constante amitié*. He told Bredin that she reminded him of the woman depicted in an engraving with that title. She was younger than Ampère and he agonized over the attention she received from other men. In September of 1811 she abruptly entered into a marriage of convenience with an elderly but wealthy provincial. Ampère was stunned and incapable of any activity for several weeks. He stayed in contact with the lady and had an emotional meeting with her in Paris in September of 1812. He confessed to her that he had been seeing another woman who had ingratiated herself with him by producing portraits of him and his son, Jean-Jacques. The friendship was repaired after another meeting and Ampère congratulated himself for choosing this path rather than the pursuit of what he called "senseless desires." The convoluted and overblown style of his letters makes it difficult to determine the extent to which these affairs were figments of his imagination. Even Bredin had a hard time deciphering his vague allusions.

But Ampère's psychological condition became far worse during 1814. This was a year in which the emotional and academic pressures of his life combined to produce almost unbearable tension. When the mathematician Bossut died in January of 1814, Ampère immediately campaigned to replace him in the Institut. Although his efforts were hampered by his heavy teaching requirements, it was during this period that he managed to write his most creative mathematics memoirs. He also tried to stay abreast of chemistry; he was bitterly disappointed when he saw other men develop what he considered to be his own insights. New romantic intrigues generated further episodes of guilt and anguish about his helplessness to overcome what he called a "superior force" that repeatedly brought suffering into his life. During March of 1814 guilt was the predominant emotion.[12]

> Through my injustice I have poisoned the entire life of the best of created beings . . . Ah, the sentiment dominating in my heart is to sacrifice my entire life, all that which depends upon my will, to attempt to soften sufferings for which I cannot atone. But what can I do? Almost nothing! Nothing at all in comparison to the evil I have done!

Whether the cause of all this soul-searching was the *constante amitié* or someone else, Ampère soon overcame his temporary inclination to "sacrifice all." Before long he was boasting to Roux that his study of chemical combining ratios and atomic structure had produced a discovery which, "after what I accomplished last summer in metaphysics, perhaps will be the one that I will conceive as the most important of my entire life." Nevertheless, he did have a dark moment later in March. He and the *constante amitié* made a painful decision to terminate their relationship. As usual, Ampère partially confided in Bredin, this time in a letter verging on hysteria.[13]

> Let this ghastly day remain engraved in your memory! Maybe one day you will know all that your friend has had to suffer! . . . I will see you again, I am sure of it at present. Do not fear at all for my life! The connection that has just been broken had attached me perhaps more than all the rest; but my two children still remain to me and will still attach me there. Besides, I have nothing with which to reproach anyone. There remains to me only the consolation of thinking that no human will could have foreseen or ordained this event; but it is an omnipotent will which has brought to bear on me the blows of its vengeance. Ah, no doubt I deserve what has happened to me since it has permitted it, since it has so brought it to pass that no one, at least I flatter myself in this respect, could have foreseen it! I do not know what I am writing to you; this is not what I wished to write you. All my organization is revolting! I can do no more! There, alone in this little room. Ah! My friend, and who knows what misfortune perhaps awaits me; I feel myself becoming faint, I can no longer write! Don't worry on my account, I will see you again!

With due allowance for Ampère's penchant for exaggeration, his allusions to suicide should be taken seriously. The deaths of Julie, Élise, and his mother, the tragicomedy of his second marriage, and now the painful denouement of his last romance, all left him sufficiently overwhelmed to attribute his fate to the wrath of an omnipotent power. His letter to Bredin concluded by describing the temporary solace he found in opening his copy of the *Imitation of Christ*. Unable to penetrate Ampère's opaque prose, Bredin could only respond with perplexed words of consolation. At the risk of oversimplifying Ampère's complex personality, it is surely true that his first decade in Paris was a painful experiment by a man who had lost his religious convictions. His attempt to savor worldly pleasures was an abysmal failure and he was left in a bitter state of despair. Although Bredin was always a sympathetic correspondent, the two men did more to amplify each other's gloomy condition than they did to improve matters. Ampère's spirits were not bolstered by reading Bredin's fatalistic reports of poor health and domestic problems.

Furthermore, throughout this period the relationship between Ampère and his son Jean-Jacques was strained. They had spent very little time together prior to their reunion in Paris in 1807. By that time Jean-Jacques was seven years old and had been cared for primarily by Ampère's mother and his sister Joséphine. He was quite bright, learning to read at an early age as his father wished. He also had violent fits of temper, behaving, as Joséphine put it, like a little devil. Ampère too was subject to bursts of anger, and he lacked the patience required to cope with his son's moods. He had no qualms about sending him off to boarding school so as to have to live with him only during vacations. Since Ampère spent a large portion of those summers traveling in his capacity as a University inspector, he had very little contact with his son. Thoroughly caught up in his own anxieties, romances, and intellectual life, he devoted little time or energy to Jean-Jacques. This bland relationship stands in sharp contrast to the close bond between Ampère and his own father. As Jean-Jacques matured, Ampère gradually paid more attention to him and encouraged his literary career; but during the childhood years there was little interaction and Ampère's battered emotions found little solace there.

Ampère did not recover from the episode of 14 March, 1814, quickly. He was oblivious of the hectic political events of that spring – the armies of the allies marching into Paris on 31 March and the abdication of Napoleon on 6 April. He worried about the effect his delirious letters might have had on Bredin and he tried to reassure him.[14]

Everything takes place around me as in a theatrical performance. I am absorbed by too many reflections about myself. But nevertheless I am aware how my poor children are still attached to me for life. May Heaven reserve a different fate for them than that of their father!

Ampère's brooding was disrupted intermittently by his campaign for admission to the Institut. Memories of his unsuccessful campaign of 1813 fueled Ampère's anxiety as he tried to concentrate on mathematics. As he mentioned in a letter to Ballanche, he could find genuine relief in doing mathematics when his attention was focused.[15]

> I am going to take up mathematics again. I have some trouble at first, but, when I have overcome the initial repugnance, I no longer want to leave the calculations. I still experience a great charm there when I can eliminate every other thought and occupy myself with it alone, absolutely alone.

He completed a significant memoir on partial differential equations and he presented parts of it to the Institut during July and September of 1814. He also profitably invested some energy in the customary social calls on influential *savants*, and he wrote a long letter to Delambre summarizing the results of his research. The election took place in November of 1814. In spite of lobbying by Laplace and Cuvier in favor of Cauchy, Ampère triumphed with 28 of the 52 votes cast. Nevertheless, after a brief period of exhilaration, he once again fell into a state of lethargy.

Shortly before the election Ampère had helped to organize an *académie psychologique* for the discussion of psychology and ethics. The members included Royer-Collard, Degérando, Maine de Biran, Georges Cuvier, and Frédéric Cuvier. Meetings were held on alternate Thursdays and consisted of discussions of papers read by the members. The format thus resembled that of the earlier *société chrétienne* in Lyon except that the focus had shifted from religion to ethics and the analysis of perception. Ampère loved the spontaneity and freedom of informal discussion. When the 1814 discussion group dissolved, he organized others in 1817 and 1819; he was always searching for a forum more congenial than the formal meetings of the Institut. Nevertheless, he was disappointed that his efforts did not result in influential publications, particularly on the part of Maine de Biran.

During 1816 Ampère passed through another religious crisis that resulted in a return to Catholicism. Ironically, it was during this period that Bredin, who had steadily admonished Ampère to return to the faith, found himself disenchanted with institutionalized religion and drawn to mystics such as Jacob Boehm. Ampère had never been comfortable with agnosticism but he had accepted it as part of his miserable condition. His renewed religious conviction was at least partially based on arguments similar to those he had found compelling in Lyon. His correspondence with Bredin contains lengthy religious discussions; unfortunately, many of their letters during the crucial year of 1816 are missing. In one of the surviving letters from January, 1817, Ampère wrote with the confidence of his renewed commitment.[16]

> What furnishes me this proof is the fact that 250 sects have been created in 100 years, sects all filled with this proud sentiment that each of them alone has the truth, all of which differ from the permanent Church only in order to bend its mysteries and miracles in such a fashion as to lose from the one and the other all their sublimity by rendering them ever more contrary to what seems to us naturally true in the state of ignorance in which we are.

Although Ampère berated himself for having introduced Bredin to the mystics he feared were leading his friend astray, he emphasized devotion more than doctrine in his own practice of Catholicism. He did insist upon a recognition of the sole authority of the Church, an act of humility that he had found difficult. The theme of the gift of faith appears frequently in the emotional personal exhortations and prayers of thanksgiving he composed. For example, the following lines were probably written in a little notebook shortly after his final conversion.[17]

> Today, 5 October, the nineteenth Sunday after Pentecost, God has revealed to me what my eternal salvation depends upon. Could I ever forget it? Great Saint Joseph, to whose intercession above all I owe this grace, Saint Mary, mother of God, whose name I received at my baptism and to whom I also have this inexpressible gift [sic], always intercede before God that he may conserve it for me and that I might make myself worthy of it!

And in another passage he expressed a new resignation to the limitations of the intellectual life, a resignation fostered by the disappointments he had experienced since his arrival in Paris.[18]

> Scorn your mind; it has deceived you so often. *How could you still rely upon it?* When you made an effort to become a philosopher, you already felt how vain this mind was that consisted in a certain facility in producing brilliant thoughts. Today, when you aspire to become Christian, do you not feel that there is no good mind save that which comes from God? The mind that separates you from God, the mind that leads you astray from the true good, however penetrating, however pleasant, however clever it may be to procure us corruptible goods, is only a mind of illusion and *aberration*.

134

He resolved to keep his intellect in check and subordinated to his spiritual goals.[19]

> My God, what are all these sciences, all these reasonings, all these ingenious discoveries and vast conceptions that the world admires and that our curiosity feasts upon so avidly? . . . In truth, nothing but pure vanity.
>
> So study, but without being over zealous! . . . Let the half-exhausted vigor of your soul be used in less frivolous pursuits! . . . Do not consume it in such vain things!
>
> Take care not to let yourself be preoccupied by the sciences as you have in the past. Work in a spirit of devotion! . . . Study the things of this world, that is the duty of your station, but consider them with only a single eye; let your other eye be steadily fixed on the eternal light! Listen to the savants, but listen to them with only one ear! . . . Let the other be ever ready to receive the soft tones of the voice of your celestial friend!
>
> Write with only one hand! Hold tightly with the other to God's raiment like a child clinging to his father's cloak! . . . Without this precaution you will surely hit your skull against a stone.

Probably written shortly before the important transition year of 1820, this passage is a remarkably poignant expression of Ampère's ambivalence toward the intellectual life. He did not always abide by the guidelines he prescribed for himself. We will see that he pursued electrodynamics with the same ambition and insistence upon due recognition that accompanied his research in chemistry. It was relatively easy for him to pledge himself to moderation and humility in 1817 or 1818, that is, immediately following his disappointing attempts to achieve major recognition as a philosopher and chemist. His ego and vanity reasserted themselves powerfully during the priority disputes and competition of the early 1820s. Nevertheless, beginning in 1817 Ampère did enjoy a new sense of emotional and spiritual stability. The Catholic revival of the Restoration period provided a sympathetic environment. His return to the religious practices of his childhood also lent some structure to his resigned acceptance of middle age. This attitude is evident in an 1818 letter to Bredin:[20] "Well dear Bredin, God wanted to prove to me that all is vain save for loving him and serving him."

In addition to his religious renewal, other developments contributed to Ampère's fortified sense of stoic perseverance. He gradually sold off his property at Poleymieux and purchased a house near the École Polytechnique at 19 Rue Fossés-Saint-Victor. He directed extensive renovations and moved in during September of 1818. He enjoyed his small garden and courtyard and readily took in scholarly boarders. The chemist Desprez was the first of many; others were Augustin Fresnel between 1822 and 1827 and Antoine-Frédéric Ozanam in 1831. Although Ampère had been loath to part with the family homestead, financial and practical considerations convinced him. Since most of his summers were taken up by his university inspection tours, he could spend very little time at Poleymieux. By renting part of his new house he supplemented his income and had an audience ready at hand to fill his insatiable need for conversation. Psychologically it did Ampère immense good to have an environment in which he could expound at length on whatever subjects crossed his mind.

He also began taking more interest in the education and ambitions of his son Jean-Jacques. The boy was well trained in languages, literature, and the sciences. Ampère initially encouraged him to find a position in the chemical industry, but Jean-Jacques insisted upon a literary career by the time he reached the age of 20. It was at that point, January of 1820, that he met Madame Récamier and began attending her salon. Although Ampère was disappointed at his son's slow start as a writer, he sincerely sympathized with Jean-Jacques' literary efforts; in 1820 he had high hopes for his son's future.

The Academic Life

Ampère's ability to produce creative scientific research in spite of the turmoil of his personal life is all the more remarkable when we factor in the demands of his professional career as a teacher and administrator. In 1808 Napoleon created the Université Impériale, and Ampère joined Poinsot as one of the original *inspecteurs*. In spite of its idiosyncratic terminology, the Université was actually a national system of education in which the country was geographically divided into 30 *académies* which in turn were subdivided into *départements*. The system included both *lycées* and more sophisticated *facultés*, the most prominent of which was the Faculté des Sciences in Paris. The task of an *inspecteur* was to supervise the curriculum and staff of the institutions within a specific *académie*. As one of six *inspecteurs générales*, Ampère was responsible for several *académies* in any given year; he thus had to devote a significant portion of his summers to lengthy inspection tours in the provinces. He tried to draw southern tours so he could visit his friends in Lyon. Although this provided some respite, he often returned to his teaching in the fall more exhausted than he had been when the summer began.

Ampère's teaching responsibilities were taxing and extremely time-consuming. In chapter 3 we noted the early history of the École Polytechnique and the militarization that was decreed in 1804, the year of Ampère's arrival. The full militarization process was delayed for one year by Napoleon's decision to move the school from the Palais Bourbon to the Collège de Navarre in the Latin Quarter. The new site combined teaching facilities, student barracks, and room for military exercises. Ampère lodged there during his early days in Paris. Located between the Sorbonne and the Jardin des Plantes on the Montagne Sainte-Geneviève, the school remained in these buildings until 1976, when it was moved to a suburban location outside Paris. Ampère's colleagues were a talented and ambitious group. Lacroix and Gaspard de Prony were *professeurs* of analysis and mechanics on Ampère's arrival in 1804; both had been active members of the Académie des Sciences and the Institut since 1799 and 1795 respectively. Lacroix was the renowned author of mathematics textbooks, books which Ampère had relied upon before coming to Paris. De Prony was the longstanding *directeur* of the École des Ponts et Chaussées, one of the most important of the *écoles des applications* attended by École Polytechnique graduates. In 1804 the young Poisson had already been a *répétiteur* for four years. Although six years younger than Ampère, he was elected to the Institut in 1812 and was welcomed into the elite membership of the Société d'Arcueil. His narrow dedication to mathematics and teaching stands in sharp contrast to Ampère's eclecticism. Outside their professional activities the two men saw little of each other. Many other men taught analysis and mechanics in less regular positions as adjoints. Poinsot, for example, taught irregularly as a *répétiteur* between 1808 and 1815; he was elected to the Institut in 1813 following the death of Lagrange. Ampère's defeat in this election was naturally disappointing; he was more assertive during his successful campaign the following year. The concentration of French scientific talent in Paris, the relatively few positions available, and the aggressive campaigning required by *le cumul* produced a competitive environment that Ampère found more vexing than exhilarating.

Prior to the Restoration, the major ideological tension within the École Polytechnique was due to Napoleon's intention that it primarily provide military technicians. Curriculum and staffing policies were directed by the school's Conseil de Perfectionnement which included Laplace, Monge, and Berthollet. Although these men were not inclined to risk a major confrontation with the Emperor whose patronage they enjoyed, they did want to ensure that the École would continue to

137

produce students capable of contributing to their own research programs. In Laplace's case, the extensive military applications of mathematics made it relatively easy to preserve a major mathematical component in the curriculum. Nevertheless, students objected that they were forced to spend a disproportionate amount of their time on mathematics that was too abstruse to be of practical significance for their careers. They were joined in this protest by senior scientists, who considered much of this mathematical training to be of little value for practical science. As one who taught analysis and mechanics, Ampère thus found himself in a position of considerable controversy. Debate about the appropriate content for the analysis course elicited strong feelings and Ampère was repeatedly called upon to defend his choice of topics and the level of mathematical sophistication he expected.

Tension increased in 1816 when Augustin-Louis Cauchy was appointed to the faculty, an appointment that took place under the new political conditions of the Restoration. Although Ampère had been Cauchy's *répétiteur* between 1805 and 1807, the two men never became close friends. To some extent this may have been due to Cauchy's irascible personality. Unlike Ampère, Cauchy was raised in Paris and was educated at the École Centrale du Panthéon and the École Polytechnique. In further contrast to the self-taught Ampère, he studied analysis under Lacroix and mechanics with de Prony. Following graduation from the École he went on to the École des Ponts et Chaussées and worked for several years as a civil engineer. His delicate health forced him to give up field work and he unsuccessfully sought election to the Institut in 1813, 1814, and 1815. Throughout this process he was supported by his politically active father, Louis-François Cauchy. Deeply antagonistic to the Revolution, Louis-François passed on his royalist political attitude to his son. Steadfastly committed to Catholicism, Cauchy suffered from none of the ambivalence that plagued Ampère. He was sincerely devout but in a cold and sanctimonious manner.

The new political atmosphere of the Restoration was at least partially responsible for Cauchy's appointment as an adjoint *professeur* of analysis and mechanics at the École Polytechnique. Hiring decisions were complicated by the lack of retirement pensions, a situation that encouraged elderly professors to retain their positions even when they were physically unable to fulfill their responsibilities. One common practice was for a younger teacher to "substitute" for an ailing professor and receive a portion of his salary. Even though he was only a *répétiteur*, Poinsot had repeatedly followed this procedure because of his poor health. Early in November of 1815 the Conseil de Perfectionnement decided to retire the professors who were no longer able to teach; this allowed Ampère and Poinsot to advance to the position of full *professeur* even though they would not receive the full salary of that position until the death of the emeritus professors. Poinsot immediately complicated matters; he declared himself incapable of teaching that term and requested his usual replacement. The administration balked at this maneuver and not only revoked Poinsot's promotion but hired Cauchy to replace him and promoted him to *professeur* the following year. It was rumored that Cauchy received the position because his religious and political views were attractive under the new political climate. Monge, Guyton-Morveau, and Lacroix all saw their careers terminated, owing to their revolutionary and Napoleonic politics. The more flexible Laplace flourished and acted as one of Cauchy's patrons. On the other hand, Cauchy's mathematical ability had become common knowledge; the only uncertainty pertained to how he would perform in the classroom.

Conditions at the École soon reached a crisis. The students were notorious for their political independence and their resistance to religious conformity. They became unruly during the spring of 1816 and they were disbanded by royal decree on 12 April. During the next few months Laplace directed a commission that reorganized the École; the school was demilitarized, fees were raised, liberal faculty members such as Poinsot were dismissed, and Cauchy's and Ampère's positions were solidified as *professeurs* of analysis and mechanics. Ampère's return to Catholicism at precisely this juncture came at an opportune moment.

During November and December of 1816 Ampère and Cauchy cooperated in an attempt to revise the analysis and mechanics curriculum. After considerable deliberation, the curriculum commission allowed them to alternate teaching analysis and mechanics to the two *promotions* in residence, that is, in even-numbered years Cauchy taught analysis to first-year students while Ampère taught mechanics to those in the second year. In odd-numbered years Cauchy taught mechanics to his *promotion* while Ampère began a new cycle by teaching analysis. This pattern continued until 1828 when Ampère's position at the Collège de France allowed him to resign from the École. But these reforms did not forestall a protracted debate about how much applied mathematics should be included in the analysis course. Ampère generally followed Cauchy's lead in these curriculum disputes. It was also indicative of Ampère's renewed sense of humility that he attended Cauchy's classes along with the students.

Unfortunately, Cauchy's teaching techniques rapidly became

notoriously highhanded. He repeatedly extended his analysis lectures at the expense of the mechanics course and indulged in mathematical discussions more appropriate to a research memoir than to an introductory class. Ampère's status was thus complicated by the extreme tactics exercised by his colleague. Teaching calculus at a time when the foundations of the subject were still uncertain was difficult under any circumstances. Ampère took a relatively conservative approach, at least in comparison to Cauchy. He jotted down some of Lacroix's observations about the analysis course in an undated note; Lacroix's assessment probably took place during the period between 1808 and 1815 when he was still an examiner for the École and Cauchy had not yet joined the faculty. Three of the eight comments are particularly interesting; no professors are named although presumably Ampère was involved as a *répétiteur*.[21]

2° they complicate their calculations too much by putting all the differential coefficients into fractions instead of making use of differentials.

3° they have not sufficiently insisted upon the passage to the limits of the infinitely small.

4° Mr. Lacroix prefers the equations to be written in differentials without fractions.

Beginning in 1816, Cauchy had no qualms about introducing a more thorough algebraic foundation for the calculus. By accurately defining the derivative as the limit of a difference quotient, Cauchy accomplished the task Ampère had only broached during his work on the subject ten years earlier. We will study these mathematical developments in more detail in chapter 6. For the moment we should note that Cauchy could not resist including his path-breaking research in what were supposedly introductory classes. These innovations were not well received by either students or other faculty members. Complaints were to no avail and in June of 1820 the Conseil d'Instruction directed Cauchy and Ampère each to publish a collection of lectures that would be open to public scrutiny and could be made available to students. Cauchy responded with his *Cours d'Analyse de l'École Royale Polytechnique*. It was entirely devoted to Cauchy's algebraic analysis of limits, continuity, and convergence; practical calculus techniques were not included. The misleadingly titled *Cours* was thus quite inappropriate as a text for courses at the École, and apparently it was never distributed to students.

On the other hand, it is ironic that just as Cauchy entered his most creative period, Ampère almost entirely lost interest in mathematics. He did complete the memoirs that ensured his election to the Institut in 1814; the second installment was finally ready for publication in 1820. This was a labor of duty rather than love. Consequently, Ampère was not tempted to be innovative in the classroom. He was content to let Cauchy wage a battle on that front while he satisfied what he took to be the requirements of his position. The 1820 decree that he publish an edition of his lectures came just prior to his obsession with electro-dynamics; he had to be repeatedly reminded of this obligation during the next few years. By 1824 he did manage to put together 152 pages of a collection he titled *Précis de Calcul différentiel et de calcul intégral*. Although Ampère never completed it, it does provide some insight into his teaching. However, for the very reason that Ampère took seriously the fact that he was teaching an introductory course, the edited version of his lectures is not particularly interesting. His pre-liminary discussion of the "derived function" was essentially the same one he had developed in a memoir published in 1806. We will in-vestigate that memoir in chapter 6; at present we should simply note that Ampère did not introduce Cauchy's procedure of limits. He defined the derivative as the value of the difference quotient when the change in the independent variable becomes zero. He concentrated on examples of how to calculate derivatives of algebraic, logarithmic, and trigono-metric functions of one and several variables. In contrast to Cauchy, he did not use his classes as a sounding board for revolutionary ideas.

Indeed, these classes became increasingly burdensome as the years passed. In 1824 Ampère successfully campaigned for a 5,000-franc position at the Collège de France with the intention of resigning from the École faculty. A position at the Collège held obvious attractions for him. There were no examinations to grade; students attended from a desire to study with specific professors who were free to design courses around their own research interests. One could hardly imagine an environment more different from the École. Ampère's campaign was something of a comedy of errors. In December of 1823 Lefèvre Gineau, the professor of experimental physics, was dismissed from the Collège for his lack of royalist enthusiasm. Six candidates immediately applied for the position, including Fresnel and Biot's protegé Pouillet. In ap-pointments of this type, the national Minister of the Interior accepted recommendations from the Collège faculty and the Académie, but ultimately he made the decision himself. Ampère committed some social gaffes during the preliminary campaigning and the faculty voted for the Minister's favorite, Beudant. Ampère was then offered the sup-port of the government if he would resign his position as Inspecteur.

141

Ampère of course wanted to retain this position and resign from the École; he dropped out of the competition at the thought of trying to teach two courses in his declining state of health. The Académie promptly voted for Fresnel, a development that induced Beudant to throw his support to Ampère. The Collège faculty now also recommended Ampère and he was granted the position of professor of experimental physics on 20 August, 1824. After lengthy negotiations he managed to retain a 2,400-franc pension from his Inspector's position, and he was eventually reinstated at full salary in 1828. It was only at that point that he could at last afford to resign from the École Polytechnique.

Ampère joined an eclectic faculty at the Collège, many of whom he had already encountered in other contexts within the hothouse atmosphere of the Parisian scientific community. Ironically, the professor of mathematical physics was Biot, primarily an experimentalist who was resolutely opposed to Ampère's theories of electrodynamics and magnetism. Cuvier taught natural history; he and Ampère engaged in some interesting debates during the last years of Ampère's life. Ampère also unsuccessfully tried to marry off his son Jean-Jacques to Cuvier's daughter. Although Ampère's health declined during this period, he was free to teach in a more extemporaneous mode than was allowed by the rigid curriculum of the École Polytechnique. Initially he concentrated on electrodynamics. During 1826–7 one of his students was Joseph Liouville, who had taken his course at the École Polytechnique the previous academic year. Liouville helped edit a set of notes for the course and they contain some interesting developments of Ampère's electrodynamics. In later years Ampère indulged his old fascination for classification schemes. His lectures on this topic must have been free-wheeling indeed; they contributed to his *Essai sur la Philosophie des Sciences* which we will consider in chapter 10.

The years in which Ampère taught at the Collège de France also constitute the period in which he established his reputation as a stereotypically absent-minded professor. Most anecdotes about his bemused behavior stem from this time. He was reported to write equations on the back of carriages only to see them carried off into traffic. He was said to have thrown his watch in the Seine while he placed a stone in his pocket. With due allowance for exaggeration, this activity was accepted more tolerantly in his new surroundings than it would have been by the wags at the École Polytechnique.

Ampère's experience with *le cumul* exemplifies the competitive professionalization of science in early nineteenth-century France. The small size of the teaching institutions meant that they could not pay salaries enough to live on to the highly specialized professors who taught the small number of classes offered in any specific discipline. Holding more than one position inevitably increased the time each man had to devote to administrative and bureaucratic responsibilities. Personal relationships became strained as the competition for each available position produced one victor and many losers. Within this narrowly technical, elite, and often snobbish community, Ampère was notorious as one of the few both willing and able to interact with a broader intellectual environment. This is the context in which we should consider his achievements as a philosopher and scientist prior to 1820. We begin with the metaphysics that he often claimed was the only science of lasting importance.

5

Metaphysics: Ampère, Kant, and Maine de Biran

Ampère never abandoned one of the optimistic convictions of his youth, the belief that scientific research would eventually reveal the true causal structure of nature. Ampère's intuition that scientific knowledge is not limited to phenomenological laws was bolstered by his study of metaphysics, the discipline he called "the only truly important science." His attempt to stipulate the objective content of scientific knowledge drove him to construct an alternative to what he understood to be Kant's "idealism." Relying upon what he knew about the function of "explicative hypotheses" in physics, he argued for an analogous hypothetico-deductive epistemology as early as 1802. Maine de Biran's criticism of this project forced Ampère to give more careful attention to the logic of scientific reasoning and particularly the concept of a *fait primitif* or "fundamental fact." During 1817 Ampère lectured on scientific methodology at the École Normale; in these lectures he explained his preference for *direct analysis* as opposed to *indirect synthesis*, a preference he would soon put into practice in his research in electrodynamics.[1]

144

Ampère's Reaction to Kant: Transcendental Realism

Ampère's complex intellectual life brought him crucial respite from the mundane details and emotional traumas of his personal life. The discovery and contemplation of mathematical truths brought him the temporary solace of systematic reasoning. Nevertheless, mathematical subtleties quickly lost their appeal unless he was convinced that they were closely tied to physical reality; research in "pure mathematics" never held his interest for very long. On the other hand, his sustained passion for the physical sciences was fueled by a conviction that he could achieve some knowledge of nature's fundamental classifications and causal structure. As we have seen, the religiously tormented Ampère never considered science an exclusive vehicle for truth. Nevertheless, he constructed philosophical arguments that science could at least reach a deeper level of reality than that described by phenomenological laws; conversely, he tried to use the explanatory and predictive success of science to support his philosophical claims.

Ampère thus set himself apart from his Laplacian colleagues by arguing that science and philosophy could be mutually supportive; as we might expect, his transitions between philosophical and scientific contexts were none too smooth. In an 1839 eulogy, Arago recalled some of the reactions Ampère provoked.[2]

Being both a mathematician and a metaphysician, Ampère lived in two distinct societies from the very time of his arrival in Paris. The sole feature they had in common was the renown of their members. On the one hand was the First Class of the old Institut, the professors and the examiners of the École Polytechnique; on the other Cabinis, Destutt de Tracy, Maine de Biran, Degérando, etc., ... The psychologists tried to determine the manner in which invention takes place; the *géomètres*, chemists, and physicists invented. Buffeted between these two schools, ... Ampère's lively imagination daily underwent some rather severe trials. I cannot state with certitude the light in which the exact sciences were considered at that time by the metaphysicians, but I do know that *géomètres* and chemists accorded little esteem to purely psychological research ... When Ampère, still highly enthused by the conversations he had just had with psychologists, interjected without warning the term *émesthèse*, for example, into a meeting of *géomètres*, physicists, or naturalists; or when, yielding to his enthusiasm, he submitted that an obscure or, at least, misunderstood word contained the most beautiful discovery of the century, was it not natural that he should encounter some unbelievers? All this would have been in order if the extreme kindheartedness of our colleague had not allowed mocking unbelievers to usurp the place of serious ones.

While Poisson and Biot may have considered philosophy an idle distraction, Ampère was invigorated by his new philosophical environment. Prior to his arrival in Paris, the *idéologues* Destutt de Tracy and Cabanis had promoted a materialism that the Kantian translator Charles Villers called the *"bella-donna* of all philosophy."[3] Nevertheless, by 1804 an idealist resurgence was afoot. Maine de Biran promoted a reassessment of the *idéologues'* psychology, and Kant's *Critique of Pure Reason* finally began to attract attention. Although Kant's treatise was not translated into French until 1835, popular 1801 commentaries by Kinker and Villers generated discussion, even if these renditions of Kant's ideas were not particularly accurate.[4] Although Ampère read Villers first, both he and Maine de Biran made more extensive use of Kinker's *Exposition*, a work far superior to Villers in both content and style. Degérando also published a long polemic against Kant's *Critique* in his 1804 *Histoire comparée des Systèmes de Philosophie.* Degérando had studied German for the express purpose of reading Kant; both Maine de Biran and Ampère profited from conversations with him. His radical interpretation of Kant's theory of experience as entailing both idealism and empiricism was very probably the source for Ampère's similar understanding.[5]

Shortly before coming to Paris in 1804, Ampère began an essay he intended to submit to the Institut in competition for a prize eventually won by Maine de Biran.[6] In his introduction he described his early dissatisfaction with idealism.[7]

> But how disappointed my expectations were when I read those works where I expected to find the key to all the sciences and the foundation of certitude. All truth vanished . . . the factual sciences, natural history, physics, and chemistry disappear. If what we observe in the objects that surround us is only the modifications of our own being, without relation to these objects, how could we draw consequences from them regarding their mutual action? The dust of the stamens would no longer fertilize the seeds hidden at the base of the calyx of a flower, and it would only seem to me to be so.

While Ampère instinctively recoiled from the solipsism he thought idealism entailed, he did not swing to the opposite extreme of positivism. He was more inclined to follow Maine de Biran's lead and argue that some causal knowledge is possible and trustworthy experience is not limited to sensations. During 1804 Maine de Biran argued that at least one direct experience of causality is obvious and takes place during every personal exertion of ego; this experience is the *fait primitif* or "fundamental fact" of psychological experience. He recorded this point of view in his *Mémoire sur la décomposition de la pensée*, and it was during the composition of this essay that he met Ampère in Paris. Subsequent friendship and intellectual exchange between the two men generated their voluminous philosophical correspondence. Letters were particularly frequent prior to Maine de Biran's permanent move to Paris late in 1812. Both Maine de Biran and Ampère also drafted long philosophical notes and essays, often in reaction to meetings of Maine de Biran's *société philosophique*. Between August of 1814 and January of 1815 *société* meetings often highlighted Ampère and Maine de Biran joining forces to defend the views they held in common. The most faithful in attendance were Royer-Collard, Degérando, Guizot, Georges Cuvier, Frédéric Cuvier, Thurot, Durivau, Christian, and Maurice.[8]

Reconstruction of Ampère's philosophical views is complicated by the fact that he never managed to organize his epistemological ideas into a publishable presentation. Nevertheless, in spite of the chaotic state of the surviving manuscripts, some persistent themes stand out and indicate long-term interest. For example, there are two salient respects in which Ampère took strong exception to Kant's theory of experience. First, Ampère argued that some aspects of the most fundamental material basis of reality are accessible to human investigation and are not hopelessly obscured by the limited human perceptions of how reality "appears." Second, Ampère held that space is not merely the form in which human sensibility presents external appearances to the interpretive faculty of understanding; his view was that space is substantial, although nonmaterial, and exists in an absolute sense independent from human thought.

Discussion of Ampère's attempts to establish these two points must be prefaced by some preliminary remarks on the Kantian distinction between phenomena and "noumena." Bluntly stated, in his *Critique of Pure Reason* Kant argued that all knowledge of the external world pertains to the "appearances" of empirical objects. This knowledge is comprised of judgments in which the concepts of the understanding are brought to bear upon the intuitions presented to it by the sensibility. Human sensibility is structured in such a way that all intuitions have a spatial or perspectival character. As Kant put it: "Space is nothing but the form of all appearances of outer sense." Kant's term "empirical object" applies not only to the common objects of direct perception but also those things of which the "grossness of our senses" prevents direct perception; magnetic fluid is one of Kant's examples. When we consider any familiar empirical object, we notice that some of its appearances vary, depending upon our spatial orientation or the condition of our sensory organs. We distinguish our variant and personal intuitions of the appearance of an object from the objective thing we conceive to be a perduring entity. This distinction is "only an empirical one" in that both of the two terms of the contrast are conceived by relying on the application of the spatial and temporal forms of human sensibility and the limited concepts of human understanding. Although Kant did sometimes refer to empirical objects as "things in themselves," in the *Critique* he avoided the term *noumena* in this context.

Kant had used the term *noumena* in the above sense in his 1770 *Dissertation*; his transition to a new critical point of view was prompted by his recognition that empirical objects not only reflect the limitations of the forms of sensibility but those of the concepts of understanding as well. In particular, the concept of a specific empirical object must itself be considered in two ways. First, it is considered "as appearance" insofar as it issues from a reflective judgment of appearances. In this sense the object is considered as a phenomenon. On the other hand, the empirical object can be "considered in itself," that is, not as appearance. That this second consideration cannot be pursued beyond its negative sense so as to consider the object as a "non-appearance" is due to the fact that our power of reflective judgment is restricted to appearances by the limited faculties of our sensibility and understanding. It is in this context that Kant introduced the term *noumena* in the *Critique*, but only while carefully stipulating that this concept does not designate any entity of which we can have the slightest knowledge.

For Kant the new and correct significance of the phenomena/noumena distinction does not lie in an ontological cleavage between a phenomenal world of human intuitions of appearances and a "real" world of noumenal entities. Rather, the philosophical value of the distinction is an epistemological one in that it brings to our attention the limited nature of the faculties with which the concept of "object" was originally schematized. From Kant's point of view, scientific progress brings us no closer to a knowledge of *noumena*. All scientific concepts rely upon the forms of sensibility and the categories of the understanding and thus are not applicable when things are considered "in themselves" rather than "as appearances."

Unfortunately, both Villers and Kinker failed to give an accurate description of Kant's transcendental reflection upon empirical objects either as appearances or "not as appearances." Instead, they conflated this distinction with the "merely empirical" one between objective things and subjective appearances. They did correctly report that Kant denied the possibility of any knowledge of "noumena," but they erred by interpreting "noumena" in the "positive" ontological sense that Kant expressly warned should be avoided. Their misleading translation of Kant's *noumena* as "things in themselves" (*choses en elles-mêmes*) failed to bring out the fact that Kant had abandoned his earlier "positive" usage of *noumena* and had given the concept a primarily restrictive role in the *Critique*.

Based on his reading of Villers and Kinker, Ampère quite naturally adopted their terminology and expressed the distinction between objective reality and perceptions of that reality by using the terms "noumena" and "phenomena." In one collection of his notes we find the following explicit definitions:[9]

> Whatever we believe to exist in itself, whether we know it or not, and whether modifying us or not, is what I call a *noumène* . . .
> In whatever way a *noumène* manifests itself to us, this manifestation is a phenomenon.

Ampère refused to accept what he understood to be Kant's argument that objective reality must be considered either as "appearances" or as Kantian *noumena*. For example, in an 1805 letter to Maine de Biran, Ampère gave the following enthusiastic interpretation of theoretical physics as an attempt to describe the "noumenal world."[10]

> Even though we only know through our impressions the phenomenal world where colors exist on objects, where the sun is a foot in diameter, where the earth is flat and immobile, where the planets move retrogradely, etc., the physicists and astronomers conceive of a hypothetical noumenal world where colors are sensations excited within a sentient being by certain rays that only exist within that being, where the sun is 307,000 leagues in diameter, where the earth is a flattened spheroid that revolves around it, where the planets always move in the same direction, etc.

Ampère's "noumenal world" is thus distinguished from the phenomenal world by eliminating the personal aspects of subjective human perceptions; on the other hand, it is still based upon a conceptual framework of spatial and temporal relations. Ampère assumed that these relations pertain to the substances constituting objective reality. From Kant's critical point of view, the error of Ampère's "transcendental realism" is the failure to subject the empirical things Ampère called "noumena" to further philosophical reflection. According to Kant, these empirical things should also be considered "in themselves," that is, they should also be considered insofar as they would be known by minds unburdened by the limitations of the faculties through which human knowledge arises, including the forms of space and time. For Ampère, when empirical objects are considered "in themselves" the result is not, in Kant's negative sense, *noumena* beyond the pale of spatial relations. On the contrary, Ampère thought that by perceiving "modes of coordination" among the appearances of empirical objects, he could infer analogous modes of coordination among the constituents of objective reality. In this sense Ampère took the "noumenal world" to be the goal of scientific research, a point he made clear in an 1810 letter to Maine de Biran.[11]

> It is not absurd *a priori* to suppose that some relations – albeit relations of which we have notions only because of our having perceived them between our own modifications, but which do not depend in any way upon the nature of those terms between which they hold – also exist between the noumena, the nature of which is absolutely unknown to us ... In this way we conceive a hypothetical noumenal world entirely different from the phenomenal or subjective world, and I repeat that it is no longer absurd to suppose that this hypothetical noumenal world is real, independent of our own existence, and that the same relations that we posit there also existed before we had the notions of them and before we existed.

From a Kantian point of view, of course, Ampère's "noumenal world," thoroughly dependent upon spatial and temporal concepts, is not "entirely different from the phenomenal or subjective world." Maine de Biran brought this point to Ampère's attention in notes composed between 1810 and 1813.[12] In response, Ampère repeatedly emphasized that he was only offering his point of view as a tentative hypothesis subject to confirmation with some degree of "probability" following successful observations of its empirical consequences. For example, during May of 1810 Ampère argued his position at philosophical discussions hosted by the mathematician Joseph Fourier in Grenoble. Ampère's efforts to include mathematical concepts within the "desubjectivized" noumenal world brought objections from Sebastian Planta, an enthusiastic convert to Kant's philosophy. Following the Grenoble meetings Planta subjected Ampère to a barrage of pro-Kantian apologia, including extensive translations from Kant and German commentators such as Reinhold, Schultz, and Wronski. The following passage by Planta is typical.[13]

> If by desubjectivized relations you mean those of number, orientation, figure, or extension taken in an abstract manner, I reply that nothing is more *subjective*, for they are nothing other than the expression of the modification of the intimate forms of our subjectivity.

Maine de Biran also remained unconvinced that an appeal to a hypothetico-deductive mode of reasoning could decide the issue in Ampère's favor. When Ampère posed the hypothesis that his "noumenal world" must include relations analogous to the spatial and numerical relations of phenomena, Maine de Biran retorted that even if Ampère could derive observable consequences from this hypothesis, that would not rule out the possibility that the same consequences might be due to a "noumenal world" structured in a manner that eludes understanding. Nor could the Kantian alternative be ruled out.[14]

To be fair, we should acknowledge that the examples Ampère cited in his early letters to Maine de Biran are misleading in that in those cases he claimed that individual "noumena" of the planets, for example, bear geometric properties such as size or shape. In general Ampère argued only that collections of "noumena" sustain mutual spatial relations. During 1813 Ampère's confidence was bolstered when he developed a geometric theory of atomic structure. As we will see in chapter 6, the Platonic foundation of this theory readily harmonized with Ampère's argument that it is only because "noumena" sustain mutual relations that they produce perceptions of geometric shapes. As Ampère put the issue in an undated note, "A body, as small as it may be, is composed of a great number of noumena juxtaposed by continuity."[15]

Another of Ampère's fundamental disagreements with Kant was his belief in the existence of absolute space as a nonmaterial substance that structures physical reality and is not simply projected upon experience by human perception or cognition. Ampère's son, Jean-Jacques, had powerful memories of how crucial the existence of absolute space was for his father's peace of mind.[16]

> In every subject, my father had a thirst for certitude. I recall with genuine remorse my having thrown him into a bitter fit of despair by denying the absolute existence of space. "What?" he said to me, "in that case there would be no real space in which the planets could describe their orbits according to Kepler's beautiful laws!"

I hope I will not be accused of lacking respect for this dear memory by recalling the suffering which today I would greatly prefer not to have caused him and which stemmed from his passionate need for his beliefs to be true and to be recognized as true.

While we will see that Ampère's conception of "noumenal matter" is clarified to some extent by his chemistry, it is more difficult to understand his alternative to Kant's theory of space. Other than the claim that noumenal space has "nothing in common with visual or tactile space," Ampère's letters provide little information. However, in a set of notes apparently written in about 1813, we find Ampère waxing Newtonian.[17]

Men knowledgeable in all the sciences, while agreeing that phenomena are not in bodies, think just as the vulgar do that the other concepts really and incontestably are there. Because phenomenal motion presumes, besides displaced intuitions, a phenomenal duration and an extended and fixed representation in which the motion takes place and where the route traversed existed in advance, they can conceive motion in bodies only by conceiving, in addition to material mobile extension, a noumenal duration and an infinite, immobile, penetrable extension in which motion takes place, and the parts of which are coordinated for all eternity according to all imaginable figures, since they can be traversed by bodies of all shapes and motions.

More tersely, another note tells us that "the noumenal displacement of a body can only be conceived within an immobile and infinite coordination of penetrable noumena, parts of the void."[18] For Ampère, objective space was thus more then a set of relations among "noumena"; for him space itself "is the true substance or the true substratum of all the impressions that refer outside."[19] Other remarks on space are to be found among notes in which Ampère recorded his earliest reactions to Kant. Once again, we find him echoing Newtonian themes.[20] "Time, space, thought, and free will – necessarily infinite – are the attributes of the being I call God." Not only did Ampère consider space to be a nonmaterial substance expressive of God's omnipresence, but he also felt that because molecules stand in geometric relationships to the parts of space, these molecules sustain objective relations to each other, relations that are ultimately responsible for our attribution of primary qualities to the bodies they compose.

By October of 1817 Ampère was hard at work on a lengthy philosophical treatise. Although he was interrupted by the task of drawing up a *résumé* of the logic course he was teaching at the École Normale, he hoped to include his epistemological ideas in an appendix to this outline and publish it as *Éléments de Logique*. The project was never completed, but Ampère's notes indicate that Maine de Biran's criticisms had brought about a more sober reappraisal of his philosophical accomplishments. For example, Ampère repeated that the confirmation of his theory is based upon the deduction of observable phenomena from hypothetical noumenal relations that are at least possible due to their independence from the secondary properties of the terms they relate.[21] In the last analysis Ampère was forced to have recourse to divine providence.[22]

> The much sought passage from the phenomenal world to the noumenal world lies in that law of our existence that the creator, wishing us to come to know the relations of causality, duration, extension, shape, motion, etc., of the noumena without knowing them in themselves as they can be known only by their author, has constructed us in such a way that these relations are necessarily established between our phenomena when they pre-exist between the corresponding noumena.

Expanding upon this conclusion, Ampère gave a modest final assessment of his contribution to epistemology.[23]

> I have added to this conviction by showing:
> (i) the possibility that the relations that make up these conceptions, although originally perceived between phenomena, are precisely the same between the noumena.
> (ii) by explicating both the physiological facts and the manner in which we have gradually formed true objective conceptions by relating some of these phenomenal conceptions.

Ampère had relied upon an inductive logic of confirmation to support his alternative to what he understood to be Kant's position. By 1817 he realized that Maine de Biran had correctly pointed out that the evidence supported other philosophical positions equally well. Furthermore, as we will see in chapter 6, it was just at this juncture that Ampère participated in the spirited arguments between advocates of the competing particle and wave theories of light. Although Ampère had been a long-standing supporter of the particle theory, he now shifted his allegiance to Fresnel's wave theory. He thus directly experienced the fickle and inconclusive nature of inductive arguments. He could only conclude that his own philosophical argument was equally indecisive.

153

Analysis, Synthesis, and "Fundamental Facts"

Maine de Biran's criticism of hypothetico-deductive reasoning inspired Ampère to study scientific methodology more systematically than he had hitherto. Maine de Biran was especially dubious about scientific explanation based upon unobservable entities such as the imponderable fluids that gradually became commonplace for Laplacian physics. For Ampère on the other hand, belief in the existence of imponderable fluids was consistent with his argument that some knowledge of the "noumenal world" was possible. In his 1802 criticisms of Coulomb he had only objected to non-gravitational forces that are postulated as being inherent in particles and acting at a distance. After his arrival in Paris he apparently still held the view that only the gravitational force is an intrinsic property of each molecule of matter; the only other "force" he considered to be fundamental was simply the repulsion between neighboring microscopic particles due to their impenetrability.

Without interjecting more clarity into Ampère's thoughts than the evidence warrants, we should at least note that he attempted to incorporate Leibniz's concept of active substance into a metaphysics compatible with his own research in chemistry and physics. He was sympathetic to Maine de Biran's praise for Leibniz's attribution of forces to "simple entities" or monads. Ampère himself drew a contrast between "substance" and "force" on the basis of the two types of change they produce. Those phenomena that involve motion he attributed to "forces"; static modifications he ascribed directly to "substances." The following passage is a relatively succinct synopsis of Ampère's argument that "the world of forces is superimposed on that of substances."[24]

> We once again take the conception of effort that we experience within ourselves and transport it to the outside under the name of force whenever we are witness to motions produced by our ego. From this there results, in a certain sense, *two noumenal worlds*, that of substance or causes passive with respect to motion, and that of active causal forces ... Furthermore, is it not the case that the world of forces is superimposed on that of substances just as the phenomenal worlds of the various senses are ... ?

Unfortunately, Ampère cited no scientific implications of his vague references to the "superposition" of two noumenal worlds of substance and force. During 1814 he made no effort to use his molecular theory of chemical affinities to exemplify the concepts of force and substance he relied upon in philosophical debate. In conjunction with the traumas of his personal life, it is not surprising that he became temporarily deranged during March of that tempestuous year and found himself wandering through Paris as if in a "représentation théâtrale."[25]

Ampère and Maine de Biran came to a more productive agreement about the importance of the *fait primitif* or "fundamental fact" for any given domain of phenomena. We have already noted how this topic contributed to the Laplacian criteria for the "principles" of a physical theory. Haüy and Biot both emphasized the central role of fundamental force laws as the fundamental facts of well-constructed theories. On the other hand, Ampère was perhaps the only member of the Parisian scientific community who attributed methodological significance to Maine de Biran's thoughts about the *fait primitif* of human psychology. Maine de Biran argued that introspection compels the conclusion that the fundamental constituent of all psychological experience is one of "effort, willed action, or volition." But while the psychological *fait primitif* is directly experienced, fundamental facts in the natural sciences are only inferred through an arduous refinement of empirical data.

In undated notes probably intended for a philosophy lecture, Ampère drew a similar but more specific distinction between the origin of fundamental facts in the *sciences mathématiques particulières*, such as geometry or statics, and in the *sciences physico-mathématiques*, such as optics or astronomy. In the first case,[26]

> These facts are so familiar that everyone grants them upon their very statement – such as the fact that space has three dimensions, or the equality of all the perpendiculars dropped between two parallels, etc., for geometry, and the equilibrium of three equal forces forming angles of 120°, etc., in statics.

In physics, on the other hand,[27]

> these facts can only be deduced from the observed phenomena by a complicated procedure; such as the attraction in inverse proportion to the square of the distance in astronomy, or the constant ratio of the sines of the angles of incidence in optics.

Ampère agreed with the Laplacians that although microscopic phenomenological laws are sometimes carelessly referred to as "fundamental facts," they are actually "fundamental" only in the sense that they have not yet been traced to the action of truly fundamental substances; subsequent determination of fundamental laws is the ultimate goal of physical theory.

This issue weighed heavily on Ampère's mind during the two decades prior to 1820. His research in chemistry and his participation in debate over the wave theory of light gave him direct experience of how methodological arguments could be influential. He attempted a systematic presentation of his views in a series of lectures delivered at the École Normale during 1817 and 1818. The broad range of material he considered relevant to this course in "logic" set him apart from colleagues faced with a similar task. For example, Joseph-Diez Gergonne taught a comparable course at the Montpellier Faculté des Sciences attended by John Stuart Mill during the winter of 1820–1.[28] Gergonne held in very low esteem the type of "metaphysics" so dear to Ampère. Ampère used his course as an occasion to expound not only on Aristotelian syllogisms and the methodology of the physical sciences, but his latest taxonomy of psychological phenomena as well. He could not have had a very receptive audience for his efforts to summarize his repeatedly revised categories of *comparatif, réflexif, physico-mathématique,* and *ontologique "systèmes"* and the *époques* of the psychological judgments that exemplify them. Nevertheless, Ampère maintained a lifelong interest in these classification schemes; detailed consideration of them is best postponed to the discussion in chapter 10 of his *Essai sur la Philosophie des Sciences*, written during the last years of his life.

On the other hand, Ampère's interest in logic and scientific method- ology was shared by many of those who wrote about mathematics in a semipopular or educational vein, a group that included Lacroix, Carnot, and Gergonne. All three, for example, wrote at great length on the distinction in proof theory between *analysis* and *synthesis*. Gergonne published an essay on this topic in June of 1817, a publication that may well have inspired the attention Ampère gave the subject in his École Normale lectures.[29] Gergonne bemoaned the fact that the term "analysis" was used so indiscriminately that it had become largely honorific and was often simply synonymous with "good methodology" as opposed to bad reasoning. He located much of the blame for this state of affairs with Condillac and his disciples. From Gergonne's point of view, analysis and synthesis were equally rigorous and the choice of one method over the other should be decided by the task at hand. His definitions bear considerable similarity to those used by Ampère.[30]

These two inverse manners of proceeding have from the greatest anti- quity received different denominations. Someone has called *Synthèse*, or the *Méthode synthètique*, the procedure by which one raises oneself, by degrees, from the most elementary truths to those which are less so; and someone has called *Analise*, or the *Méthode analitique*, the method that consists, on the contrary, in coming back down from the highest truths to more elementary ones, with the intention to show that the former basically reduce to the latter. These two methods thus traverse the same route but in precisely opposite directions; furthermore, neither has absolutely any advantage over the other, either with respect to rigor or to brevity.

Gergonne provided exemplary logical forms for the two methods; the synthetic "ascent" to prove a contingent theorem E is represented as follows, where A, B, C, and D are more "elementary" propositions and A is known to be true.[31]

> If A is true, B will also be true.
> If B is true, C will also be true.
> If C is true, D will also be true.
> If D is true, E will also be true.
> Now, A is true.
> Therefore, E also is true.

On the other hand, the "analytic" form begins by a consideration of the contingent theorem E and "descends" to more elementary propositions until a true proposition A is found that ensures the truth of the entire set.[32]

> E would be true if D were true.
> D would be true if C were true.
> C would be true if B were true.
> B would be true if A were true.
> Now A is true.
> Therefore, E also is true.

Two points should be noted about Gergonne's distinction. First, he is primarily interested in the structure of proofs in mathematics. Neither of his logical forms is an example of hypothetico-deductive reasoning from a tentative hypothesis to experimentally confirmed consequences. Although his "analysis" includes hypothetical statements, these statements do not take the contingent theorem E as hypothetically true; for example, the first step of the analytic proof simply states that if another, more "elementary" statement D were true, then E would be true. Second, the analytic "proof" assumes that a synthetic proof will follow and simply reverse the steps of analysis. In this respect, Gergonne

157

pointed out that analysis is generally a more valuable exploratory procedure than synthesis in the search for new proofs; on the other hand, synthesis can subsequently be employed for teaching purposes to demonstrate succinctly the proof of the discovered theorem.

Although Ampère shared Gergonne's interest in logic, he took a much broader view of the subject. In his notes for Lesson 15 of his École Normale lectures, he commented that[33]

> ordinarily logic is divided into the examination of ideas, judgments, arguments, and methods. The two latter are generally reduced to judgments, that is, arguments are reduced to apodictic judgments that such and such conclusions follow from such and such premises, and method is reduced to judgments that prescribe the procedure that should be followed in the search for truth.

Ampère included both mathematics and the physical sciences within the "search for truth." Above all, he was concerned about the logical structure of arguments intended to link scientific theories to experimental evidence. He distinguished four different argumentative modes based upon whether they involve analysis or synthesis and whether they are "direct" or "indirect."

In terms that bring to mind Gergonne's discussion, Ampère defined synthesis as a mode of reasoning that begins from relatively "simple" premises to reach conclusions that are more "complicated." Analysis, on the other hand, arrives at conclusions that are more "simple" than the original premises.[34]

> There is *synthesis* when, in combining therein judgments that are made known to us from simpler relations, one deduces judgments from them relative to more complicated relations.
> There is *analysis* when from a complicated truth one deduces more simple truths.

The relative simplicity or complexity of a statement is dependent upon the state of affairs or "relations" (*rapports*) it describes. Conditions are more complicated insofar as they are composite results of many individual causal agents acting under attenuating circumstances. For example, in the context of an imponderable fluid theory of electricity, a very "simple" statement would be the claim that the force between two particles of electric fluid is inversely proportional to the square of the distance between them. On the other hand, a claim that a large collection of these particles has reached a specific state of static equilibrium on the surface of a conductor would describe much more complicated circumstances resulting from the mutual interactions of many electric fluid particles.

Ampère drew his second distinction between "direct" and "indirect" reasoning. Direct arguments proceed from premises that are acknowledged to be true; they establish a conclusion that had previously been contingent. Indirect reasoning takes a contingent hypothesis as its major premise and derives a conclusion that can be straightforwardly recognized as true or false. Combining this dichotomy with that of analysis and synthesis results in four possible modes of reasoning: direct analysis, indirect analysis, direct synthesis, and indirect synthesis.

The deduction of new geometrical theorems from more simple, previously established theorems is Ampère's prime example of direct synthesis; in his notes he cites this as the typical method of discovery in geometry.[35]

> Either one or the other [analysis or synthesis] may be direct or indirect. The direct procedure is when the point of departure is known – direct synthesis in the elements of geometry. By combining at random simple truths with each other, more complicated ones are deduced from them. This is the method of discovery, the special method of inventions, contrary to popular opinion.

In the physical sciences, Ampère took Newton's presentation of the universal law of gravitation as an example of direct analysis. One stage in Newton's argument involved taking as a "known" premise the approximate truth of the relatively complicated statement of Kepler's third law in order to derive the more "simple" inverse square law for the attraction of each planet to the sun. Ampère emphasized the fact that the initial experimental data encountered by a working scientist seldom consist of the most "simple" facts about the domain. Much of the creative task of scientific discovery is the determination of the fundamental processes responsible for the complex phenomena observed. Only after a preliminary stage of exploratory discovery can textbook writers use direct synthesis to give an axiomatic presentation of the fundamental laws of a theory for didactic purposes.[36]

> It is commonly said that one must always proceed from the simple to the composite. Here we have one of those banal claims that is repeated without examining what is involved. If it is a matter of studying an entire science, and its truths are already consigned to books, it seems preferable, in general, to begin with that which is most simple in this

science. However, this is only to say that direct synthesis is to be preferred in the study or teaching of the sciences . . .
   But is the study of a science the only use of our intellectual faculties?
Is it never necessary to demonstrate a truth that cannot be directly deduced from more simple principles? Does one never need to resolve
specific questions, and was it never necessary to begin by discovering
the principles of those very sciences which today can be studied or
taught through direct synthesis by taking these principles as known?

As these rhetorical questions imply, Ampère was very interested in
the process Biot called the "divination" of the principles or fundamental laws of a theory. Ampère did concede that direct synthesis of
phenomenological laws from fundamental laws provided an impressive
teaching device; nevertheless, even when teaching science he preferred
to follow the procedure of indirect synthesis, the process he claimed
to be more representative of actual scientific discovery.[37]

   I think that indirect synthesis is also better for education, but it is not
   indispensable. If you have listeners or readers originally disposed to
   take you at your word and to wait until becoming acquainted with an
   entire science before verifying through observations the consequences
   of the principles which will have to be accepted on trust, you can proceed by direct synthesis.
      Although I regard this procedure as very little suited for giving clear
   ideas and enlightening the minds of those who are studying a science,
   it might be very difficult to banish it entirely from education. Nothing
   is more convenient for those who teach, and nothing condenses more
   the exposition of the principles of a science.

   In these lecture notes Ampère did not distinguish between microscopic phenomenological laws and fundamental laws, a distinction
that would become of primary importance during his research in electrodynamics after 1820. He did comment at length that, in contrast to
the direct synthesis utilized in textbooks, both direct analysis and indirect synthesis provide tools for the creative scientist confronted by
the complexity of actual phenomena. Ampère thus concentrated on
these two modes and made some interesting remarks about his
preferences. He pointed out that only in very rare cases might it be
possible to emulate Newton's impressive use of direct analysis to
derive the microscopic phenomenological law of gravitation. More
commonly, recourse must be had to indirect synthesis, that is, a law
must be adopted tentatively as a hypothesis from which observable
consequences can be derived and compared to experience.[38]

This need to change method is no less evident when it is a matter of . . . finding the explication of a phenomenon which nature offers to us in all its complexity. Here, the data being by their very nature more complicated than the conclusion being sought, direct synthesis becomes inapplicable, and one must have recourse either to direct analysis, if possible, or else to indirect synthesis, trial and error, and explicative hypotheses . . .

In the explicative sciences such as astronomy, chemistry, etc., the point of departure given by observation is very complicated. The direct procedure would have been an analysis through which, without hypotheses, the retrograde motion of the planets would have been traced to the motion of the earth, and the phenomena presented by the mutual action of bodies would have been traced to the nature and properties of their elements. But this analysis was impossible. Indirect synthesis replaced it, and in this way Copernicus and Lavoisier changed the visage of science. This method, determined by the very nature of the questions with which they were occupied, was the only one capable of leading them to their brilliant discoveries.

These passages are a clear statement of an important methodological preference Ampère adopted prior to his research in electrodynamics. From Ampère's point of view, only *de facto* mathematical and physical complexity prevents scientists from following Newton's exemplary use of direct analysis. Although the inductive confirmation of a hypothesis is the typical pragmatic procedure in physics, it is less trustworthy than direct analysis.[39]

Recall that in order to trace the cause of a collection of observed facts, there are two procedures – direct analysis and indirect synthesis.

Admittedly, the former would be more natural if it could be used – useless efforts have been made by even some very great men to make people believe that they followed it.

Successful observations of empirical consequences of a hypothesis, no matter how numerous, are not equivalent to a proof; true consequences can be derived from false premises. In his 1817 notes Ampère made this point in language that brings to mind the falsification methodology of the twentieth-century philosopher Karl Popper:[40] "It follows that in the two indirect procedures, a single false conclusion implies the negation of the hypothesis, while, in general, a true conclusion establishes only a probability in favor of the hypothesis." Given that the experimental confirmation of a hypothesis only generates some degree of "probability" in favor of the hypothesis, Ampère gave some thought to how various types of evidence should be assessed. For example, he warned against an excessive insistence on quantitative evidence.[41]

> There is immense value in the exact verification of a calculated conse-
> quence – providing more probability than two non-quantitative ones.
> Nevertheless, some *savants* are excessive in their rejection of this latter
> genre of proofs – example: for the theory of light.

Aware that simple tabulation of confirmatory evidence often cannot decide the issue between two rival theories, Ampère was inclined to give more confirmatory weight to the successful prediction of new phenomena, that is, phenomena unknown prior to prediction based upon the hypothesis in question. Ampère realized that he lacked any logical grounds upon which to distinguish this type of evidence from confirmation through explanation of phenomena known prior to the creation of the hypothesis. As justification, he relied primarily on the history of progress in physics. For example, in the early electricity manuscript he composed during 1801, Ampère posed a prediction that he hoped subsequent observations would transform into con-firmatory evidence. He motivated his procedure by pointing out the futility of any attempt to capture all details of the known facts within any scientific domain, and then appealed to historical precedent.[42]

> I thus believe that it behooves all those who know the limits of their
> conceptions to . . . found a physical system only on some . . . of the facts
> that they are supposed to coordinate; it is luck alone that can then
> discover the truth. However, *one has all the more means of verifying it as
> one has used fewer in discovering it.* One has only to imagine some new
> experiments and predict their outcome; it then suffices to try them in
> order to see if the conjectures one has made are confirmed or not. This
> is roughly the procedure that has been followed in physics. Lavoisier
> and Newton himself did not foresee the new observations and new
> experiments that have demonstrated their sublime theories.

At this very early stage in his scientific career, Ampère wrote as if he wanted to restrict confirmatory evidence to the prediction of new phenomena that had not contributed to the formulation of the hypo-thesis in question. But there is ample indication that he did not persist in this rather extreme view; later he allowed the explanation of previ-ously known phenomena to complement prediction in the confirma-tion process. For example, in 1814 he commented as follows concerning his hypothesis that all gases at a given temperature and pressure have the same inter-particle separation.[43]

162

Whatever may be the theoretical reasons that seem to me to support it, one can only consider it as a hypothesis. However, when comparing the consequences, which are the necessary implications of it, with the phenomena or the properties that we observe, if it is in agreement with all the known results of experiment, and *if one deduces consequences from it that are found to be confirmed by later experiments,* it will acquire a degree of probability that will approach what, in physics, is called *certitude.*

Although in this passage Ampère did not provide a relative assessment of the two types of confirmatory evidence he cited, he clearly continued to give predictions of new phenomena an important status. A more detailed account is provided by an introductory passage Ampère wrote in 1816 for a draft entitled *traité de psychologie.* Ampère characterized scientific method according to three normative guidelines.[44]

(i)   to observe, confirm, and class the greatest possible number of facts with respect to the questions concerned, before trying to explicate any of them;

(ii) renounce all the explanations that originally present themselves – reject all those that are found to be in contradiction with some of the observed facts, even when they seem to account for the majority of the others in a satisfactory manner;

(iii) definitely adopt the one explanation that seems the most probable only after having verified it by drawing some consequences from it ... that are *confirmed by some facts that one did not have recourse to during the research that led to it and which these consequences themselves indicate to the observer in the case where he would not yet have noticed them.*

Although the last phrase of this passage is obscure, Ampère seems to take the view that decisive evidence is provided by all entailed facts that were not used to create the theory in question; temporal prediction of new phenomena does not appear to be essential.[45]

Bearing in mind that there is often considerable disparity between a scientist's methodological statements and actual scientific practice, there can be no doubt that Ampère took these issues seriously. His emphasis on successful prediction as an essential element in the confirmation of a hypothesis gave an interesting twist to his application of indirect synthesis. This aspect of Ampère's methodology was an expression of his impetuous personality. Confronted by a new field of scientific research, he could rarely resist the temptation to pose startling predictions and conjectures on the basis of limited experimental support. This was particularly the case during the early years of his research in electrodynamics between 1820 and 1822. On the other hand, he was quite capable of retaining a belief in a hypothesis even when confronted by evidence that, according to his own methodological norms, should have required him to "renounce" or "reject" it.

Before we turn to Ampère's accomplishments as a mathematician and scientist prior to 1820, we should recapitulate the central philosophical and methodological attitudes he advocated. First, throughout the two decades between 1800 and 1820, Ampère repeatedly argued that physical theories could provide a limited but significant knowledge of objective reality. He was convinced that this knowledge must be expressed in terms of the mutual relations that coordinate "noumenal" entities, namely, the motions and vibrations of the most fundamental "molecules" of noumenal substance. He thought of these motions taking place within an immobile "noumenal space" of penetrable substance. His desire to rebut what he understood to be Kant's "idealism" made him eager to illustrate his own position through concrete examples. Fresnel's wave theory of light made Ampère receptive to the possibility that an elastic ether might provide the "noumenal" basis for the phenomena the Laplacian school associated with forces acting at a distance.

Second, Ampère was in full agreement with Maine de Biran that physical theories should be based on a prior determination of the "fundamental fact" of the domain in question. More specifically, Ampère advocated an initial description of these fundamental facts through microscopic laws derived from macroscopic phenomenological laws, a process he referred to as direct analysis. Given that this direct analysis is seldom possible, it is to be expected that a fundamental fact would emerge from a "divination" of a tentative "explicative hypothesis" followed by the experimental confirmation Ampère called indirect synthesis. This confirmation should include and emphasize the successful prediction of new phenomena that might not have been noticed otherwise. Although Ampère put these ideas into practice most fully in electrodynamics, they also contributed to his scientific work prior to 1820; indeed, as we would expect from Ampère, his philosophy and his science developed symbiotically.

6

Mathematics, Chemistry, and Physics (1804–1820)

During the same period in which Ampère labored over his metaphysics, he also maintained an active role in the French scientific community. It is indicative of his innate intellectual ability that he could advance to the forefront of research in all three fields of mathematics, chemistry, and physics. His accomplishments are all the more impressive when we consider his tumultuous personal life and the political and social upheaval during the Napoleonic wars and the Restoration of Louis XVIII. Ironically, during this period Ampère's highest professional recognition came in mathematics, the field for which he felt the least attraction. He wrote mathematics memoirs with the explicit design of earning election to the Institut; once that goal was achieved he put his energy elsewhere. Ampère's penchant for classification schemes had an impact on both his mathematics and his chemistry; it diverted his attention to problems that were not central to the research of his contemporaries. While mathematicians of his generation explored the foundations of the calculus and Fourier series, Ampère devised classification schemes for partial differential equations. While chemists analyzed compounds and worried about atomic weights, Ampère speculated about the geometry of atomic structure as the key to a "natural" classification scheme for the elements. His outsider status was equally pronounced in physics. His long-standing aversion to

165

fundamental forces that act through empty space made him receptive to Fresnel's wave optics and the hypothesis of an etherial medium. Ampère's interaction with Fresnel began just a few years before 1820 and served as an important prelude to the electrodynamics that followed.

Calculus and Its Algebraic Foundations

Writing to Élise in 1805, Ampère bemoaned the fact that while his attention was absorbed by metaphysics, he was nevertheless expected to function as a mathematician.[1]

> I have only a single pleasure, quite futile, quite artificial, and one which I rarely taste. It is that of arguing over questions of metaphysics with those who occupy themselves with this science in Paris, and who show me yet more friendship than the mathematicians. Nevertheless, my position requires me to work to the taste of the latter, which does not help amuse me because I no longer like mathematics at all. Since I have been here, I have nevertheless completed two calculus memoirs that are going to be published in the *Journal de l'École Polytechnique*. It is scarcely only on Sundays that I can see the metaphysicians, such as Maine Biran with whom I am closely linked, and Senator Tracy, to whom I sometimes go to dine at Auteuil where he lives; it is almost the sole place near Paris where the countryside reminds one of the banks of the Saône.

Although Ampère's heart and mind were elsewhere, one of the memoirs he mentioned in this characteristically forlorn passage became the only publication in which he made an effort to establish rigorous foundations for the calculus.[2] It probably contains the ideas Ampère obliquely alluded to in his earlier paper on the calculus of variations. Published in 1806, it lacks the systematic organizational style of definitions, axioms, and theorems later championed by Cauchy. Ampère's memoir preserves the tentative and exploratory cadence of a preliminary draft.

The paper is typical of Ampère's early mathematics in that it was inspired by a desire to extend or amend insights by Lagrange; in this respect the memoir is an exemplar of the algebraic approach to foundational studies of the calculus prior to Cauchy. As the historian of mathematics, Judith Grabiner, has pointed out,[3] Ampère relied upon techniques and arguments Lagrange advocated in his 1799 lectures at the École Polytechnique, lectures Lagrange published as *Leçons sur le Calcul des fonctions*. Lagrange defined the "derived functions" of a given function as the coefficients in its Taylor's series expansion. Second, he calculated an expression for the remainder that results when the function is approximated by terminating its Taylor's series at any given term. Although Lagrange determined "limits" within which the remainder falls, he did not attempt to define the derivative as a limit procedure.

Ampère sought two extensions of the Lagrangian approach. He gave a new definition of the derivative in terms of inequalities similar to those Lagrange had used in his analysis of Taylor's series remainders; second, he gave a new proof of the Taylor's series expansion formula, including the remainder expression. He began by exploring some properties of the difference quotient:

$$\frac{f(x + i) - f(x)}{i}$$

Ampère did not anticipate Cauchy by defining the derivative of f(x) as the limit of this quantity as i approaches zero; instead, Ampère considered the case where i is *equal to zero* and then presented a proof that the resulting ratio is not always zero or infinite within any interval of values of x. He considered this to be equivalent to showing that the difference quotient was only zero or infinite at a finite number of "particular values of x."[4] He also referred to his proof as a demonstration of the "existence" of the difference quotient function. This claim is best interpreted as an indication of what Ampère meant by the existence of a function rather than as an invitation to scrutinize Ampère's proof according to criteria for existence proofs that were not yet operative in 1806.

Ampère was intrigued by the difference quotient when i is zero because Lagrange had proved that his own definition of the derivative implied that $I(x,0) = 0$ where:

$$f(x + i) = f(x) + i[f'(x) + I(x,i)] \tag{1}$$

Ampère thought that if he could show that

$$\frac{f(x + i) - f(x)}{i}$$

is a reasonably well-behaved function when i vanishes, then it might eventually become a convenient way to *characterize* f'(x); he did not preempt Cauchy by *defining* f'(x) as the accurately formulated limit of the difference quotient as i approaches zero.

Ampère began his study of

$$\frac{f(x+i)-f(x)}{i}$$

by presenting an indirect proof; his goal was to show that a contradiction results from assuming that when i is zero the difference quotient is either zero or infinite throughout an interval [a,b] representing a range of values of x. He first showed that within any such interval [a,b], there are points P and Q such that:[5]

$$\frac{f(P+i)-f(P)}{i} \leq \frac{K-A}{k-a} \leq \frac{f(Q+i)-f(Q)}{i} \qquad (2)$$

where A and K are the values of f(x) when x takes on the values a and k respectively. Second, he used the Intermediate Value Theorem for continuous functions to conclude that there is a point X between P and Q such that:

$$\frac{f(X+i)-f(X)}{i} = \frac{K-A}{k-a} \qquad (3)$$

But this contradicts Ampère's assumption that all values of the difference quotient within the interval are zero or infinite; since the assumption led to a contradiction, it could not be correct. As Judith Grabiner has explained, Ampère's argument is an example of mathematical reasoning during the transitional period prior to the introduction of the concept of uniform convergence. In this respect both Ampère and Cauchy were prone to offer proofs which in retrospect were seen to implicitly assume uniform convergence.

With his proof about the well-behaved nature of the difference quotient in hand, Ampère next reconsidered Lagrange's theorem concerning f′(x) as stated in (1) above. Temporarily using this relationship to characterize f′(x), Ampère used an argument similar to the preceding one to show that there are points x_0 and x_1 within any interval [a,b] such that:

$$f'(x_0) \leq \frac{K-A}{k-a} \leq f'(x_1) \qquad (4)$$

After showing that the function f′(x) satisfying (4) is unique, Ampère then used this inequality to *define* f′(x), that is, by reconsidering the inequality (4) and letting x and (x + i) replace a and k respectively, Ampère gave the following definition:[6]

The derived function of f(x) is a function of x such that $\dfrac{f(x+i) - f(x)}{i}$ is always included between two of the values that this derived function takes between x and x + i, whatever x and i may be.

Ampère next showed that this definition agrees with the existing procedures for calculating derivatives of trigonometric functions, areas, tangents to curves, and arc lengths. Furthermore, it also preserves standard applications to velocities and accelerations in mechanics. In effect this means that the old techniques can still be used but with the understanding that the derivative has been defined in a manner Ampère preferred.

Heavily indebted to Lagrange, Ampère's efforts to define the derivative in terms of algebraic inequalities place him squarely within the algebraic tradition. Nevertheless, he also hoped to do without his mentor's dependence upon Taylor's series as a foundation. Instead, Ampère used his discussion of the derivative to provide a new proof of Taylor's theorem. This transition in Ampère's memoir begins with some introductory remarks that illustrate his algebraic procedure. He recalls that one property satisfied by the derivative is equation (1):

$$f(x + i) = f(x) + i[f'(x) + I(x,i)], \text{ where } I(x,0) = 0.$$

Ampère had shown that if a function f'(x) satisfies this equation, then it also satisfies the inequalities he used to define the derivative; furthermore, since he had shown that such a function is unique, the above expression can be used to explore additional properties of the derivative rather than relying upon the more unwieldy definition. Ampère let the variable x be augmented by the small quantity i and wrote the resulting power series expansion:

$$f(x + i) = f(x) + p(x)i + q(x)i^2 + r(x)i^3 + \ldots \tag{5}$$

Equating the two expressions for f(x + i), eliminating common terms, and dividing through by the factor i resulted in:

$$f'(x) + I(x,i) = p(x) + q(x)i + r(x)i^2 + \ldots \tag{6}$$

Equating the parts of this equation that are functions only of x and those that are dependent on both x and i gives:

$$f'(x) = p(x) \quad \text{and} \quad I(x,i) = q(x)i + r(x)i^2 + \ldots \tag{7}$$

Ampère thus had shown that the first coefficient in a power series expansion of $f(x + i)$ in powers of i is equal to $f'(x)$.[7] He concluded these introductory remarks as follows:[8]

> This procedure is evidently that of the differential calculus, where one represents by dx what we here call i, and by dy the term $if'(x)$, y being the function of x that we have expressed as $f(x)$. This calculus can thus in this manner be disengaged, not only from the consideration of infinitesimals (*infiniment petits*), but also from that of the formula of Taylor, of which it then furnishes a very simple demonstration.

Ampère then proceeded to provide his proof of Taylor's theorem in the form:

$$f(z) = f(x) + f'(x)(z - x) + \frac{f''(x)(z - x)^2}{2} + \frac{f'''(x)}{2 \cdot 3}(z - x)^3 +$$

$$+ \ldots + \frac{p^{(n)}}{1 \cdot 2 \cdot 3 \ldots n} i^{n+1} \tag{8}$$

where z is $(x + i)$ and $p^{(n)}$ represents the nth derivative of the expression

$$\frac{f(x + i) - f(x)}{i}.$$

The proof is based on a manipulation of algebraic relationships analogous to Ampère's preliminary remarks. Ampère thus felt that he had properly adopted Lagrange's aversion to infinitesimals and limits and had also eliminated the need to rely upon Lagrange's definition for the derived functions. Ampère concluded by briefly addressing two other topics. He commented on Lagrange's determination of the limits within which the remainder term falls,[9] and he added a note on the extension to functions of two variables.[10]

It is not clear how Ampère expected other mathematicians to react to his new definition for the derivative. Even within the context of his own paper he did not attempt to apply it to see what implications it might have. On the contrary, he had recourse to the characterization of the derivative as the value of the difference quotient

$$\frac{f(x + i) - f(x)}{i}$$

when i takes the value zero. Further indication that this is how Ampère thought of the derivative is provided by his abstract of a paper

170

presented by Binet.[11] Ampère's introductory statement of the theorem at issue closely resembles the statement with which he had introduced his own memoir.[12]

> M. Binet proposes to demonstrate, in a manner more simple than has been done up to the present, the following theorem, upon which rests the entire theory of differential calculus. With f(x) representing an arbitrary function of x, if one considers the quantity $\dfrac{f(x+h)-f(x)}{h}$, which is evidently a function of x and h, and if one supposes that one substitutes into it smaller and smaller values of h, the corresponding values of this function, except for particular and isolated values of x, will not be able to decrease or increase in such a manner as to become smaller or greater than any given magnitude; but they will tend in general toward a determined limit (*limite déterminée*), which one ought to consider as the value that this quantity takes when one sets h = 0, and where it is presented under the undetermined form 0/0. This value will necessarily be a function of x, since that of $\dfrac{f(x+h)-f(x)}{h}$, depending in general only on x and h, can only depend upon x when h is determined there by making h = 0. It is, as is known, this function that one calls the derived function or differential coefficient of the first order of the function designated by f(x).

This passage is a remarkable illustration of the limited extent to which Ampère thought of the derivative as the result of a limiting procedure. Preceding Cauchy by over a decade, he lacked a precise stipulation of what it means to say that the value of h approaches zero as a limit; as a result, he was content to say that the derivative was the value of the difference quotient when h became zero. A severe conceptual gap separated Ampère and his contemporaries from the revolutionary breakthrough carried out by Cauchy during the 1820s.

Meanwhile, Binet's proof was slightly more straightforward than Ampère's and, unlike Ampère, he did not present it as a prelude to a new definition of the derivative. Binet assumed that an interval (a, a + b) could be found such that f(x) is monotonically increasing or decreasing. He then chose h small enough so that within the interval the difference quotient is smaller than

$$\frac{f(a+b)-f(a)}{b}.$$

He divided the interval into n equal segments and showed that a contradiction resulted from the assumption that the difference quotient is not restricted to finite non-zero values.

If we assess the impact of Ampère's work on the foundations of the calculus by the published record, the result is meager. While Binet's proof was repeated approvingly by Lacroix,[13] Ampère's attempt to redefine the derivative was quietly ignored. Ampère did not retain his interest in this topic during the years subsequent to 1806. As was the case with his early thoughts on the calculus of variations, Ampère was capable of moving to the forefront of a research topic and then dropping it without making a significant contribution to future developments.

On the other hand, we should not ignore Ampère's efforts as a teacher, particularly in the light of the fact that his students included Augustin-Louis Cauchy and Joseph Liouville. There is ample manuscript evidence of Ampère's diligence as a professor of analysis and mechanics. For the decade beginning in the academic year 1815–16, Ampère and Cauchy taught analysis and mechanics in alternate years. Students who entered the École in even-numbered years would have Cauchy for both courses and those entering in odd years would have Ampère. Ampère took an active role in debates over curriculum reform, a source of considerable controversy during the period in which Cauchy accomplished the rigorous reformulation of calculus based upon limits. Indeed, Cauchy became a much more controversial teacher than Ampère.[14] Cauchy customarily devoted more than the allotted time to his analysis lectures and shortened the number of mechanics lectures. He did so to dwell on what for him were the most exciting aspects of the new analysis; the students were often baffled and repeatedly protested to the administration.

In contrast to Cauchy, Ampère was genuinely concerned about the progress of his students and he organized his lectures accordingly. Under continued pressure from the administration, he finally published some of his analysis lectures in 1824 under the title *Précis de Calcul différentiel et de calcul intégral*.[15] The lectures begin with a discussion of continuous functions that bears the stamp of Cauchy's approach.[16]

> When, in making an independent variable increase or decrease by insensible degrees, a function of this variable also increases or decreases by insensible degrees in such a manner that, in taking at will in the interval between these two values, two other values of the independent variable, the difference of which is as small as one wishes, the difference of the corresponding values of the function similarly become as small as one wishes, then one says that the function is continuous in this same interval.

On the other hand, Ampère did not introduce limit procedures in his definition of the derived function. He simply repeated his earlier 1806 approach and defined the derived function as the value of the difference quotient when the denominator vanishes. He used the term "differential of the first order" to refer to the quantity df(x) or f'(x)(X – x), where X – x represents the difference in the independent variable. He felt that students were unlikely to benefit from Cauchy's methods; having lost interest in the issue long ago, he was content to teach practical techniques in a fairly mechanical manner.

A significant portion of Ampère's teaching was devoted to the solution of differential equations. The subject caught his attention during 1814 and 1815, especially partial differential equations. His motivation was a frank desire for admission to the Institut. His precarious financial situation and his hope for some release from the burdens of repetitious teaching and examining inspired his most creative contribution to pure mathematics.

Partial Differential Equations and the Institut

It is hardly surprising that historical interest in the theory of partial differential equations has given high priority to problems motivated by developments in mathematical physics: wave equations, diffusion equations, hydrodynamics, elasticity, and potential theory. Due to the importance of this research for physics, most modern mathematicians, physicists, and engineers are familiar with the names Laplace, Cauchy, Poisson, and Fourier. These men took physical problems as the starting point for their analysis; specific boundary conditions and initial conditions played a central role in their approach. They also completed the eighteenth-century debate over the acceptability of the series solutions championed by Fourier. The triumph of Fourier's approach evolved into one of the most fruitful topics in nineteenth-century mathematics.

Generally speaking, Ampère remained aloof from these issues. With the exception of mechanics and minor applications in optics, his study of partial differential equations was not related to physical problems; nor did he contribute to the debate over Fourier series. In contrast to most of his contemporaries, Ampère's memoirs on partial differential equations can be read as exercises in "pure mathematics." In particular, Ampère put more attention on the classification of different types of equations than he did on solution techniques for equations known to have physical significance. In this respect he represents an exception to the generality posed by the historian of mathematics, Morris Kline.[17]

We are accustomed today to classifying partial differential equations according to types. At the beginning of the nineteenth century, so little was known about the subject that the idea of distinguishing the various types could not have occurred. The physical problems dictated which equations were to be pursued and the mathematicians passed freely from one type to another without recognizing some differences among them that we now consider fundamental.

It is true that Ampère's classificatory efforts were thoroughly surpassed by those of Du Bois-Reymond in 1889. Modern mathematicians who rarely cite Ampère continue to use Du Bois-Reymond's terminology to distinguish elliptic, hyperbolic, and parabolic linear second-order homogeneous partial differential equations. Nor did Ampère's contemporaries find easy reading in his memoirs. His definitions lack precision, his notation is cumbersome, theorems are not presented in a format that would indicate their importance, and examples are often only partially developed. Nevertheless, Ampère's encyclopedic penchant for classification resulted in pioneering contributions to the field. His peers appreciated his creative navigation of uncharted territory and he was awarded an appointment to the Institut in 1814.

Following a verbal presentation to the Institut on 11 July, 1814, Ampère published his general results in 1815 and later added a lengthy 1820 supplement primarily concerned with applications to specific cases.[18] The major conclusions of the 1815 paper fall into two categories. Following his extensive introductory remarks, Ampère proved that the number of arbitrary functions appearing in a "general integral" solution to a partial differential equation must be as large as the order of the equation. Second, he investigated conditions and properties that hold when the general integral of a partial differential equation contains no partial integrations, or "partial quadratures" as they were subsequently termed. He referred to equations with integrals of this type as the "first class," and he developed techniques for determining and classifying the arguments of the arbitrary functions that appear in general integrals of the first class. This investigation of integrals of the first class was easily extended into a procedure for the discovery of general integrals, although Ampère did not emphasize that aspect of the topic.

Organizing his memoir into four sections, Ampère opened with an introductory discussion of definitions illustrated by examples. By 1814 the domain of partial differential equations had become entangled by a multiplicity of alternative definitions.[19] As usual, Ampère began by introducing new terminology based upon a disagreement with

174

Lagrange. Ampère's innovations were guided by his hope that they would clarify the distinctions at stake. In general, the task of solving or "integrating" a partial differential equation requires a discovery of how the variables are directly related to each other. There are different degrees of generality in these solutions. Ampère decided to limit himself to situations in which there are two independent variables, x and y. A solution to a partial differential equation then would be a stipulation of a function $z(x,y)$ representing how the dependent variable z depends on x and y. When the partial differential equation is of the second order, it involves terms customarily abbreviated as follows:

$$p = \partial z/\partial x \quad q = \partial z/\partial y \quad r = \partial^2 z/\partial x^2 \quad s = \partial^2 z/\partial x \partial y \quad t = \partial^2 z/\partial y^2$$

The equation can thus be written in the form:

$$f(x,y,z,p,q,r,s,t) = 0$$

In general, Ampère called relations that stipulate z as a function of x and y "primitive integrals" (*intégrales primitives*) and used the term "primitive particular solutions" (*solutions particulières primitives*) to refer to less general solutions. He tried to spell out this distinction more carefully as follows.[20]

> In order that an integral be *general*, it is necessary that there results from it, among the variables that one considers and their derivatives, to whatever degree (*à l'infini*), only the relations expressed by the given equation and by the equations that are derived from it by differentiating it . . .
>
> One distinguishes *particular integrals* from *particular solutions* in that the former can be completed in such a manner as to produce a general integral in which they are included, and in that this does not take place with respect to the latter which express relations incompatible with those represented by general integrals.

The definitions suffer from Ampère's customary lack of precision. To clarify the distinction he had in mind, Ampère considered as an example the following relatively simple first-order partial differential equation:

$$z = px + qy \tag{1}$$

One solution is $z = Ax + By$, where A and B are constants. However, according to Ampère's definition this is not a "general integral"

175

because differentiation of the initial equation yields the second-order equations:

$$rx + sy = 0$$
$$sx + ty = 0 \qquad (2)$$

On the other hand, differentiation of the solution in question yields r = s = t = 0. While it seems too strong to say as Ampère does that this result is "incompatible" with the second-order equations (2), it is true that the stipulation that r, s, and t vanish is only a specific particularly simple way of satisfying those equations. Furthermore, the relations r = s = t = 0 cannot de derived by differentiating the initial equation (1).

Ampère argued that two other solutions to equation (1) should appropriately be called "general integrals". One is given by

$$z = x\psi(y/x) \qquad (3)$$

where ψ is an arbitrary function of (y/x). Another general integral is provided by the two equations:

$$z = \alpha x + y\phi(\alpha)$$
$$x + y\phi'(\alpha) = 0 \qquad (4)$$

where α is another constant and $\o$ is another arbitrary function. Differentiation of either solution (3) or solution (4) results in expressions for r, s, and t that satisfy conditions (2). Furthermore, in Ampère's terminology, the "particular integral," $z = Ax + By$, results from choosing specific functions ψ and ϕ for the solutions (3) and (4) respectively, namely, let

$$\psi(y/x) = A + B(y/x)$$
$$\phi(\alpha) = B + K(\alpha - A) \qquad (5)$$

where K is an additional constant. In this respect Ampère preferred the term "particular integrals" to Lagrange's misleading "complete integrals" (*intégrales complètes*).[21]

In a second introductory discussion, Ampère commented on the "arbitrary quantities" to be expected in a general integral, for example, the arbitrary functions ψ and ϕ and the argument α in equation (4). Given that the partial differential equation is of order m, Ampère assumed that differentiating the general integral in all possible manners up to the nth order results in h independent equations and that

176

all differentiations of the initial equation up to order (n – m) result in h' independent equations. It follows from his definition of a general integral that the h equations must contain at least (h – h') arbitrary quantities in order that by eliminating these quantities the h equations can be reduced to h' equations. When arbitrary functions are differentiated, it is possible that new arbitrary functions are produced. Ampère called integrals in which this is the case "heterogeneous" with respect to a particular variable of differentiation; an integral can also be "homogeneous" with respect to other variables if the relevant differentiations up to any specific order produce no new arbitrary function.

Ampère also distinguished two types of "arbitrary" functions according to whether they do or do not include "partial integrations," that is, integrations over one of the independent variables with z and the other independent variable treated as constants. He called "integrals of the first class" those that do not include any partial integrations. After deriving a general formula for differentiating partial integrals, Ampère stipulated the parametric formulation he would assume general integrals to have prior to further analysis. Suppose that there are k arbitrary functions disengaged from partial integrations, and k' partial integration variables. The general integral then should be characterized by (k + 1) "principal equations of the integral" from which the arbitrary functions can be eliminated; 2k' "accessory equations of the integral" stipulate the limits of partial integration.

In his second section Ampère addressed the issue of how new "arbitrary functions" are produced by differentiation of the integral of a partial differential equation. He proved several "theorems", although he did not refer to them as such. He began by introducing his own cumbersome but accurate notation for partial derivatives. When α is a variable that acts as the argument of an arbitrary function, α can in turn be a function of x and y, such as the function (y/x) in equation (3). Arbitrary functions thus can be differentiated with respect to x, y, or α. In Ampère's notation the variable that is held constant during a partial differentiation is placed behind a parenthesis. For example, the partial derivative of a function u with respect to x with α held constant is written:

$$\frac{du}{dx\,(\alpha}$$

Suppose now that a new function of α, produced by successive differentiations of the primitive integral, appears for the first time in one of the nth-order derivatives of z. Call u and v the two corresponding derivatives of order (n − 1); the new function is thus produced by the differentiation represented equally well by either of the two expressions:

$$\frac{du}{dy\ (x} = \frac{dv}{dx\ (y}$$

Ampère proved that a new function appearing for the first time in one of the derivatives of z of order n will also appear in the two other derivatives of order n preceding and following it:

$$\frac{du}{dx\ (y} \ /\ \text{and}\ \frac{dv}{dy\ (x}.$$

He then showed that it must also appear in all derivatives of order n.[22]

With these extensive preliminaries in place, Ampère devoted the third section of his paper to the first major topic of the memoir, the demonstration that a partial differential equation of order m has a general integral with at least m arbitrary functions.[23] He based his proof on his definition of a general integral, namely, one such that differentiations produce a set of equations that can be reduced to those formed by differentiations of the initial partial differential equation.

Given that the arbitrary functions in the integral contain k independent parameters, the integral can be put in the form of k + 1 parametric equations; derivatives of these equations taken up to and including the nth degree in x and y then produce a total of $\frac{1}{2}(n + 1)(n + 2)(k + 1)$ equations. Ampère next determined how many parameters and arbitrary functions must be available in order to reduce these equations to the $\frac{1}{2}(n − m + 1)(n − m + 2)$ equations derived by differentiating the initial partial differential equation of order m. The k parameters and their derivatives with respect to x and y result in $\frac{1}{2}(n + 1)(n + 2)$ quantities. He let g represent the number of arbitrary functions in the general integral and g′ the number of additional new functions produced by first derivatives. In cases where the initial differentiation of one of the g functions produces no new arbitrary function, he let l, l′, . . . represent the additional number of differentiations required before new functions do result; each of the primed parameters corresponded to one of the g arbitrary functions. In each of the g different cases, subsequent additional differentiations produce additional functions until the maximum order of differentiation is reached at n. Since there are g different functions to be considered, the total number of new functions produced in this manner is given by:

$$(n − l) + (n − l′) + . . . = ng − l − l′ − . . .$$

The total number of parameters and arbitrary functions to be eliminated is thus:

$$\tfrac{1}{2}(n + 1)(n + 2)k + g + g' + ng - 1 - 1' - \ldots$$

The result of these eliminations from $\tfrac{1}{2}(n + 1)(n + 2)(k + 1)$ equations is a new set of equations of number:

$$\tfrac{1}{2}(n + 1)(n + 2) - g - g' - ng + 1 + 1' + \ldots$$

Regardless of the value of n, this number must not exceed $\tfrac{1}{2}(n - m + 1)(n - m + 2)$ if the integral is to be general in Ampère's sense. Through algebraic manipulation of this inequality Ampère then concluded that g must be no smaller than m, the order of the partial differential equation.

Since integrals of the first class are far simpler than those that involve partial integrations, Ampère devoted the initial pages of the concluding section of his memoir to an investigation of integrals of the first class.[24] Given a partial differential equation of order n with independent variables x and y,

$$f(x,y,z,p,q,r,s,t, \ldots) = 0 \qquad (6)$$

Ampère carried out a transformation to variables x and α, where α is the argument of any arbitrary function in the general integral of the equation. The independent variables now are x and α, and all derivatives are carried out with y considered to be a function of x and α. The highest nth-order derivative of z will thus involve a derivative with respect to α that is one order higher than any other derivative. For example, in the case of a second-order equation, this condition would apply to t, given by:

$$t = (\partial q / \partial \alpha)/(\partial y / \partial \alpha)$$

Ampère next rewrote the differential equation using x and α as independent variables and then put it in the form of a power series with respect to the highest order derivative in α. Once again using the second order as an example, the transformed equation would be of the form:

$$P + Qt + Rt^2 + \ldots = 0$$

179

Now a general integral must make this equation an identity regardless of which arbitrary function has the argument α. Various cases can be considered. Ampère assumed that integrals of the first class are under consideration; the partial integrations involved in integrals not of the first class would complicate analysis of the transformed equation in such a way that no general conclusions could be drawn. On the other hand, an integral of the first class may involve arbitrary functions that are either homogeneous or heterogeneous with respect to the argument α. In the heterogeneous case, t will involve a new function produced by the final differentiation with respect to α; the transformed equation can thus only be an identity if all coefficients of t vanish:

$$P = Q = R = \ldots = 0 \tag{7}$$

In the homogeneous case, the equation could be satisfied without all coefficients vanishing. Thus, if in a specific case it is found to be impossible for all the coefficient terms to vanish, the argument α cannot appear in an arbitrary function that is heterogeneous. Satisfaction of the condition that all the coefficients in the transformed equation vanish thus acts as a necessary condition for an integral to be of the first class and heterogeneous with respect to a specific argument. Analysis of equation (7) can be used to calculate possible values of the argument α in terms of x and y.

Ampère next carried out a second analysis to find consequences applicable to the more general case where an integral of the first class includes some homogeneous arbitrary functions.[25] By repeatedly differentiating equation (6), and transforming to independent variables x and α, he found that a common factor must vanish. For example, in the case of second-order equations, this condition is:

$$\partial f/\partial t - \partial y/\partial x \, \partial f/\partial s + (\partial y/\partial x)^2 \, \partial f/\partial r = 0 \tag{8}$$

This is a quadratic equation in powers of $\partial y/\partial x$, where the partial derivative is taken with α held constant. The roots of the equation can thus be used to determine possible values of the argument α. Used in conjunction with equation (7), equation (8) thus provided a means to determine both heterogeneous and homogeneous arguments. Ampère also noted the more specific cases that arise when the roots of equation (8) are identical or vanish.

Ampère concluded the 1815 memoir with two brief illustrations of his techniques. For example, he considered the second-order equation:

$$st + x(rt - s^2)^2 = 0 \qquad (9)$$

Using the relevant versions of equations (7) and (8), he showed that they were compatible with equation (9) and that each could be solved for a value of

$$\frac{dy}{dx\,(\alpha}$$

in terms of x, r, t, and s. Equation (9) thus has one heterogeneous and one homogeneous arbitrary function. This was as far as Ampère pursued his examples. The formal classification of the type of integral at stake was of far more interest to him than the specific details of the solution itself.[26]

More detailed applications were reserved for a lengthy second memoir that did not reach publication until early in 1820. Relevant versions of equation (7) were applied to a variety of first- and second-order equations, and the resulting calculations filled 188 pages. The first-order cases were relatively straightforward and Ampère's impact on subsequent developments of that topic was not significant. Second-order equations were his strong point, and he asserted that his approach was far more "direct" than the earlier efforts to find intermediary integrals of the first order. It also could be applied to hitherto untractable nonlinear equations.

For second-order equations, Ampère began by distinguishing four different cases depending upon whether the arbitrary functions are heterogeneous or homogeneous and whether they are functions of either x or y alone or of both x and y. He developed one example in full detail, the equation:[27]

$$(r - pt)^2 = q^2 rt$$

He then gave extensive attention to equations containing the term $(rt - s^2)$. He wrote these equations in the form:

$$Hr + 2Ks + Lt + M + N(rt - s^2) = 0$$

where H, K, L, M, and N are functions of x, y, z, p, and q. His usual application of equation (7) reduced to a set of first-order differential equations. This was a major simplification and Ampère explored its consequences for a variety of special cases. When $N = 0$, the reduced equations contain a term $K^2 - HL$ which Ampère abbreviates by G. He then applied the technique of Lagrange multipliers to the set of

equations made up of the reduced equations and the expression for
dz/dx. He laboriously analyzed the resulting simultaneous equations
for cases where G is identically zero, such as:[28]

$$x^4r - 4x^2qs + 4q^2t + 2px^3 = 0$$

and where G does not vanish, such as:

$$r + 2qs + (q^2 - x^2)t - q = 0$$

Seldom do any references to physical applications arise. A rare exception is Ampère's last example:

$$r + 2qs + (q^2 - b^2)t = 0$$

Ampère cites this as an equation pertinent to the movement of an
elastic fluid and refers to Poisson's treatment of it in terms of definite
intervals. In the manuscript records of his presentations to the Institut,
Ampère generally began with references to Poisson's work as a motivation for his own research.

Several aspects of Ampère's mathematical disposition are clear in
this, the most significant of his publications in pure mathematics. We
know from his correspondence that he generally found lengthy calculations to be "arid" and destructive of his sense of creativity and *joie
de vivre*. Nevertheless, when his concentration was intact, he could
lose himself in the details of these calculations; the temporary escape
from his customary depression was a welcome relief. Furthermore,
he was proud of his ability to carry out extended applications of the
clever insight expressed in equation (7). There was a sense of satisfaction in the knowledge that he had made a creative contribution to
mathematics, albeit one that was less widely acknowledged than those
of more celebrated contemporaries such as Cauchy and Fourier.

Poisson's review of Ampère's presentation during July of 1814 was
measured in tone but fairly accurate.[29] He noted Ampère's proofs about
arbitrary functions and alluded to the implications of equation (7). He
concluded by drawing a connection to Monge's study of characteristics that Ampère had not mentioned.[30]

These questions are those that M. Monge has given to determine the
curves that he calls *characteristics*, and which are, as is known, the very
remarkable lines following which two different surfaces which correspond to a given equation in partial differences can have, without coin-
ciding, a contact of as high an order as is desired.

Although Ampère often mentioned Monge in his notes, his own preference was for a more algebraic approach, as was the case in his study of the calculus. The techniques Ampère developed were more general than Monge's and the slightly more direct methods of George Boole. They were not significantly extended until Darboux did so in the 1870s. As Andrew Forsyth noted in his exhaustive 1906 history of the theory of partial differential equations, in the case of second-order equations, Ampère's method was "still of fundamental importance"[31] and could easily be extended to three independent variables. Nevertheless, it is also true that Ampère did not participate in the accelerated progress that was taking place in topics that most French mathematicians considered to be central to their discipline. Here the most relevant contrast is between Ampère's approach and those of his contemporaries Poisson, Fourier, and, slightly later, Cauchy. While Ampère classified and sought general criteria for integration, these men were arguing over the acceptability of Fourier series solutions to specific equations of interest to mathematical physicists. Equations for heat diffusion, vibrations of elastic surfaces, and surface waves all claimed their extended attention. By 1821 Cauchy was even replacing earlier power series solutions with more sophisticated applications of complex analysis. There may be no simple answer to the question of why Ampère did not share these interests. It is unlikely that his characteristic demand for "certainty" made him opposed to power series solutions as such. A more relevant factor would be Ampère's eclectic interests and his refusal to dedicate himself exclusively to mathematics. Poisson, Fourier, and Cauchy were single-minded in their mathematical investigations of carefully posed physical problems. They were unconcerned about the issues that attracted Ampère to experimental chemistry, and even more indifferent to his metaphysics.

At any rate, once the agonizing campaign for the Institut came to an end with the election of 28 November, 1814, Ampère would never again devote extensive energy to pure mathematics. He did not even complete the publication of his 1814 research until 1820, the year in which he resolutely turned to electrodynamics. In the meantime, chemistry and physics dominated his scientific activity.

The Eighteenth-Century Revolution in Chemistry

Historians often cite developments in chemistry during the half-century prior to 1789 as an example of a full-blown scientific revolution.

Although argument continues about the relative importance of one or another revolutionary innovation, several factors are clearly relevant for an understanding of Ampère's contributions to chemistry early in the nineteenth century. First, as might be expected from an empirically minded post-Newtonian milieu, eighteenth-century chemists eventually insisted that all substances cited in chemical reactions must be attributed measurable weight. The old alchemical "spirits" or essences were gradually banished from chemical language in favor of material substances. Alchemical "principles," although they might bear such seemingly substantial names as "mercury" or "sulphur," were alleged to be responsible for observable properties without themselves being material in the manner that ordinary physical substances are. In this respect some alchemists shared a perspective of the Aristotelian tradition by claiming that nonmaterial alchemical principles "informed" matter and thus acted as formal causes of observable qualities. This mode of reasoning gradually lost its appeal in the face of a more positivistic insistence that the study and description of chemical substances must be restricted to measurable properties. Measurements of mass and density became central to eighteenth-century chemical analysis. Furthermore, the insistence that matter could not be created or destroyed by chemical reactions was gradually formalized as the principle of conservation of mass. For example, Lavoisier gave a famous statement of the principle in his influential 1789 *Traité élémentaire de Chimie*.[32]

> We may lay it down as an incontestable axiom, that, in all the operations of art and nature, nothing is created; an equal quantity of matter exists both before and after the experiment; the quantity and quality of the elements remain precisely the same; and nothing takes place beyond changes and modifications of these elements.

Lavoisier's allusion to "elements" brings to mind a second aspect of eighteenth-century chemistry, the recognition that matter can be classified into a large number of pure chemical substances or elements. As late as the middle of the eighteenth century, chemistry texts still referred to the Aristotelian elements of earth, air, water, and fire as the four fundamental elements. These abstractions became increasingly irrelevant as chemists devised more elaborate classifications for practical purposes. For example, considerable attention was devoted to the formation of "neutral salts" through the combination of an acid and a base; the bases in turn were generally classified as alkalis, metals, or earths. By 1782 so many materials had been analyzed into components

184

that Louis-Bernard Guyton de Morveau could publish a table of no-menclature that included over 500 substances. Affinity tables, pub-lished for the first time in 1718, also provided lists of materials according to the strength of their "affinity" for a given substance. The tables were intended to be empirical descriptions of observed displacements of one component of a compound when that compound is exposed to a material with greater affinity than the displaced component. Acids, bases, and salts made up many of the affinity table entries; little con-sideration was given to the issue of which, if any, of these substances were elementary.

In common with contemporaries such as Guyton de Morveau and Torbern Bergman, Lavoisier adopted a working operational definition of "element" as a substance that has not yielded to chemical analysis into constituent components. This would become the basis of his famous table of simple substances in the 1789 *Traité*. In the meantime, the analytic process was applied to atmospheric air and water, sub-stances that were at least nominally associated with two of the Aris-totelian elements. Cavendish, Joseph Priestley, and Lavoisier all successfully isolated hydrogen, oxygen, carbon dioxide (fixed air), and nitrogen (*azote*) within a relatively short period of time prior to 1775. Similarly, Lavoisier demonstrated that water is not a simple substance but a compound of hydrogen and oxygen. Chemists also isolated a wide variety of apparently pure and elementary "earths" that would not yield to further chemical analysis.

The old Aristotelian element of fire posed difficulties of a different sort. Lavoisier's insistence that chemical substances have measurable mass was one motivation for his famous oxygen theory of combus-tion. At the beginning of the eighteenth century, Georg Stahl had proposed that the alchemical principles of sulphur and mercury might actually be a single principle of combustibility, a principle he referred to as phlogiston. Combustion then was said to involve a release of phlogiston from the burning substance. But experiments on the com-bustion of metals, some of which had been performed as early as the seventeenth century, showed that the solid remnants of the process, the metallic calx, had a larger mass than the initial metallic sample. Guided by the principle of conservation of matter, Lavoisier and Priestley performed the delicate experiments that Lavoisier interpreted as an indication that the increase in mass is due to a chemical reaction that draws oxygen out of the atmosphere to form a metal oxide. Al-though resisted by Priestley, Lavoisier's interpretation gradually tri-umphed. On the other hand, Lavoisier retained an allegiance to the caloric theory of heat. At the turn of the nineteenth century there was still debate among chemists about whether caloric should be considered to be chemically bonded to material particles or whether it was simply a material responsible for repulsion between particles but did not itself act as a chemical agent.

Classification and Nomenclature

Lavoisier was quick to promote his own interpretation of new developments by introducing new terminology. In particular, by 1779 he was convinced that acids owe their properties to the same gas that is responsible for the combustion processes he had been studying; he thus chose the name "oxygen" with its Greek connotation of "acid former." Encouraged by his analysis of water into hydrogen and oxygen, Lavoisier advocated a thorough reform of chemical terminology. His collaboration with Guyton de Morveau, Antoine-François de Fourcroy, and Claude-Louis Berthollet, all recent converts to the oxygen theory, resulted in their 1787 publication, *Méthode de Nomenclature chimique*. The new terminology represented a concerted effort to purge chemistry of the arbitrary names that had haphazardly accumulated over centuries of alchemical and chemical research. Little accurate information was conveyed by antiquated names such as "flowers of zinc" or "oil of tartar." In the case of simple gases such as oxygen, hydrogen, and azote (nitrogen), the Greek name was chosen to indicate the primary property involved. Oxygen, at least according to Lavoisier, is the source of acids, hydrogen is one of the components of water, and azote is a gas that does not support life. Compounds were represented by a binomial term that indicated the components, for example, zinc oxide replaced the old and less informative term pomphilix.

Although the primary goal of the *Méthode de Nomenclature chimique* was not classification as such, the authors did provide a system of five classes of 55 allegedly simple substances; this scheme formed the basis for Lavoisier's revised classification in his *Traité* two years later. The first class was made up of light, caloric, oxygen, hydrogen, and nitrogen (*azote*). These were all widely recognized simple substances; the inclusion of light and caloric strikes the modern eye as anomalous. Although eventually both would be dropped from tables of chemical substances, they were retained by Lavoisier in the *Traité*. The second class contained "acidifiable bases," such as sulphur, phosphorus, and carbon, and included the unknown bases of other acids, such as *acide muriatique*, muriatic acid. Lavoisier and his colleagues incorrectly

186

expected that all acids would eventually be analyzed into oxygen and an "acidifiable base"; the oxygen theory of acids was thus incorporated into the 1787 classification scheme. The third, fourth, and fifth classes were made up of the metals, earths, and alkalis respectively.

The authors of the *Méthode de Nomenclature* included a dictionary of synonyms for the translation of old terminology into the new usages; chemistry students of Ampère's generation were the first to learn chemistry without the burden of what had become an unwieldy and unsystematic terminology. Interest in systematics and a philosophical predilection for accurate language were central themes of Enlightenment culture. The penchant for classification directed research within a broad spectrum of disciplines: botany, mineralogy, chemistry, mathematics, medicine, and natural history. Furthermore, debate over chemical terminology in some respects followed patterns established in botany and natural history. The middle of the eighteenth century was dominated by Linnaeus' efforts to create an admittedly artificial classification scheme for botany based solely upon the reproductive systems of plants. His plant and animal taxonomies of class, order, genus, species, and variety became a characteristic fixture of Enlightenment rationalization. The widespread acceptance of his binomial system of genus and species represented the level of success Lavoisier and Guyton de Morveau hoped to achieve with their revision of chemical terminology.

Furthermore, the controversy about the "artificiality" of Linnaeus' system was duplicated by chemists. Following Linnaeus' death in 1778, the French botanists Michel Adanson and Antoine-Laurent de Jussieu advocated a "natural" method of classification based upon a holistic study of the full scope of a plant's parts. The publication of de Jussieu's *Genera plantarum* in 1789 represented the full arrival of an alternative to the Linnaean system. By calling attention to the fact that plants fall into groups in which various properties vary in degree or intensity, the natural method generated predictions about hitherto unknown members of the group. We have already noted Ampère's youthful sympathy for Rousseau's vigorous objections to Linnaeus. His preference for "natural" classifications carried over into chemistry and reached expression in his 1816 memoir on classification.[33]

Lavoisier's Traité

The single most important factor in Ampère's chemical education was his reading of Lavoisier's *Traité* during his relatively carefree days in Lyon. Ampère was young, uncommitted to any specific chemical theories, and eager to learn; his reading of the *Traité* left a lasting impression. In later years he would often add Lavoisier to the names of Copernicus and Newton when he felt called upon to cite scientists of revolutionary impact. Lavoisier would have been particularly pleased to see Ampère responding favorably to both the general methodology and the specific content of the *Traité*. In his introduction, Lavoisier cited Condillac in support of the view that scientific progress is dependent on linguistic accuracy. He also insisted upon "direct" reasoning from known experimental facts to less obvious consequences. It is tempting to discount these admonitions as so much pedagogical preaching far removed from actual scientific practice. In point of fact, Lavoisier's own text was a complex composite of theory, observational reports, and interpretation. At any rate, Ampère did develop a preference for the direct reasoning Lavoisier advocated.

Lavoisier also emphasized a reductionistic approach to chemical research which Ampère took thoroughly to heart. In particular, Lavoisier argued that the determination of the chemical constituents of complex substances must be based upon both analysis into components and synthesis out of those components. The tandem concepts of analysis and synthesis arose so frequently in Ampère's early reading in mathematics, chemistry, and philosophy that they became an essential component of his scientific mentality. His interest in chemistry during the two decades prior to 1820 ensured that analysis would remain central to his conception of scientific methodology.

Turning to the theoretical content of the *Traité*, we find a three-part organization. Part I is Lavoisier at his most original. Here he vigorously advocated his caloric theory of gases and his oxygen theories of combustion and acids. Part II is devoted to an extension of the binomial nomenclature proposed by Lavoisier and his colleagues in 1787; Part III describes apparatus and procedures. Ampère was deeply affected by several aspects of Lavoisier's text. In Part I, the presentation of the caloric theory of gases emphasized the "expansibility" of the caloric fluid. An increase in the number of caloric particles between the particles of ordinary matter increases the average distance between those material particles and thus brings about phenomena such as expansion, melting, evaporation, and the gaseous state. Lavoisier drew heavily upon Turgot's notions that the *expansibilité* of vapors is due to the subtle fluid that constitutes heat, a view Turgot had expressed in an anonymous article for the *Encyclopédie*.[34] Thus Ampère had probably already encountered these ideas and was prone to accept and amplify them. We have already seen how his first excursion into scientific theorizing in 1800 and 1801 resulted in an imaginative reduction of electric and magnetic phenomena to the "springiness" of electric and magnetic "atmospheres" surrounding the particles of ordinary matter. In conjunction with his reading of Lavoisier, Ampère became committed to the material nature of heat; he would not abandon it until many years later. Furthermore, his early thoughts about the composition of gases and the interaction between caloric and ponderable matter were later transformed into analogous reasoning about the luminiferous ether. The ether eventually became the most important unifying concept in Ampère's efforts to integrate theories of light, heat, electricity, and magnetism.

Lavoisier's oxygen theories of acids and combustion were major motivations for his classification of the "simple substances," the classification that resulted in the famous chart of 33 elements at the beginning of Part II of the *Traité*. Believing that oxygen is a component of all acids, he suspected that the bases which form neutral salts when combined with acids might also have oxygen as a component and thus might be metallic oxides.[35] Oxygen functioned as both a conceptual and a material unifying principle for Lavoisier. The extent of this conviction resulted in some inconsistency in his choice of "simple substances." While operationally defining elementary substances as presently nonsusceptible to chemical analysis, he nevertheless listed *radical muriatique, radical fluorique,* and *radical boracique* as simple substances in expectation that the three acids concerned would eventually be analyzed into oxygen and these three radicals. On the other hand, he had less clear theoretical reasons for not including the alkalis in his list. It is not clear whether Lavoisier suspected that nitrogen or oxygen was a component of the alkalis.[36] At any rate, by his own definition of an element, he should not have allowed these theoretical suspicions to interfere with his stipulation of the list of elements as presently constituted. Perhaps to some degree he shared the scruples of his critics, who accused him of making chemical terminology excessively contingent upon day by day developments in laboratory analysis.

As represented in plate 2, Lavoisier's full list was divided into four categories. Reminiscent of his earlier 1787 contribution, the first group is made up of light, heat, oxygen, nitrogen, and hydrogen, substances having little in common save perhaps their respective roles as sources of important phenomena or compounds. The second group contains the bases of well understood acids and it reserves positions for the three bases expected to result from future analysis. Third are the metals, a class Lavoisier takes to be easily recognizable; as we will see, Ampère would disagree. Finally, simple "earthy substances" make up

TABLEAU DES SUBSTANCES SIMPLES

	Noms nouveaux	Noms anciens correspondans
Substances simples qui appartiennent aux trois règnes & qu'on peut-regarder comme les élémens des corps	Lumière	Lumière.
	Calorique	Chaleur. Principe de la chaleur. Fluide igné. Feu. Matière du feu & de la chaleur.
	Oxygène . . . . : . . .	Air déphlogistiqué. Air empiréal. Air vital. Base de l'air vital.
	Azote.	Gaz phlogistiqué. Mofète. Base de la mofete.
	Hydrogène	Gaz inflammable. Base du gaz inflammable.
Substances simples non métalliques oxidables & acidifiables	Soufre	Soufre.
	Phosphore	Phosphore.
	Carbone	Charbon pur.
	Radical muriatique	Inconnu.
	Radical fluorique	Inconnu.
	Radical boracique	Inconnu.
Substances simple métalliques oxidables & acidifiables	Antimoine	Antimoine.
	Argent.	Argent.
	Arsenic	Arsenic.
	Bismuth	Bismuth.
	Cobolt	Cobolt.
	Cuivre	Cuivre.
	Etain	Etain.
	Fer.	Fer.
	Manganèse	Manganèse.
	Mercure.	Mercure.
	Molybdène	Molybdène.
	Nickel	Nickel.
	Or	Or.
	Platine	Platine.
	Plomb	Plomb.
	Tungstène.	Tungstene.
	Zinc.	Zinc.
Substances simples salifiables terreuses	Chaux	Terre calcaire, chaux.
	Magnésie.	Magnése, base du sel d'Epsom.
	Baryte.	Barote, terre pesante.
	Alumine	Argile, terre de l'alun, base de l'alun.
	Silice	Terre siliceuse, terre vitrifiable.

2 Lavoisier's Table of Elements, first published in his *Traité élémentaire de chimie* (Paris, 1789) and in Robert Kerr's translation, *Elements of Chemistry* (Edinburgh, 1790, pp. 175–6).

Lavoisier's last class; he included them in spite of his suspicion that they were actually metallic oxides.

Lavoisier's legacy to early nineteenth-century chemistry was complex. His untimely death during the Terror in 1794 deprived his colleagues of both his theoretical insights and his leadership within the Académie. His insistence on careful analysis of compounds into more elementary constituents helped make the study of neutral salts and acids a major empirical topic for chemists during the early decades of the nineteenth century. Although the study of chemical composition, molecular formulas, combining weights, and the classification of elementary substances could in principle proceed independent of debates over the possible atomic components of molecules, atomic theories constituted another aspect of chemistry that appealed to Ampère. We should note how Dalton's ideas were absorbed into the French context.

Atomic Theory and Atomic Debates

Berthollet, the most prominent French chemist, encouraged empirical investigation and took a skeptical attitude toward what he called "arbitrary hypotheses" about the microstructure of chemical compounds.[37] This conservative attitude naturally kept him from supporting Dalton's atomic hypothesis. French debates over atomism thus took place under a stern eye of disapproval from the acknowledged master and premier source of patronage for the field. It was much safer to concentrate on the phenomenological laws of stoichiometry, laws pertaining to the relative weights of the elements that make up compounds. Although based upon empirical data, phenomenological laws such as Gay-Lussac's law of combining volumes for gases generated controversy in their own right. Nevertheless, this phenomenological research was not directly inspired by theoretical interest in atomic hypotheses. Some chemists realized that it was possible to account for the stoichiometric laws by adopting a chemical atomic theory; that is, they claimed that chemical reactions take place between chemically indivisible units pertaining to each element, each unit with a fixed atomic weight characteristic of its element. In this respect it is worthwhile to follow the historian Alan Rocke and distinguish "chemical atomism" from the more controversial "physical atomism."[38] Although John Dalton was the most influential proponent of chemical atomism, he also advocated a more extreme view by arguing that elements are made up of *physically* indivisible atoms, the ultimate building blocks of all matter. The reception of Dalton's views and the resulting atomic debates in

France were thus complicated by the uncertain status of phenomenological laws, Dalton's combination of chemical and physical
atomism, and Berthollet's general aversion to all forms of atomism.
Nevertheless, in spite of the complications generated by the Napoleonic wars, by 1810 debate over Dalton's ideas was a major component of the French scientific milieu.

Dalton's atomic theory was rooted in his interest in the physics of
gases, particularly with applications to meteorology in mind. In 1801
he proposed his phenomenological law of partial pressures; a mixture
of gases produces a pressure that is the sum of the partial pressures
produced by the individual gases. To explain why this law holds,
Dalton speculated that the atoms of each gas repel only other atoms
of that gas and not those of the other gases of the mixture. To determine the relative weights of different atoms, he adopted his "rule of
greatest simplicity," that is, unless there is good reason to do otherwise, one should assume the simplest possible formula for the elements in a compound. For compounds made up of two elements, this
meant assuming a binary formula; for example, water initially was
assumed to have the formula HO.

In view of Berthollet's resistance to atomic theory and with the interpretation of so much empirical research still at issue, it is not surprising that the majority of chemists placed little importance on the
paper Amedeo Avagadro published in the *Journal de Physique* in 1811.
Half a century would pass before Avagadro's hypothesis was recognized as an important contribution to atomic theory. Ampère was one
of the few French scientists to take the hypothesis seriously; he actually argued for it independently before reading Avagadro. This was
made possible in part by Ampère's peripheral status with respect to
the elite circle of chemists under Berthollet's patronage. Not sharing
the agenda of that group, Ampère was free to cultivate his own ideas.
To some extent Avagadro was ignored by the French for reasons that
also applied to Oersted during the same period. Although both men
published in prestigious French journals, they were eyed with suspicion due to their outsider status as foreigners and their willingness to
draw sweeping conclusions from preliminary data.

Born in Turin in the kingdom of Sardinia, Avagadro was forced
to respond from a distance to French and English developments in
chemistry and the physics of gases. He was unsupported, but also
unburdened, by the constraints of Berthollet's patronage. In an 1811
paper Avagadro introduced his famous hypothesis that equal volumes
of gases contain equal numbers of particles. However, Avagadro made
this assumption only as one stage in a far more complex argument.

192

His real goal was to determine chemical affinities by linking them to calculations of affinity for caloric. His equal numbers hypothesis simply allowed him to use density measurements in place of molecular weights to determine molecular formulas. He pursued his more general project over a subsequent period of three decades. By 1824 he had argued that the "attractive power of heat" could be written as a function of specific heat and then transformed into an "affinity number" by dividing by density and normalizing with respect to oxygen. There were too many unjustified steps in Avagadro's argument and his efforts generated little interest, particularly following Dulong's publication of more accurate specific heats in 1829. French indifference to Avagadro's equal numbers in equal volumes hypothesis was only in part due to the resistance to atomic theory advocated by Berthollet's school. That resistance might have weakened had Avagadro more successfully determined affinities using atomic theory and his hypothesis.

Furthermore, Avagadro complicated atomic theory by introducing a second hypothesis concerning the divisibility of molecules. Recognizing that an atomic model provided insight into Gay-Lussac's law of combining volumes, Avagadro also realized that serious anomalies had to be resolved. For example, one volume of oxygen combines with two volumes of hydrogen to form two volumes of water. In modern terminology, Avagadro in effect argued that oxygen molecules must be divisible into two oxygen atoms, each of which combines with two hydrogen atoms. However, Avagadro himself did not use these terms and his 1811 terminology did not convey a clear conception of molecular structure. Rather than use the term "atom" (*atome*) to refer to the simplest parts of a molecule, he used the terms *intégrante molécule*, *constituante molécule*, and *élémentaire molécule* without being entirely consistent. It is only with historically misleading hindsight that these terms can be translated respectively as compound molecule, elementary molecule, and atom. Furthermore, Avagadro's contemporary Berzelius produced the most accurate available determination of atomic weights without using Avagadro's assumptions. Berzelius' electrochemical theory actually ruled out the possibility of polyatomic molecules.

While the fate of Avagadro's research program is interesting in its own right, he did not have a significant personal impact on Ampère; Avagadro's hasty applications of his equal numbers in equal volumes hypothesis may even have contributed to the lack of positive response to Ampère's own argument for the hypothesis. On the other hand, there were other more reputable sources for Ampère's interest in atomic theory and molecular structure. For example, we have already noted the importance of René Just Haüy's 1803 *Traité élémentaire de physique* as an early manifesto of Laplacian physics. In his earlier study of crystal structure and mineralogy, Haüy had already devised the concept of the *molécule intégrante* to refer to both the simplest unit of crystal structure and the chemical molecule, the smallest particle that retains the chemical properties of a given substance.[39] He distinguished this concept from the basic units of crystal structures, the "primitive forms" that were directly observable in crystal cleavage fragments. Haüy stipulated six of these forms: parallelepiped, regular tetrahedron, octahedron with triangular faces, hexagonal prism, rhombic dodecahedron, and dodecahedron with isosceles triangles for faces. He also thought of these forms as constructed from *molécules intégrantes* of three forms: tetrahedron, parallelepiped, and triangular prism. Although he included a new discussion of affinity theory in the 1806 edition of his *Traité*, he did not frame it in terms of molecular shape. Instead, he conformed to the program of Laplace and Berthollet by alluding to "spheres of effective affinities" for particles that were left structurally unspecified. This conservative stance is not surprising for a man with little interest in chemistry as such. For Ampère, on the other hand, the primitive forms of Haüy's crystallography provided the key to a mathematical basis for chemical composition.

Ampère's Platonic Atomism

Ampère maintained his interest in chemistry throughout his first decade in Paris. Although he performed virtually no experimental work himself, his encyclopedic memory was well suited for arguments about chemical composition and the related disputes over nomenclature. At the École Polytechnique he was in close touch with active researchers; Thenard was joined there by Gay-Lussac in 1809. Ampère became quite excited about Davy's isolation of potassium and sodium in 1807, and he began a predominantly one-sided correspondence with the English chemist in 1810. Ampère's letters reveal both his strong and weak points as an outsider to the established group of professional chemists. He obviously had the ability to contemplate a large set of chemical reactions simultaneously and to draw creative conclusions consistent with the lot. On the other hand, many of the experiments that bolstered his arguments were wholly hypothetical, a characteristic that gave his conclusions an equally tenuous nature. So while in some cases he reasoned his way to results that were indeed confirmed by later experimentation, such as his intuition that iodine and fluorine would be recognized as elements, these insights were accompanied by scores of less promising suggestions and premature appeals for revisions in terminology.

Perhaps the most famous of Ampère's letters to Davy was the very first he wrote in November of 1810. At that point Davy was investigating the properties of fluoric acid. Ampère proposed that the acid is similar to muriatic acid (HCl) in that it contains no oxygen and is simply a compound of hydrogen and a radical Ampère called *oxy-fluorique*.[40] Davy's formal reply was brief and far less effusive. He thanked Ampère for his "very instructive" comments but shared no details about his own research. This pattern continued throughout the subsequent decade. Ampère's enthusiastic speculations and very specific suggestions for experimental tests met with polite acknowledgement but little else. Davy probably did not intend to be disparaging; he simply adopted the practice typical of his contemporaries by keeping a guarded silence about research plans until the results reached publication. Ampère could not maintain this restraint; he broadcast ideas as soon as they occurred to him and hoped that his colleagues would follow suit. He was sharply disappointed in this hope during the peak years of his interest in chemistry.

During the Napoleonic wars competition between English and French chemists was even more vigorous than usual, for example, between Davy and the Gay-Lussac and Thenard team. The discovery of new elements was especially coveted since it represented a fairly direct route to some degree of immortality. In 1813 Davy did write to Ampère to acknowledge that his intuition about *oxy-fluorique* had been correct.[41] In print, however, he simply remarked that "during the period that I was engaged in these investigations, I received two letters from M. Ampère, from Paris, containing many ingenious and original arguments in favour of the analogy between the muriatic and fluoric compounds." He also proposed to call the new element *fluorine*, "a name suggested to me by M. Ampère."[42] Ironically, Ampère's own choice for this name was *phtore*, a name he preferred because its Greek root implied a substance that is destructive or ruinous. Davy's aloofness may in part have been due to the gulf separating the experimentalist from the theorist; he had almost been deprived of an eye while experimenting on a compound of nitrogen and chlorine casually suggested to him by Ampère.

In another famous episode, Ampère was one of those who provided Davy with his first sample of the element iodine.[43] Between October and December of 1813 Davy visited Paris with his wife and Faraday to receive the Napoleonic prize he had been awarded by the Institut for his work on electrochemistry several years earlier. The presence of an Englishman in France was an unusual event during the Napoleonic wars and a high level of mutual suspicion prevailed. Although Davy met Laplace and attended a meeting of the Société d'Arcueil with Lady Davy, the English chemist felt more at ease and more welcome with less exalted chemists such as Ampère, Nicolas Clément, and Charles Bernard Desormes. On 23 November these three called on Davy and presented him with a sample of an intriguing substance that had been isolated by Bernard Courtois about two years earlier. The general opinion in France was that it was a compound of chlorine and oxygen, and on 29 November Clément and Desormes presented a memoir to that effect before the Institut. Like Ampère, Clément was struggling for a position in the Institut and hoped to be elected as a corresponding member. Meanwhile, experimenting with his portable equipment, Davy quickly concluded that the mysterious substance was a new element and he established its name as iodine. By 3 or 4 December Davy had confided his conclusion to Baron Cuvier and at the 6 December meeting he was elected to the position Clément had hoped for. At that same meeting Gay-Lussac read his own paper on *iode*, thereby igniting a rather harsh priority dispute.

The hapless Ampère thus found himself in the midst of an unusually vigorous priority dispute. Although Davy may have been privately appreciative of Ampère's efforts, his departure for Italy and England left Ampère with little consolation. Obviously Davy was not interested in any significant collaboration. Although the two men were nearly the same age and shared romantic literary sensibilities, they were too different culturally and in social station to become close colleagues. Ampère subsequently harbored considerable bitterness about the treatment he received. He had shared his ideas spontaneously and then saw them transformed into discoveries by other men. His resentment was compounded by the realization that he had not had the time or facilities to do the necessary research himself. Later he also sensed that the cool reception of his ideas by some of his French colleagues was partially due to his relations with their English rival Davy.

Between 1813 and 1816, an interval of truly chaotic personal life and his most creative work in philosophy and mathematics, Ampère managed to publish three chemistry memoirs, an output that he claimed represented just a small sample of a much more thorough and empirically supported theory of chemical composition and classification. He presented the first of these essays to the Institut during January of 1814; here he cautiously limited himself to a derivation of the gas law known in France as Mariotte's law, the inverse proportionality between pressure and volume. A decade in Paris had taught Ampère to proceed more carefully than he had in the early manuscripts of 1801. A preliminary paragraph of the published version of the memoir reveals his respect for the Laplacian analysis of short-range forces and his retreat from his earlier optimism about his model for electric and magnetic forces.[44]

> In the current state of our knowledge, and leaving out of account the forces which produce the phenomena of electricity and the magnet, which are too little known to allow their effects to be related by general laws to those of the other forces of nature, the distance and the relative position of the particles of bodies are determined by three kinds of forces: the pressure that they support, the repulsion between their particles produced by caloric, and the attractive and repulsive forces pertaining to each of these particles that depend upon their nature and bestow various qualities upon bodies of different species.

Ampère argued that the third of these forces, that of chemical affinity, was of much shorter range than the distance between gas particles, a claim he justified by noting that when gases that do not combine chemically are mixed, there is no change in the total volume. Ampère thus adopted as a working principle that "the particles in all gases are at a sufficient distance that the forces which are specific to them no longer have any influence on their mutual distances."[45] We can see here a prelude to the hypothesis that would be explicitly adopted in his next paper, the equal particles in equal volumes hypothesis. If the caloric atmosphere around each gas particle makes chemical affinities irrelevant, then equilibrium requires simply a neutralization of the repulsive effect of caloric by the force due to the pressure exerted by a restraining barrier. Ampère then considered the pressure exerted by the gas on a movable wall or piston. The repulsion between a gas particle and the wall is written as a function $\phi(t,z)$, where z is the distance to the wall and the temperature is t. The pressure or force per unit area then would be determined by integrating this expression over the variable z and introducing a constant factor n, the density of gas particles. Regardless of what the function ϕ is, the result will be proportional to n and thus inversely proportional to volume.

Ampère's derivation is interesting in that it does not introduce a kinetic theory of gases; the gas particles are imagined to have fixed positions, and the temperature of the gas is determined by the quantity of caloric that separates the particles. His procedure was thoroughly in keeping with Laplacian physics and in this respect Ampère's memoir represents his most conventional work in physics. Nevertheless, it was not well received and Ampère vented his frustration in a gloomy letter to Bredin.[46]

> I certainly see that I will run aground at the Institut, but I cannot resign myself to take the necessary steps and not work to attain positions there. I have read there and my memoir is condemned without examining it. The one of the members whose friendship ought to be the most assured, for whom I have sacrificed a place which might presently be my resource, indeed you know who, has reproached me, to the point of the gravest insults, about my correspondence with Mr. Davy as a crime. I find myself the butt of scorn from those to whom I have never done any harm, upset to the highest degree about the present and the future, seeing myself perhaps soon without any resource to subsist here; and all this makes up only a very small part of my troubles. They are of all kinds, but to whom do I dare complain? Heaven's curse seems affixed to me, but it is always through my fault that I am precipitated into misfortune. Always, I know not what fate puts me in a situation which enlightens me, which gives me time to reflect, and then a sudden event sweeps me away just at the moment when I seem to have nothing to fear.

Ampère's biographer de Launay cited Thenard as Ampère's major detractor, a man known for his aggressive personality. Thenard had been elected to the Institut in 1810, and in Ampère's paranoid and self-pitying state of mind that event might now have been perceived as a "sacrifice" on Ampère's part. The more mild-mannered Gay-Lussac harbored no grudge and had Ampère proof-read his iodine memoir. Ampère was pleasantly surprised by this request and was delighted to find, as he put it, "all my predictions of five years ago verified and generally admitted."[47]

Oscillating violently between depression and exhilaration, Ampère began a much more extended and theoretical chemistry memoir in January of 1814. Early in the project he shared his high expectations with Roux-Bordier.[48]

> For about two months I occupied myself with a project the upshot of which seemed to me should inaugurate a new era in this science and provide the means of predicting *a priori* the fixed ratios according to which bodies combine with each other by relating their various combinations to principles which would be the expression of a law of nature; this discovery, second only to the one I made last summer in metaphysics, will possibly be the most important thing I will have

conceived in my entire life. I say "second to what I have done in meta-physics," because the latter science is the only truly important one; for the theory of chemical combinations is indeed particularly clear and indisputable and will become something as common in the physical sciences as the other theories generally admitted.

Ironically, Ampère presented his atomic theory in the form of an open letter to Berthollet. Ampère was of course an outsider to the elite circle of French chemists; even if he had been given the opportunity, we can hardly imagine him making a favorable presentation to the sophisticates of Berthollet's Société at Arcueil. Instead he wrote a rambling forty-three page "abstract" of what he claimed was a much more detailed theory. A passage from a letter to Bredin during the spring of 1814 captures the tense state of mind that was typical for Ampère during this time. His frustrated ambition and exasperation with conditions that were beyond his control combined to exert a pressure that he found almost unbearable.[49]

Here is what has happened to me! I was working on a memoir which I was supposed to read to the Institut on partial differentials. I was not very pleased with it myself although there indeed might have been some new things; but I know that they will not please the Bonaparts of mathematics and they alone will be judges of it. Suddenly I was told that Mr. Dalton is occupying himself in England with the manner in which molecules of bodies are arranged in chemical combinations. I know that last January I had written a memoir on that. Suddenly the fear came over me that he would find and publish before me part of what I had done. I speak of my fear. I am advised to make an abstract, in the form of a letter to Mr. Berthollet, of the January memoir and it will be printed in the *Annales de Chimie*. Mr. Berthollet is told and he strongly agrees. I begin the abstract believing that there would be two days of work, maybe three. This memoir was a formless chaos. I no longer saw anything there, having lost sight of its ideas. Finally I gave it up. The editor of the *Annales de Chimie* complained to Mr. Berthollet, telling him that he had counted on it and that he finds himself short. Mr. Berthollet finds me at the Institut and tells me.

I return, I hire a copyist to write under my dictation. I lodge him quite near me so as to work very late in the evening and early in the morning. But I am forgetting the mathematics memoir, I lose almost all hope of attaining the Institut, and after three weeks of dictating I give the meas-ure to the press; half is at the press and the remainder is finally almost finished, but this abstract is as long as the memoir from which it is thought to be drawn. No one will read it. None of it will be understood.

And all of this constitutes just so many events arranged to upset, on every point, all the plans to which I have successively held in my life.

We can hardly imagine a scenario more removed from the polished procedure followed by an insider such as Gay-Lussac. Ampère's fears about the insignificant impact of his memoir were well taken. The link between his theoretical propositions and empirical evidence was far too weak to impress working chemists as the inauguration of a "new era" for their science. Ampère based his theory upon several fundamental assumptions and related inferences. As had been the case with Avagadro in 1811, Ampère did not use the term "atom" to refer to components of larger assemblages. Instead he cited the Gay-Lussac law of combining volumes as inspiration for a distinction between the "particles" that make up an element or compound, and the assemblage of punctal "molecules" that determine the shape of these particles. The particles are assumed to be three-dimensional and are thus made up of a minimum of four "molecules." The molecules are thought of as indivisible, punctal, and located at positions that determine the apices of a polyhedron, a shape specific to each element or compound and which Ampère called the "representative form" of the particle. The polyhedron-shaped particles are thus mostly empty space with all mass located in the point molecules at the apices of the polyhedron.

As had been the case in the earlier memoir on gas theory, Ampère retained his long-standing commitment to the caloric theory of gases and made the "basic assumption" that caloric keeps particles of a gas far enough apart so that cohesion and affinity do not operate and the distance between gas particles depends only upon temperature and pressure. He then concluded that this means that at any specific temperature and pressure the inter-particle distance is the same for all gases and thus the number of particles in any two equal volumes is the same. This is of course the hypothesis adopted by Avagadro in his 1811 memoir. Ampère added a footnote saying that he had not known of Avagadro's proposal when composing his own argument. He then applied the equal numbers in equal volumes hypothesis to explain the law of combining volumes. For example, Ampère argued that since a volume of oxygen combines with two volumes of hydrogen to form two volumes of steam, the resulting water "particle" must be made up of one hydrogen particle and one half of an oxygen particle.

But how many molecules make up any specific particle? Here Ampère introduced his second basic assumption, namely, that the common elements hydrogen, oxygen, and nitrogen (azote) have particles made up of four molecules. This assumption Ampère claimed was "the simplest" and was justified by agreement between its implications and observations. How are these molecules arranged? Ampère claimed that in general they should be located at the apices of one of five "primitive forms" recognized in crystallography: the tetrahedron, the octahedron, the parallelepiped, the hexahedral prism, and the rhomboidal dodecahedron. These polyhedra require four, six, eight, twelve, and fourteen molecules respectively. The particles of oxygen, hydrogen, and nitrogen are thus tetrahedra made up of four molecules each. On the other hand, each chlorine particle is an parallelepiped formed by eight molecules, an assumption Ampère claimed was required to account for observed combining ratios of gases.

The basis of a simple chemical compound is the interpenetration of pairs of particles, one from each element, so that they have a common center of gravity and also produce another regular polyhedron from the total assemblage of the two sets of molecules. On the other hand, although Ampère does not explicitly say so, he implies that when half of a particle combines with a particle of another element, the total collection of molecules adopts a new configuration corresponding to one of the acceptable structures. Water particles, for example, are octahedra made up of six molecules of which four come from a hydrogen tetrahedron and two from one half of an oxygen tetrahedron.

Setting aside cases involving half particles, Ampère coined names for the 23 polyhedra produced by permitted combinations of two of his representative forms, the tetrahedron and the octahedron. Of course he was not the first scientist to appreciate the geometric beauty of these shapes. Among others, Plato and Kepler had introduced aesthetically pleasing polyhedra into their theoretical attempts to explain observable phenomena. In Plato's case, tiny right triangles are the "atoms" that make up "molecular" polyhedra, an atomism with a beauty that Plato felt made it a far more "likely story" than the chaotic atomic world imagined by predecessors such as Democritus. Early in his career, Kepler argued that the six known planets held the orbits they did because they were located at distances from the sun determined by a nested system of spheres and regular polyhedra. He only abandoned this aesthetically appealing model through its imperfect agreement with Tycho Brahe's data. Similarly, Ampère felt that the beauty of his theory was a major factor in its favor; furthermore, the theory did have an empirical basis in the sense that his primitive forms are those recognized in crystallography and are indeed regular polyhedra.

But how was the theory to be evaluated in more detail? Ampère's answer appears early in the 1814 memoir; characteristically, his methodological comments come in the midst of a theoretical discussion. Rather than list all the assumptions of his theory and then propose a testing procedure, Ampère paused after stating the equal numbers in equal volumes hypothesis and made the following remark.[50]

> Whatever may be the theoretical reasons which seem to me to support it, one can only consider it as a hypothesis; however, in comparing the consequences, which are the necessary implications of it, with the phenomena or the properties that we observe, if it is in agreement with all the known results of experiment, and if one deduces consequences from it that are found to be confirmed by later experiments, it will acquire a degree of probability which will approach what, in physics, is called *certitude.*

Now Ampère of course did not draw consequences from the Avagadro hypothesis alone; presumably, his comment about predictive success applied to the full set of his theoretical assumptions. At any rate, it is clear that in 1814 he retained his earlier emphasis on successful predictions as a particularly valuable form of evidence. In point of fact, Ampère devoted a far larger portion of his "abstract" to the geometric description of acceptable particle combinations than he did to the comparison of theoretical implications with empirical data. For chemists who struggled with actual measurements, Ampère's elegant geometry was easy to ignore until he demonstrated that it satisfied the predictive criterion he himself had stated. At the very least, to be compelling the theory had to resolve some of the disputes over how existing data was to be interpreted.

Ampère's failure to attract interest in his theory was primarily due to the tenuous links that connected it to empirical research. After all, chemistry in France was passing through an empirical phase in which speculative theorizing was not well received. For example, in 1814, the same year Ampère published his "abstract," Gay-Lussac published a 155-page *Annales de Chimie* memoir simply on the properties of iodine. Furthermore, Ampère's own attempts to apply his theory were disappointing. He presented an elaborate table showing his 23 permitted polyhedra and the number of tetrahedra and octahedra from which they are constructed through superposition. Ampère's claim was that this table is the key to possible combinations of elements and the proportions in which they combine. But in practice the table gave little guidance. For example, there are 21 imaginable combinations of tetrahedra and octahedra up to and including the case of six tetrahedra combined to form a hexa-tetrahedron. Of these 21 possibilities,

Ampère's system recognizes 14 as feasible chemical compounds. The system was thus not a very efficient mechanism for predicting actual compounds for the simple elements represented by tetrahedra such as oxygen, hydrogen, and nitrogen. Furthermore, the theory provided little guidance even in relatively simple cases such as water. Ampère knew that the combining ratio for water was two volumes of hydrogen with one volume of oxygen to produce one volume of water vapor. He thus claimed that the molecular process involved was the combination of one tetrahedral hydrogen particle with one half of a tetrahedral oxygen molecule. If we let the symbols H and O represent Ampère's hydrogen and oxygen molecules, or hydrogen and oxygen atoms in subsequent terminology, the formula for a particle of water would presumably be written H_4O_2. But many other possible combinations of hydrogen and oxygen are also permitted by Ampère's theory; even without introducing half particle cases, the theory allows 14 different formulas just for combinations resulting in two, four, five, or six tetrahedra. Although Ampère could pick a formula compatible with the known combining ratio, his theory could never have made a prediction of that ratio prior to actual measurement. More generally of course, the theory ruled out the possibility that any particle could have less than four molecules.

Nevertheless, as was his wont when he felt inspired, Ampère quickly passed over the simple examples and plunged headlong into cases that were at the center of major controversies. The six examples he cited as conclusive proof of his theory all depended upon empirical research that was still in a state of flux and was not uniformly interpreted by the chemists Ampère hoped to persuade. For example, he claimed that his theory made predictions that agreed with data on the hydrates of potassium and the ratios of oxygen found in various forms of potasse. This subject had been a source of controversy between Davy, Gay-Lussac, and Thenard ever since Davy had used an electric discharge to produce potassium in 1807. Gay-Lussac and Thenard devoted 50 pages to the subject in their 1811 *Recherches physico-chimiques*. For Ampère to select one piece of data from this morass and claim that it confirmed a wildly speculative theory must have struck his readers as presumptuous to say the least.

For Ampère it was natural to think that "molecules" are held within their respective "particles" by electrical forces; he described them in a brief letter to Pierre Prévost late in 1814.[51] He thus joined Davy and Berzelius, among many others, in trying to reduce chemical phenomena to electrostatics. This project would be absorbed into a broader context following his electrodynamics research beginning in 1820.

Restricting our attention for the moment to chemistry, an interesting assessment of Ampère as a chemist appeared in the lectures of Jean-Baptiste Dumas during 1836. Significantly, Dumas did not even mention Ampère's geometric atomism; instead, he concentrated on the electric theory of chemical forces. He summarized Ampère's theory as follows, using the term "molecules" to refer to what Ampère had called "particles," a change in terminology that Ampère himself also adopted in publications during the 1830s.[52]

> For him, molecules of substances would have a constant electricity from which they could not be separated, and a shell of opposite electricity would form around each of them, neutralized at a distance by that of the molecule. Each molecule of hydrogen, for example, would include a certain quantity of positive electricity that would be specific to it, and it would be surrounded by an electrical atmosphere of the negative type; molecules of oxygen, on the contrary, would be found to be negative on the interior and positive on the exterior.
>
> With the aid of this fundamental hypothesis, Mr. Ampère finds himself in a position to explain many facts. Bring sufficiently close together two particles constituted in this way and oppositely electrified and their atmospheres will combine; from this, heat and light. Then the molecules themselves, in virtue of their opposite electrical states, will combine and remain closely united; from this, permanent combination.

Ampère's attempt to unify optics, heat theory, and chemical affinity was based on the wave theory of light and his conception of the luminiferous ether as a superposition of positive and negative electric fluid particles. But Dumas was not favorably impressed by these speculations. He used sulphur as an example of the anomaly that some elements can combine with either a negative or a positive element; he then rendered his general verdict on Ampère the chemist, a judgment that is probably a fair indication of the general opinion held by his colleagues.[53]

> It must be concluded that the hypothesis of Mr. Ampère, as ingenious as it may be, is absolutely inadmissible. This is the fate, and this circumstance is to be noted, of systems of affinity and systems of molecular grouping presented by physicists. Even when, as is the case with Mr. Ampère, they possess exact notions of the phenomena and the laws of chemistry, the lack of experience in the practice of this science can always be detected among them . . . if one wishes to provide a theory of chemical action, let the physicists come to us, let them march in step with chemists, and let them be quite convinced that the least details of our science are to be considered.

Dumas' reference to Ampère as a physicist was written after the electrodynamics research that established Ampère's reputation. In 1814 the situation was quite different; at that point Ampère was even more suspect as a mathematician and philosopher who dabbled in chemistry without dirtying his hands in the laboratory. Of the few chemists who took Ampère's geometric model seriously, Marc Antoine Gaudin was the most energetic. He attended Ampère's lectures at the Collège de France in 1827 and thereafter attempted to use Ampère's ideas to determine atomic weights and the arrangement of atoms within molecules.[54] Unfortunately, he became excessively committed to ideas that struck other chemists as unjustified; he became increasingly isolated and had little impact on either chemistry or crystallography.

In retrospect, Ampère's distinction between "particles" and "molecules" represents a rare intuition into the important concept of polyatomic molecules. But Ampère did not have the time or energy to seriously confront his elegant theory with empirical data. The valuable concepts of polyatomic molecules and the equal numbers in equal volumes hypothesis thus were not significantly furthered by the efforts of Ampère himself. Chemists were too skeptical of Ampère's system as a whole to isolate these ideas and subject them to independent consideration. His geometric formalism and his speculations about inter-molecular forces were too far out of step with the research problems of his contemporaries to have a significant impact. Concrete chemical analysis and the discovery of empirical laws constituted the accepted agenda.

The indifference that greeted Ampère's atomism memoir when it was published in June of 1814 left him depressed and temporarily disenchanted with chemistry. He devoted the summer to mathematics and wrote to Ballanche in September.[55]

> Happy are those who cultivate a science in the period in which it is not complete, but when its last revolution is mature! That is entirely accomplished by Gay-Lussac who is completing the rough sketch created by the genius of Mr. Davy, but which I would infallibly have done, which I actually did the first, but which, unfortunately, I did not publish when there was time. But what does it matter? The time of suffering is short on earth; that is what should console me for everything; but it is no less painful.

Convinced that his teaching responsibilities and his lack of laboratory facilities had deprived him of the recognition he deserved, Ampère could only vent his spleen in letters to Lyon. The blow to his ambition and sense of justice did not heal quickly. It left a scar in the form of an increased cynicism about the contingent nature of scientific fame and the self-serving maneuvering that was commonplace in a competitive environment where ideas were thought of as commodities to be bargained for. He may well have plunged into a far deeper depression had he not won election to the mathematics section of the Institut a few months later in November of 1814. During 1815 he roused himself for one more attempt to contribute to chemistry, an effort that resulted in a lengthy publication in the new *Annales de Chimie et de Physique*, edited by Gay-Lussac and Arago.

Ampère's 1816 Classification of the Elements

Lack of response to his atomism memoir and his exhausting campaign for admission to the Institut left Ampère in an ambivalent state of mind. Uncertain about the course he should take intellectually and despondent about the past, he still harbored ambition. He wrote to Bredin in February 1815.[56]

> I dream alternately about chemistry and psychology while always planning to occupy myself exclusively with mathematics. You know that my mind, full of disgust with all that I see investigated with such ardor by other men, can only attach itself to what I imagine as necessarily bearing light to other centuries . . . in the sciences I can take a vivid interest only in works which are presented as being able to change the face of it in some respects.

During 1815 Ampère reverted to his old interest in classification schemes. Between May and July of the following year he made the final revisions to a long memoir on the subject published in four installments in the *Annales de Chimie et de Physique*. Although this subject was less controversial than his earlier espousal of atomism, Ampère did not expect an enthusiastic reception. Writing to Bredin, he remarked wearily that "there will result from all this only an imperfect work the publication of which will not do me very much honor."[57] This assessment was fairly accurate; few chemists took any interest in a memoir that did not introduce new elements or compounds and primarily argued for a "natural" classification scheme for known elements. Nevertheless, the subject was a fitting conclusion to Ampère's chemistry publications; by exploring it he temporarily satisfied his encyclopedic urge to classify, the tendency he had developed in his youth while reading the *Encyclopédie* and Buffon.

From a broad perspective, the two most significant modern efforts to classify the elements are probably Lavoisier's list in his 1789 *Traité* and Dmitri Mendeleyev's first publication of an early version of the modern periodic table in 1869. Ampère's effort thus falls within an intermediate stage in which empirical information increased rapidly. The determination of accurate atomic weights was still a source of great controversy; it would have to be resolved, and a sufficient number of elements would have to be discovered, before chemists could recognize the cyclic recurrence of chemical properties essential to Mendeleev's table. Due to widespread interest in the isolation of new substances, Ampère had a list of 48 elements to ponder in 1815. Lavoisier's list of 33 had been augmented and light and caloric were no longer recognized as chemical elements. As had been the case for Lavoisier himself, the injunction to define an element as a substance that resists all attempts to resolve it into simpler parts was invoked selectively by the subsequent generation of chemists. For example, fluorine and several metals such as aluminum, uranium, zirconium, yttrium, and beryllium were accepted as elements although they had not yet been produced in pure form. Terminology was not entirely uniform. Ampère used the older terms glucinium and colombium for beryllium and niobium respectively, and he did not list tantalum. Also, although niobium and tantalum had initially been recognized as distinct elements, following Wollaston's 1809 analysis, they were incorrectly thought to be identical and were not recognized as distinct until 1846. Nor did Ampère list vanadium, initially cited as an element by Andrés Manuel del Rio in 1801 before he retracted and incorrectly identified it with chromium.[58]

As usual, Ampère was undeterred by these empirical uncertainties; he was convinced that he could stipulate a "natural" classification scheme that expresses the real order of materials. This attitude had its origins in his early days of herborizing at Poleymieux. An avid student of Buffon in his youth, Ampère's reading of Rousseau's *Lettres sur la Botanique* had been a major factor in his recovery from his father's death. For the rest of his life he shared with these two botanists an avid opposition to "artificial" systems such as Linnaeus's which relied upon a single character as the basis for categories. He opened the 1816 memoir by defending the same strategy for chemistry:[59]

> it seemed to me that one should make an effort to banish artificial classifications from chemistry and begin to assign to each simple substance the place it must occupy in the natural order by comparing it in succession to all the others and combining it with those which are related to

it by the greatest number of common characters and above all by the importance of these characters.

Ampère expected the result of these natural associations to be a set of *"genres"* or groups of elements that can be arranged in an order such that similar groups are adjacent to each other. When properly ordered, this sequence looped back upon itself to form a closed cycle, as shown in plate 3.

The first step was to follow the example of the life sciences and draw a fundamental distinction corresponding to that between animals and plants; this could be followed by further resolution into families and *genres*. Furthermore, to avoid "artificiality," the crucial characteristics that formally define these distinctions must only be chosen after assessment of the groups generated by a study of the full ensemble of chemical properties.

Dubious about any straightforward distinction between metals and non-metals, Ampère noted that among elements widely recognized as non-metallic, such as silicon, tellurium, and arsenic, there is a common property in that they can all form "with other substances of the same class, permanent gases which can subsist without decomposing when they are mixed with atmospheric air."[60] Ampère called these non-metallic elements *gazolytes* and he arranged them in five *genres* in the sequence: silicon, boron, carbon, hydrogen, nitrogen, oxygen, sulphur, chlorine, fluorine (*phtore*), iodine, tellurium, phosphorus, and arsenic.

Second, he divided the metals into two groups, *leucolytes* and *chroicolytes*. *Leucolytes* have melting points below 25 degrees, as measured by Wedgwood's pyrometer, and produce no colors when dissolved in colorless acids. *Chroicolytes* have higher melting points and do produce colors when dissolved in colorless acids. To distinguish *genres* within these major divisions, Ampère employed more specific "third-order" properties, such as an affinity for oxygen, expressed either in the formation of acids or compounds that either are or are not decomposable by carbon or chlorine. Ampère used these third-order properties to order the elements within his *genres* in such a way that each *genre* leads "naturally" to the next. He noted other "fourth-order" properties such as the formation of oxides, malleability, and the property of decomposing water, and warned that an exclusive reliance on one of these properties would result in an artificial system analogous to Linnaeus' botany. He completed the second installment of the memoir with a flourish.[61]

> The irresistible progress of the sciences toward their true goal, the most complete possible knowledge of the laws of nature, will lead necessarily

Tableau des quinze genres et des quarante-huit espèces des Corps simples pondérables, rangés dans l'ordre naturel.

Carbone.	Bore.
Hydrogène.	Sicilium.
Azote.	Colombium.
Oxigène.	Molybdène.
Soufre.	Chrôme.
	Tungstène.
Chlore.	
Phtore.	Titane.
Iode.	Osmium.
Tellure.	Rhodium.
Phosphore.	Iridium.
Arsenic.	Or.
	Platine.
Antimoine.	Palladium.
Étain.	
Zinc.	
	Cuivre.
	Nickel.
Bismuth.	Fer.
Mercure.	Cobalt.
Argent.	Urane.
Plomb.	
Sodium.	Manganèse.
Potassium.	Cérium.
Barium.	Zirconium.
Strontium.	Aluminium.
Calcium.	Glucynium.
Magnesium.	Yttrium.

3   Ampère's Table of Elements, first published in 1816 in *Annales de Chimie et de Physique*, volume 2, p. 116.

to the adoption of a natural classification in chemistry, as it has led to it in the other branches of the physical sciences.

If we think of the modern periodic table as a "natural" classification scheme based on knowledge of the internal structure of atoms, Ampère's prediction has indeed been fulfilled. Although his own attempt fell far short of his expectations, it included some interesting insights. Ampère's 15 *genres* are the analogue to the 15 groups of elements found in columns of the periodic table when we omit the noble gases, a group unknown in Ampère's time. Although this exact agreement in number is largely coincidence, some of Ampère's *genres* do represent the same sets of elements as modern groups. His fourth *genre* is made up of chlorine, fluorine, and iodine, the three halogens known in 1816. His eighth *genre* is the two alkali metals then known, sodium and potassium; *genre* nine is the four known alkali earth metals: barium, strontium, calcium, and magnesium. On the other hand, many of the other *genres* were chosen less perspicaciously and did not survive in later classification schemes. To take just one example of many that could be mentioned, copper, silver, and gold all fall within a single column of the periodic table; in Ampère's system they fall in three different *genres* and silver is a *leucolyte* while copper and gold are *chroicolytes*.

In general, Ampère's choice of distinguishing characteristics and his terminological innovations failed to elicit a favorable response from chemists. The fact that he made no references to atomic masses may have contributed to the lack of enthusiasm, since this was a major research interest of his contemporaries. Ironically, even Ampère's term *phtore* was rejected in favor of fluorine, the term Davy said was suggested to him by Ampère, who devoted several pages of his memoir to a discussion of his correspondence with Davy. The feeling that chemists had not adequately recognized Ampère's contribution to the discovery of fluorine obviously still weighed heavily on his mind. Referring to the compound hydrogen fluoride, he remarked that "I am the one who first made known the composition of this hydracid and established the existence of the material that is at question here; consequently, I believe that I should give it a name appropriate to designate it." After noting that "the work that I did on this subject not having been published at the time, I can prove the existence of it only by citing two letters from the famous English chemist," Ampère included the full texts of Davy's short letters from 1811 and 1813.[62] While Ampère's complaints may have been partially justified, few chemists took them seriously; the highly specialized nature of French science made it easy to discount his priority claim as just so much grumbling by an outsider who had done little actual empirical research.

Ampère included discussion of the properties of each of his 48 elements, particularly his *gazolytes*. These discussions actually provoked more interest than his central topic of classification. For example, Pierre Louis Dulong was one of the few established chemists who responded to Ampère's memoir. Ten years younger than Ampère, Dulong was a respected member of the Société d'Arcueil and had prepared the nitrogen trichloride that Ampère described for Davy, which almost cost Davy an eye when his sample exploded. Dulong had been injured twice by similar explosions and gladly left further research on this compound to Davy. During 1816 Dulong was studying phosphorus compounds and had discovered hypophosphorus acid. He noticed an interesting idea in Ampère's long discussion of sulphur; Ampère argued that some sulphur compounds with metals contained twice as much sulphur as there was oxygen in the corresponding oxides. Dulong wrote to him in August of 1816.[63]

> Your hypothesis as to the composition of the thiosulfates is applicable to the hypophosphites, but in order for it to acquire any credence it will be necessary at least to acquire by itself the phosphoretted hydrogen gas which is combined with phosphorous acid in hypophosphorous acid.

Dulong concluded that further experimentation was called for, a conclusion that usually applied to Ampère's proposals. This was a major source of Ampère's frustration in 1816; his professional status as a teacher and *Académicien* in mathematics made it impossible for him to pursue his ideas experimentally. Although Dulong at least gave due consideration to Ampère's ideas, this was an exception to the more general lack of interest and provided little solace.

Ampère was not entirely alone in looking for relations among the elements that would provide a "natural" classification scheme. The German chemist J. W. Döbereiner worked on a theory of "triads" between 1816 and 1829. He noted that sets of elements with similar properties sometimes included triads in which the central element had an atomic weight that is the mean of those of the other two elements. Additional numerical regularities in atomic weights were noticed by others, but with little import, owing to the frequent revisions in atomic weight data. In France Dumas proposed a triad theory during the 1850s and argued that transmutations might be possible within triads. Ampère thus did not contribute significantly to the tradition that led to the modern periodic table. By trying to produce a classification

211

scheme without relying upon atomic weight he omitted what would later be recognized as a crucial ordering factor.

Ampère's detailed interest in chemistry abated after 1816. The last significant development in his career prior to 1820 was his adoption of the wave theory of optics espoused by his friend Fresnel. This took place within the context of Ampère's general interest in mathematical physics.

Mechanics and Optics:
The Wave Theory of Light

The relatively minor role of empirical observations in mechanics made it the most mathematical branch of physics during the eighteenth century. The statement of a small number of more or less self-evident principles was a typical prelude for the exploration of mathematical consequences. For example, although Lagrange relied heavily on the virtual velocities principle in his 1788 *Méchanique analitique,* he did not provide a proof at that time; he was content to state that it was as self-evident as the principle of the lever and the composition of velocities. Fourier, Laplace, Poinsot, and Prony were among those who sought to provide a "demonstration" that the principle follows from other assumptions that are more closely linked to observation. In 1798 Lagrange himself gave a proof that the principle follows from the principle of pulleys. Discussion of the subject was lively among the talented faculty at the École Polytechnique and Ampère decided to make a contribution.

The 1806 paper on the principle of virtual velocities was typical of his publications in physics prior to 1820 in that it was primarily pedagogical in nature. His proof was also idiosyncratic in several respects, for example, his statement of the principle in question was far more inclusive than might be expected. Most authors stated the principle as the requirement that, in a system with no acceleration, the sum of the moments of the active forces vanishes. Ampère's statement of the proposition included the technique of Lagrange multipliers when they are applied to equations of force moments and conditions of constraint to derive a set of equilibrium equations. What he actually proved was the more restricted claim that the sum of moments of force vanishes. He also defined his "moments" in an unusual manner, claiming that he thereby avoided the distressing use of infinitesimals that was so confusing for students. Third, his proof was based upon the transmission of forces by incompressible rods, a technique he and

212

Félix Savary would rely upon years later to analyze electrodynamic forces.

Ampère considered a special case in which the localized masses that make up a body are constrained so that they can only move along a one-dimensional curve. He then constructed a triangle determined by a point M with coordinates (x,y,z) at which a force is applied, a closely neighboring point N along the tangent to the curve at the initial point, and a third point P taken along the direction of the force. The point N is given coordinates differing from those of M by $x'i$, $y'i$ and $z'i$, where the primes indicate differentiation with respect to a new independent variable, and i is the infinitesimal quantity Ampère sets out to avoid. He does so by defining the moment of the force as the projection of the tangential side of the triangle, MN, onto the line of direction of the force and then dropping the resulting factor i since only the relative value of moments will ever be significant. He notes that the resulting quantity is the derivative of the function representing the length of the MP side of his triangle, the side that falls along the direction of the force. Ampère claims that by not relying on the notion of a derivative, his definition of force moment is more accessible to novices. But his claim that he avoided the introduction of infinitesimals in the definition is considerably overblown. The triangle he used to carry out his own definition was defined in terms of derivatives and the infinitesimal i. Furthermore, his definition is artificial and formal in the sense that it encourages little intuitive understanding of physical significance. The usual definition of moment in terms of a small displacement along the direction of the force was at least clearly related to the force itself.

At any rate, with his definition in hand, Ampère applied it in an analysis of force transmission. He imagined a rigid rod acting as a link between a mass driven along a curve by a force and a second mass that can move along a second curve. If the second mass is assumed to be driven by a second force with the same effect as the one produced by the rigid rod connected to the first mass, an algebraic analysis showed that the moments of the two forces must be equal. Second, if a collection of masses are imagined being driven along a straight line by forces linked to an initial set of forces, equilibrium of the second set of masses requires the sum of the forces to vanish, and Ampère showed that this is equivalent to the requirement that the sum of the moments of the first set of forces must also vanish.

The proof was primarily intended to serve as an algebraic exercise for students rather than as a significant revision of mechanics. Ampère did take his teaching responsibilities seriously, even if he regretted losing so much time from his own research. From a pedagogical point of view he was not satisfied with a purely formal presentation of mechanics as a mathematical analysis of forces. In his notes for mechanics lectures he organized the subject according to his insistence that motion is an epistemologically more fundamental concept than force.[64] Ampère's classroom presentations of mechanics generally began by showing that acceleration is a directly measurable quantity that is proportional to the operative force, and that the principle of the combination of forces results from the single supposition that "if a force produces the same effect as two or several others in any particular motion of a mobile point, it also produces the same effect on any other motion."[65] Once this result had been reached, Ampère could make the customary application of Lagrange's methods to derive the equations of equilibrium and motion.

As a chemist and a physicist, Ampère was deeply concerned about the material basis of the forces that appear as abstract quantities in the context of mechanics. While alternative foundations for mechanics could be judged primarily on the basis of logical consistency and in consideration of how infinitesimal quantities were to be introduced or avoided, other branches of physics required other commitments. We have seen that, as a chemist, Ampère was willing to attribute chemical affinities to electric forces originating from positively or negatively charged particles surrounded by oppositely charged electric atmospheres. But this step opened a new set of questions about the origins of electric force. Given his long-standing aversion to the thought of forces acting at a distance through empty space, it is hardly surprising that Ampère became interested in ether theories.

Optics was the domain in which ether models first became acceptable in French physics. The transition from the Laplacian particle theory of light to Fresnel's wave theory was contested and prolonged; it was accomplished only in the teeth of aggressive opposition, particularly by Biot. It was preceded by a variety of speculative theories often inspired by romantic longings for insight into nature's "dynamism" or the "polarity" of the German Nature Philosophers. For example, consider a representative passage from Lorenz Oken (1779–1851).[66]

A line, one extremity whereof strives toward the centre, the other to the periphery, the one to identity, the other to duality, will exhibit itself in the world as a *line of Light*, in the planet as a *Magnetic line*. Magnetism is centroperipheric antagonism, a radial line, $0 - \pm$, the action of the line being cleft at one extremity. Magnetism has its root in the beginning of creation. It is prophesied with time.

214

This was *not* the mentality that produced the physics of Fresnel and Ampère. While they were willing to speculate about the micro-world of unseen causes, they also insisted that hypotheses be limited to clearly stated models in direct analogies to observable processes.

Although hardly acceptable to Laplacians, Hans Christian Oersted (1777–1851) represented an alternative that avoided the extremes of *Naturphilosophie*. The historian Kenneth Caneva has argued that Oersted may well have been an incentive for Ampère's receptivity to ether theories and his subsequent enthusiastic response to electromagnetism as well.[67] Oersted earned a degree in pharmaceutical science at the University of Copenhagen in 1797 and he wrote a doctoral dissertation on Kant in 1799. He maintained an interest in philosophy throughout a scientific career in which he did research on electro-chemistry, thermo-electricity, Chladni's sound figures, fluid and gas compressibility, diamagnetism and electromagnetism. From his study of Kant he brought to scientific research a skepticism about all forms of atomic theory and a conviction of the essential unity of the apparently disparate variety of natural forces. He also became fascinated by the poetic language of *Naturphilosophie* during a period of travel in Europe between 1801 and 1804. Chastened by an encounter with the stricter standards of French chemistry in Paris, Oersted retained his Kantian presuppositions but now he also realized that as a scientist he could not allow them to act as a justification for ignoring empirical research.

Oersted made his ideas known in France through an 1806 publication in the *Journal de Physique* and in much more detail in his 1813 *Recherches sur l'Identité de forces chimiques et électriques*, a revised and translated version of the German treatise he had published the previous year.[68] Here Oersted argued that not only chemical affinities, but thermal, optical, electric, and magnetic phenomena as well, are all due to the action of two fundamental forces (*forces primitives*) of positive and negative electricity existing within all matter. It is hardly surprising that Oersted's work was ignored by the Laplacian school; his theory conflicted with virtually all their methodological norms and conceptual presuppositions. His "forces" were not expressed mathematically; he rejected both the caloric theory of heat and corpuscular optics; by trying to reduce all phenomena to the action of electric "forces," he conflated not only the distinct Laplacian disciplines of optics, heat, electricity, and magnetism, but the science of chemistry as well. Furthermore, Oersted did not accept the Laplacian interpretation of fundamental electric forces as the unexplained causes of attractions and repulsions between particles of imponderable electric fluids. Oersted's view was that[69] when we speak of primitive or fundamental forces, we only intend to designate the most simple activity our experiences can give us an idea of. Thus, not wanting to become engaged in metaphysical discussions, we will not make any decisions as to whether these forces are distributed within various molecules of bodies . . . or whether these forces are spread in space without being fixed at such points.

Oersted's concept of electric "force" was so vague that it could be imagined to apply either to the Laplacian fundamental forces mathematically investigated by Poisson or to the particles of electric fluid that Laplacians treated as the material source of these forces. In fact, Oersted probably had nothing so specific in mind. Uninhibited by Laplacian scruples that demanded that scientific concepts be expressed in mathematical language, Oersted was content to describe an electric "current" in more colorful language.[70]

Electricity thus does not flow through conductors like a liquid through a pipe; it is propagated by a kind of continual decomposition and recomposition, that is, an action which disturbs equilibrium at each instant and re-establishes it during the following instant. One might describe this series of opposed forces which exist in the transmission of electricity by saying that *electricity is always propagated in an undulatory manner*.

Oersted did not attempt to link this "undulatory" motion to an ethereal substratum. He simply described the propagation as the passage of a temporary and localized disruption of electrostatic equilibrium, so in one respect it was misleading for Ampère to cite Oersted as follows in 1822.[71]

Oersted considers the compositions and decompositions of electricity, which I have designated under the name of electric currents, as the sole cause of heat and light, that is, of the vibrations of a fluid spread throughout space and which, under the generally adopted hypothesis of two electric fluids, can hardly be considered otherwise than as the combination of these two fluids in the proportion where they mutually saturate each other.

Everything in this passage following the phrase "that is" pertains only to the ether model Ampère himself created after becoming familiar with Fresnel's optics. On the other hand, Ampère may have felt indebted to Oersted in a more general sense, thanks to the inspiration he derived from the idea of "undulatory" propagation as a single

216

mode of transmission of effects through both solid matter and empty space.

Be this as it may, Ampère adhered to the corpuscular or emissionist theory of light throughout his first decade in Paris. Several important consequences followed from what Fresnel called Ampère's "conversion" to wave optics in 1816. First, it encouraged him to think of an etherial medium as a material basis for optical, thermal, electrical, and magnetic phenomena that was compatible with his metaphysical conception of absolute space. Second, it allowed him for the first time to both observe and participate in a major revolution in physics. He recognized that a significant component of the debate involved alternative interpretations of phenomenological laws on the basis of conflicting material models and fundamental forces. He also noted the severe rivalry and even enmity that accompanied the transition from one theory to the other, particularly between Biot and Arago. Thus, when he initiated his own break with Laplacian orthodoxy four years later, Ampère was fully aware that his rivals would resist him both by trying to incorporate his empirical discoveries into their program and by exercising their powerful status within the Académie and other institutions.

Between 1805 and 1815 Laplace and Biot promoted a vigorous research program in optics, based upon the assumption that light beams are made up of particles of an imponderable fluid that are acted upon by short-range forces as they interact with matter. In practice, the mathematical development of the program proceeded by resolving a beam into rays which could be thought of as being individually asymmetric with respect to rotation about the direction of their motion.[72] This property provided an intuitively appealing explanation for phenomena such as polarization, an effect that received extensive attention beginning in 1809.

It was work on the problem of double refraction by Laplace and Malus that drew a response from Ampère in 1815. It had been known since the seventeenth century that when a light beam enters a crystal such as Iceland spar it splits into two beams at the surface. One of these beams follows the path predicted by Snel's law of refraction, that is, the path such that $\sin \theta / \sin \theta' = n$, where θ and θ' are the angles of incidence and refraction respectively and n is the ratio of the respective velocities of light in the two media involved. In contrast to this "ordinary" beam, the other "extraordinary" beam follows another direction not accounted for by Snel's law. In 1690 Christiaan Huygens developed a geometrical construction to determine the direction of extraordinary rays.[73] From Huygens' point of view the method

217

succeeded because light is propagated by a series of ether particle pulses that form a wave front and secondary pulses are emitted at each point of a medium encountered by the front. Due to the close linkage between Huygens' procedure and his pulse theory, eighteenth-century particle theorists either took little interest in it or were skeptical. Malus began working on the problem from a Laplacian perspective in 1806; he was encouraged to continue when, in January of 1808, the mathematics section of the First Class of the Institut chose the topic for a prize competition to be awarded in 1810. The Laplacian group within the Institut clearly expected Malus to provide an explanation in corpuscularian terms. It was in the initial stage of his research on this problem that Malus discovered that light rays reflected from glass at 54.5 degrees have the same properties as ordinary rays produced by double refraction. He coined the term "polarization" for these phenomena, thinking that perhaps light particles are analogous to little bar magnets with two poles. Laplace and Haüy reported his discovery to the Institut in December of 1808 and also revealed that Malus had translated Huygens' construction into algebraic relations for the direction of extraordinary rays. From the point of view of Malus and Laplace, a successful contestant for the Institut prize would provide a derivation of these phenomenological relations based upon fundamental laws of mechanics. Malus accomplished a derivation of this type within a few weeks. Laplace rather churlishly rushed ahead with the publication of his own version of a proof without crediting Malus with the work that had made it possible. In both cases the "proofs" of Huygens' phenomenological law involved a stipulation of velocity variation in addition to the principles of mechanics. Although this velocity function could be thought of as part of a model to which the principles were applied, the velocity was stipulated using an ellipsoidal construction that had no physical meaning for particle theorists. Nevertheless, Laplace proclaimed the derivation to be a great victory for the program of short-range forces and even claimed that it made Huygens' method fully trustworthy for the first time.

Although ridiculed in England by Thomas Young, Laplace's decree was not questioned in France. Fresnel shared some alternative ideas with Ampère during the fall of 1814 but they made little impact at that time. Aside from the fact that Ampère was preoccupied with chemistry, metaphysics, and his candidacy for the Institut, he was still thoroughly comfortable with Laplacian optics. In March of 1815 he presented an optics memoir to the Institut with an opening line that ensured a receptive Laplacian audience.[74]

"Count Laplace has reduced all the phenomena of ordinary and extraordinary refraction to a single principle, that of least action." Citing as inspiration Biot's discoveries of more complicated polarization phenomena, Ampère devised a generalization of Huygens' construction. As had been the case with Laplace and Malus, the proof is an analytic exercise in which no physical interpretation is given to the geometry; the directions of refracted rays are simply attributed to a "force which emanates from the axis of polarization."[75] By allowing the velocity of the light in each medium to vary with direction, Ampère constructed two ellipsoidal surfaces; in this respect his technique is more general than Huygens' construction, in which the light enters a refracting material from an isotropic medium and thus generates a spherical velocity surface for that medium. Ampère's method relies upon an application of the principle of least action to the path of an arbitrary light ray; that is, the ray follows a path such that the integral of its velocity with respect to distance is a minimum. Following the proof of a preliminary theorem, for any incident ray Ampère could construct the refracted ray by reversing the sequence of his proof; that is, his two surfaces are constructed and the incident ray is extended until it meets the relevant surface. A tangent plane is drawn at that point and from the intersection of this plane with the boundary plane between the two media a second tangent plane can be drawn; the point of intersection with the surface of refracted rays determines the direction of the refracted ray in question.

While Ampère's memoir was of some interest from a mathematical point of view, it contributed little to the subsequent history of optics. He did not follow Malus' example by putting his results into an algebraic form that would have encouraged experimental testing. Nor did he give his geometric construction any physical significance. Furthermore, within a few months Augustin Jean Fresnel presented his first wave-optics memoir to the Académie; during the following year Fresnel and Arago recruited Ampère in support of a wave theory of light.

Born in Broglie in 1788, Fresnel spent most of his childhood in the village of Mathieu near Caen where he attended the École Centrale prior to acceptance at the École Polytechnique in 1804, the year Ampère joined the faculty there. Two years later he began training as a civil engineer at the École des Ponts et Chaussées and was then assigned to a variety of engineering projects in the provinces. His maternal uncle, Léonor Mérimée, taught design at the École and he introduced Fresnel to Ampère and Arago. In 1811 Arago had discovered additional polarization phenomena using thin layers of mica. Biot was performing similar research and the subsequent collision of these two ambitious men quickly degenerated into quite a nasty priority dispute.

Biot and Arago became permanently estranged, setting the stage for Arago's championing of Fresnel's wave optics a few years later.

Brought up in a strict Jansenist household, Fresnel became a rare friend with whom Ampère could share both scientific and spiritual interests. Fresnel only started taking an interest in optics during 1814. Although at that point he had little knowledge of the subject, he already suspected that an ether theory might reveal a link between optical and electrical phenomena. He recorded his first thoughts on the subject in a paper he called his *Rêveries*. Unfortunately it has not survived; at Mérimée's suggestion it was conveyed to Ampère during the fall of 1814. At that point Ampère was thoroughly occupied with chemistry, metaphysics, and mathematics and he repeatedly postponed writing any response. Knowledge of Fresnel's ideas did not prevent Ampère from making his Laplacian presentation to the Institut in March. With Arago as his primary confidant, Fresnel began an experimental study of diffraction; he gradually developed an increasingly accurate application of the concept of interference. Encouraged by Arago, he presented the Institut with his first memoir on the subject in October of 1815. Arago and Poinsot made a very favorable report the following March. This announcement generated immediate controversy and an energetic discussion developed concerning the extent to which the Institut should appear to lend support to Fresnel's theoretical interpretation of his experimentation. A version of the conversation was published in the *Bibliothèque universelle*; it provides an interesting indication of how the new optical theory had become a threat to the Laplacian domination of French physics.

Laplace and Biot were particularly hostile to Arago's report. Laplace was quoted as saying that in view of the past success of "Newton's theory," he regretted that now there were attempts to[76]

> substitute for it another purely hypothetical one, and which, so to speak, can be arranged at will, that of Huygens' undulations. He [Laplace] believes that one must limit oneself to repeating and varying experiments and in concluding laws from them, that is, coordinated facts, and avoid any undemonstrated hypothesis.

Laplace of course had no qualms about ignoring this advice in the interests of his own "Newtonian" theory. Ampère was more conciliatory; his comments were recorded in the following slightly garbled report.[77]

> It was by a single application of his system of attraction to the moon that Newton verified the three laws of Kepler. Subsequently, it was seen

that this cause could explain the parabolic motion of the comets, the ebb and flow of the tides, etc., and it has become more and more probable. Mr. Fresnel's construction, using intersecting circles, is also deduced from a single phenomenon alone. However, if he succeeds in extending it to others, he would reduce them all to a single law. Thus, although I have always accepted the system of emission, the conclusions of the Report seem good to me.

Two months later Ampère had, as Fresnel put it, fully "converted" to the wave theory. Fresnel's first memoir on diffraction was the deciding factor. Ampère did not interpret Fresnel's clever manipulation of wave equations simply as a convenient calculating device. For Ampère the theory was "true" in the sense that it correctly attributed optical phenomena to undulations of the ether.

In July of 1816 Fresnel wrote to his brother Léonor that "you see that the vibrations party is strengthened every day (for I believe I have informed you of the conversion of Ampère)."[78] Coming from a man of Fresnel's strong religious convictions, "conversion" aptly described the situation. In 1816 there was no compelling reason to side with Fresnel and Arago against Laplace and Biot. A vibratory study of diffraction was hardly a full-blown optical theory, particularly given the problems posed by polarization effects, the strong point of the entrenched corpuscular theory. Ampère's willingness to take this leap of faith was as much due to his prior aversion to Laplacian fundamental forces acting through empty space as it was to the independent merits of Fresnel's work up to that point.

Biot and Laplace were in no mood to concede. In March of 1817 they arranged for an Académie prize competition on the subject of diffraction, a prize ultimately won by Fresnel, much to their dismay. Léonor Mérimée relayed to Fresnel Ampère's assessment of the situation.[79]

Yesterday I saw Ampère, who asked me for news of you and strongly enlisted me to write to you to put yourself in the ranks and to send your memoir to the contest, with the new observations that you have made and that you may yet make. "He will assuredly win the prize," he said to me; "for himself and for the cause he must compete."

I made some objections, based on the partiality of the commissioners if they were chosen from the sect of the *Biotistes*. – Ampère replied that there was nothing to fear, that when the commissioners were nominated General Arago would not fail to make known the impropriety of nominating party men, and that what will happen is what always happens when the Republic is warned that citizen Laplace wishes to dominate.

221

Arago rose to the occasion; he and Gay-Lussac joined Laplace, Biot, and Poisson on the commission that judged the two entries for the diffraction prize; Fresnel was declared the winner. The *Biotistes* were upstaged not only by Fresnel's strong showing but also because no one from their own ranks had come forward. The lesson to be learned by the Arago-Fresnel alliance was not lost on Ampère. His disappointing experiences with chemistry had taught him to strike swiftly to ensure proper credit. The optics debate provided experience with Académie politics and reiterated the importance of cooperative editors.

Between 1816 and 1820 Ampère also contributed to Fresnel's ideas about the ether and the nature of optical vibrations.[80] Initially Fresnel thought of light waves as longitudinal vibrations analogous to sound waves. Longitudinal waves involve vibrations of the medium in the direction of the wave motion, as in the case of a compression or shock wave. But it is difficult to imagine how polarization effects could be produced by this type of wave, since it is symmetric with respect to the direction of wave motion. Fresnel credited Ampère with the suggestion that transverse vibrations might be involved. These vibrations are perpendicular to the direction of wave propagation, as in the case of a water wave. Transverse waves can be due to vibrations restricted to any plane perpendicular to the direction of propagation and thus have an asymmetry suggestive of polarization. Ampère and Fresnel initially thought that light must be a combination of longitudinal and transverse vibrations. In a memoir on polarization published early in 1820, Fresnel recalled that in 1816 he and Ampère[81]

> had both felt that the phenomena could be explained with the greatest simplicity if the oscillatory movements of polarized waves took place only in the very plane of those waves. But what would become of the longitudinal oscillations along the rays?

Fresnel's hesitation stemmed from his conception of the ether as a diffuse fluid analogous to the air that transmits longitudinal sound waves. However, during 1820 Fresnel revised his ether model to include repulsive inter-particle forces; he now argued that all light waves are transverse vibrations in an ether in which the particles are arranged in uniform layers which "by sliding between two others should communicate motion to them."[82] Due to the impact these ideas would have on Ampère, Fresnel's description of vibrations propagating in an ethereal lattice is worth quoting in more detail.[83]

> Imagine in a fluid three consecutive long parallel rows of material points arranged in this way. If a certain law of repulsion is assumed between

222

these molecules, then in the state of equilibrium and absolute rest they will take a regular arrangement according to which they will be equally separated on the three rows, and those of the intermediary file will correspond, I imagine, to the center of the intervals falling between the molecules of the other two ... if one displaces the intermediary file slightly by making it slide along itself ... and then leaves it free, each of its material points will return toward its original situation ... and will oscillate from one side to the other like a pendulum that has been displaced from the vertical.

Needless to say, Ampère and Fresnel did not achieve a full-blown theory of ether mechanics in 1820. Nevertheless, they continued to exchange thoughts on the subject throughout the years prior to Fresnel's terribly early death in 1827 at the age of 39. During the last five of these years Fresnel rented a room in Ampère's house near the École Polytechnique and became one of his few close scientific friends. Although Ampère emphasized phenomenological laws in his electrodynamic publications, there is ample evidence that he maintained a hope that Fresnel's luminiferous ether might also be shown to act as the medium for transmission of electrodynamic effects. In this respect, Fresnel's ether replaced the caloric medium Ampère had relied upon in his 1801 manuscripts on electricity and magnetism. Metaphysically, the ether also played the role of a substratum responsible for the existence of absolute space. Ampère later introduced it into his lectures on electric currents at the Collège de France and his final publications on the transmission of light and heat.

As the year 1820 unfolded, Ampère was in a position to take full advantage of Oersted's discovery of electromagnetism. His mathematical labors had won him access to the Académie where he had powerful associates in Arago and the *secrétaire perpétuel* Delambre. He enjoyed relatively good health and emotional and spiritual stability. Oersted's discovery probably could not have come at a better time. As we take up the chronicle of the golden years of Ampère's career as a physicist, we might keep in mind the admonition he recorded in lecture notes for his 1817 class at the École Normale.[84]

Recall that in order to trace the cause of a collection of observed facts, there are two procedures – *direct analysis* and *indirect synthesis*.

Admittedly, the former would be more natural if it could be used: useless efforts have been made by even some very great men to make people believe that they followed it. Candour of Copernicus – the procedure of genius – spontaneity of the explicative conception.

Part III

Electrodynamics

7

Ampère's Response to the Discovery of Electromagnetism (1820)

Ampère's renown as a scientist is primarily due to his investigation of electrodynamics during the 1820s. In the present format it is impossible to provide a complete treatment of his experimental and theoretical accomplishments during that decade. To provide a view of both the forest and some of the trees, I have chosen themes that were particularly important to Ampère and they will be illustrated with detailed descriptions of some significant episodes. The focus of chapter 7 is Ampère's invention of the equilibrium demonstration technique during his initial search for a phenomenological law of electrodynamic forces. Experimental innovation also dominates chapter 8, with Ampère's theory of magnetism and his rivalry with Biot as major incentives. Chapter 9 highlights the culmination of Ampère's experimentation and subsequent mathematical developments related to the Biot-Savart law and Poisson's theory of magnetism. Chapter 9 also contains a discussion of Ampère's "Newtonian" methodology and his thoughts about ether mechanics as an explanatory basis for both optics and electrodynamics.

As might be expected, the initial months of Ampère's new project included many periods of frustration. Always eager to speculate and generalize, for the first time he tried to maintain a reasonably disciplined experimental investigation of an uncharted new domain. He quickly became enmeshed in the recalcitrant complexity of the new phenomena. Historians have only recently begun to investigate this exciting period; in general, there has been excessive reliance on the famous expository memoir he published several years later in 1826.[1] The methodological comments in the introductory pages of that text have usually been glossed as Ampère's attempt to imitate Newton's presentation of gravitation theory as a "deduction from the phenomena."[2] Indeed, this is a fairly accurate rendition of the image Ampère wanted to present in 1826; it places the emphasis on his electrodynamic force law as an analogue to Newton's law of gravitation. It will be worthwhile to sketch the general contour of this image before we scrutinize its origins.

Two provocative experimental discoveries provided a lasting foundation for Ampère's efforts to create a new physics of "electrodynamics," the term he coined for the dynamics of electricity in motion. In July, 1820, Oersted demonstrated that an electric current can change the normal orientation of a suspended bar magnet. Second, during the following September and October, Ampère produced attractions and repulsions between wires conducting electric currents. He attributed these phenomena to "electrodynamic" forces acting between the currents, and he also argued that internal electric currents within all magnetized bodies are responsible for magnetic forces. In keeping with the norms practiced by his Laplacian colleagues, Ampère assumed that an adequate theory of electrodynamic forces must include a microscopic phenomenological law to express the attractive or repulsive force between any two segments of an electric current small enough to be considered "infinitesimal" in comparison to the distance separating them. By 1826 he had assembled an elegant and succinct derivation of a force law of this type, based upon conceptual and mathematical analysis of four carefully chosen experimental demonstrations of apparatus held in a state of equilibrium by electrodynamic forces.

Each of these demonstrations was interpreted by attributing the static equilibrium of a potentially mobile circuit component to a set of balanced electrodynamic forces. Assuming that an electrodynamic force acts along the line joining any pair of circuit elements, Ampère applied the relevant laws of statics to formulate mathematical statements of the observed states of equilibrium. Analysis of these equilibrium conditions in terms of infinitesimal circuit elements then provided the desired derivation of an electrodynamic force law, the force being a function of the mutual orientation of any two elements and the distance between them. In 1826 Ampère presented this polished argument as what he hoped would be accepted as a model exercise in "Newtonian" methodology.

There has never been any doubt that Ampère must have done considerable spadework prior to 1826. As the English physicist James Clerk Maxwell remarked in 1879:[3]

> We can scarcely believe that Ampère really discovered the law of action by means of the experiments which he describes. We are led to suspect, what, indeed, he tells us himself, that he discovered the law by some process which he has not shown us, and that when he had afterwards built up a perfect demonstration he removed all traces of the scaffolding by which he had raised it.

Historians have taken up the task of reconstructing Ampère's discarded scaffolding. L. Pearce Williams and Christine Blondel have tried to unravel the chronology of the early events leading up to Ampère's discovery of electrodynamic forces between electric currents.[4] Blondel has provided an accurate survey of the full scope of Ampère's work in electrodynamics and Grattan-Guinness has given detailed analyses of the mathematics of some of Ampère's arguments.[5]

My own interest lies more in Ampère's early experimental methods than with any specific discovery as such.[6] In particular, it is interesting that some of Ampère's most creative experimentation during 1820 set an important precedent for his subsequent adoption of the equilibrium technique. Prior to the middle of 1822 he did not anticipate that this method could provide a basis for a full derivation of his force law; his first full set of equilibrium demonstrations was not assembled until November, 1825, and he repeatedly revised them thereafter. His famous 1826 memoir provides little information about how his initial responses to experimental anomalies initiated his gradual implementation of a new experimental method. It is true that Ampère did begin his research with a preference for the primarily deductive and "direct" mode of theory construction he eventually relied upon. Nevertheless, at the outset there was no reason for him to suspect that direct analysis would ever be applicable to the bewildering new domain of electrodynamics.

Ampère's experimental activities under these circumstances are intriguing in several respects. First, following his discovery of electrodynamic forces in 1820, Ampère's experimentation was almost always guided by predetermined goals; his initial agenda was to specify a mathematical expression for a phenomenological force law and to determine the circuits followed by the electric currents he believed existed within magnets. Thus, with rare but important exceptions, Ampère did not experiment with an exploratory mentality dedicated to the pursuit of novelty for its own sake. Instead, Ampère's attitude was that his task as an experimenter was, quite literally, to *construct* new phenomena and then to present them in a suggestive, clear, and easily replicable manner so as to foster the acceptance of his theoretical interpretations. Nevertheless, during Ampère's search for illustrative demonstrations, he repeatedly encountered novel or anomalous experimental circumstances which he had to acknowledge as unexpected but genuine experimental discoveries. Furthermore, the alacrity with which Ampère transformed initially anomalous phenomena into incentives for further theoretical progress was at the heart of his creative invention of the equilibrium technique.

We have already noted Ampère's preference for what he called direct analysis, the argumentative form that derives simple, unexpected conclusions from complicated but well established premises. In his lectures at the École Normale he argued that only *de facto* mathematical and physical complexity prevents scientists from following Newton's exemplary use of direct analysis; the inductive confirmation of a hypothesis by successful observations of its implications is a less trustworthy procedure. Nevertheless, Ampère recognized that this indirect synthesis was often a pragmatic necessity. Ultimately, his invention of the equilibrium technique was in full keeping with his preference for direct analysis. By deriving simple theoretical conclusions from complicated but clearly presented experimental demonstrations, Ampère could claim that he had given the argumentative structure of his electrodynamics the closest possible correlation with Newton's theory of gravitation. Prior to his invention of this technique, Ampère was forced to rely upon indirect synthesis to confirm tentative hypotheses. A close study of Ampère's early research reveals that his recognition that equilibrium experiments might be a foundation for direct analysis took place after his encounter with anomalous test results generated by the hypothetico-deductive method of indirect synthesis.

Oersted's Discovery of Electromagnetism and the Biot-Savart Response

Immediately following his initial discovery in the spring of 1820, Oersted carried out additional observations and summarized his results in a short Latin publication dated 21 July, 1820. Arago learned of the discovery while in Geneva and made a report to the Académie at the meeting of 4 September, 1820; the following week he repeated Oersted's experiment for the benefit of his skeptical French audience. Oersted's text was quickly translated and published in all the important French journals.

Oersted's observations seem to reveal a glaring exception to the alleged independence between electric and magnetic phenomena; furthermore, from the point of view of the Laplacian school the force acted peculiarly. A magnet suspended with its midpoint on the circumference of a circle drawn around the axis of a conducting wire tended to take a new orientation tangent to the circle. In contrast to the attractive and repulsive central forces manifested in gravitational and purely electric and magnetic phenomena, the current in the wire apparently had a "transverse" or tangential effect on the magnet.

As we have seen in our study of Oersted's 1813 publication, although his Kantian metaphysics made him receptive to the possible unity of "forces," he had no interest in the Laplacian agenda to reduce phenomena to *mathematically formulated* fundamental forces. Oersted was content to report how he had observed the new effect under a variety of initial conditions. He did, however, draw some tentative explanatory comments to which Ampère gave due consideration.[7]

> It appears, according to the reported facts, that the electric conflict is not restricted to the conducting wire, but that it has a rather extended sphere of activity around it . . . the nature of the circular action is such that the movements that it produces take place in directions precisely contrary to the two extremities of a given diameter. Furthermore, it seems that the circular movement, combined with the progressive movement in the direction of the length of the conjunctive wire, should form a mode of action which is exerted as a helix around this wire as an axis.

Oersted concluded his report by applying a two-fluid theory of electricity to claim that the fluids traverse a helical path in opposite directions and that each fluid only affects one pole of the magnet respectively. Without giving these comments more importance than they merit, it is interesting that helices figure so prominently in Oersted's initial interpretation. On the other hand, Oersted was not clear about the magnitude of the reorientation of the magnet; one of Ampère's earliest investigations was the discovery that, when the effect of terrestrial magnetism is neutralized, the magnetic needle takes an orientation normal to a line drawn from its center to the conducting

231

wire. But before turning to Ampère's efforts in detail, we should note how the Laplacian school reacted.

Biot took an immediate interest in the phenomenon and recruited a protégé, Félix Savart, to help him make measurements. Biot's approach was precisely what would be expected from a graduate of the École Polytechnique and an elite member of Laplace's circle. As he wrote retrospectively in 1821:[8] "The first thing that had to be discovered was the law according to which the force emanating from the connecting wire decreased in strength at various distances from its axis." To determine how the magnitude of Oersted's effect varied with changes in the experimental conditions, Biot and Savart magnetized a small steel needle and suspended it in a horizontal plane. An appropriately located bar magnet neutralized the magnetic effect of the earth, and the magnetized needle was then exposed to a long vertical conducting wire located at a carefully measured distance. It was observed that the needle always came to equilibrium with an orientation perpendicular both to the wire and to the line joining the needle to the wire. To study this effect quantitatively, Biot noted that just as a pendulum can be used to measure the strength of the earth's gravitational force, either the frequencies or the periods of the oscillations of the needle, when displaced from its position of equilibrium, can be used to compare the strengths of the force bringing the needle back to equilibrium for various distances of separation between the wire and the needle. The operative force is proportional to the square of the frequency or inversely proportional to the square of the period. In their first short publication, Biot and Savart reported their results as follows:[9]

> By the aid of these procedures, Biot and Savart were led to the following result which rigorously expresses the action experienced by a molecule of astral or boreal magnetism located at an arbitrary distance from an extended, very fine cylindrical wire magnetized by the voltaic current. Draw a perpendicular to the axis of the wire from the point where this molecule resides: the force which influences the molecule is perpendicular to the line and to the axis of the wire. Its intensity is inversely proportional to the simple distance.

Biot and Savart had obviously carried out quite a complex set of inferences in order to offer these statements as the "result" of their observations. Most important, they claimed that they had discovered that the wire exerts a force on individual particles of magnetic fluid; their actual measurements, of course, only pertained to the effect produced on the entire magnetized needle. As resolute champions of the two-fluid theory of magnetism, Biot and Savart assumed that, although the two types of magnetic fluid particles were constrained within individual molecular regions, the net effect of their polarized separation within those regions would be the same as that which would be produced by a concentration of one fluid near each pole of the magnet. By approximating the distance from the wire to each of these poles by the distance measured from the center of the magnet, the experimentally determined action of the wire on the magnet could be interpreted in terms of action exerted on magnetic fluid particles located at each pole. The observed inverse proportionality to distance was thus said to apply to a force acting on individual magnetic fluid particles.

Biot's reasoning illustrates the ease with which the Laplacian framework of imponderable fluids lent itself to a mathematical analysis of forces. When the particles are imagined as being concentrated within very small regions, such as the poles of a magnet, all the properties of a force acting on the pole of the magnet are readily conceived to apply to forces acting on individual fluid particles so as to produce the observed phenomena as their cumulative effect. Although the pole is treated as a mathematical point for calculational purposes, the magnitude of the total force can still be presumed to be proportional to the density of magnetic fluid concentrated there.

Following their presentation of these conclusions to the Académie on 30 October, 1820, Biot and Savart carried out a second set of measurements using wires which, although still positioned in a vertical plane, also had a single bend located at the height of the suspended magnet. Based upon the slightly retrospective account Biot presented in 1821, the new conclusions he reported to the Académie on 18 December, 1820, can be summarized as follows. First, the initial set of data with straight wires was now more specifically interpreted as measurements of a *résultat composé* brought about by the action of the entire wire on a particle of magnetic fluid. Second, the inverse dependency of this total force with respect to distance from the wire could be shown to be a mathematical consequence of an inverse square distance dependency for more elementary forces assumed to act upon the magnetic fluid particle from each "infinitesimal" segment of the wire. The second set of measurements with bent wires was then interpreted as evidence that these more elementary segmental forces were also proportional to the sine of the angle formed at the point where a line drawn from the magnetic particle to the wire segment meets the wire.

Biot's justification for these claims was an interesting combination of rhetoric and error. First, he claimed that Laplace had "deduced" the inverse square dependency of the segmental force from Biot's first set of data. More accurately stated, Laplace's claim was simply a hypothesis confirmed by Biot's data; in Ampère's terms this was indirect synthesis rather than direct analysis. It is easy to construct counterexamples to show that other forces, which are not inverse square functions of distance, are also compatible with Biot's data. Second, the stated angular dependency of the segmental forces did not really account for Biot's second set of data; he corrected this flaw several years later by altering the phenomenological law he used to summarize his measurements.[10] In any event, in spite of his claim that his conclusion about the angular dependency of his segmental forces had been "analyzed by calculation,"[11] Biot had employed once again what Ampère called indirect synthesis; he assumed that the segmental force was proportional to the sine of the relevant angle and derived a measurable implication from this hypothesis. The second set of data was then cited as confirmatory evidence. Strictly speaking, this data pertained directly only to the total composite action of conducting wires on magnets. Biot's references to "deduction" and "analysis" of more elementary forces were simply rhetorical attempts to lend an aura of certainty to what were actually inductive confirmations of hypotheses. The confirmatory procedure was precisely the indirect synthesis Ampère expected to be necessary, at least initially, for a new domain like electrodynamics.

Before turning to Ampère's own investigations, Biot's methodological commitments should be reiterated. Although he claimed to have discovered a mathematical expression for forces assumed to act upon magnetic fluid particles from each tiny section of a conducting wire, Biot considered this conclusion to be only a preliminary step toward a full explanation based on an understanding of "a true molecular magnetization impressed upon the particles of metallic bodies by the voltaic current that traverses them."[12] For Biot, the genuine fundamental forces were assumed to act between magnetic fluid particles in the wire and in the magnet. Not surprisingly, Biot never explained how "molecular magnetism" was produced in a conducting wire without violating the conventional Laplacian prohibition of any interaction between the electric and magnetic fluids. Instead, he argued that his segmental phenomenological force law could at least account for the observed interactions between magnets and conducting wires.

Ampère's fledgling attempts to create his own physics of "electrodynamics" thus took place in competition with a powerful research program headed by one of France's most prestigious scientists. Rivalry and competition between Ampère and Biot was inevitable; they advocated incompatible reductionistic programs – one based upon magnetic fluid and the other on electric fluid. Since the issue could not be resolved experimentally during 1820, a choice between the two amounted to a preference for one of two alternative phenomenological laws as a temporary expedient. With this context in mind, we should expect that, during Ampère's initial experimental investigations, he would be alert for any opportunity to convert initially perplexing observations into demonstrations readily interpreted in a manner that supported his own research program. This indeed became the scenario during the fall of 1820 when he made his first halting steps toward a direct analysis of an electrodynamic force law.

Ampère's Discovery of Electrodynamic Forces and his First Force Law Hypothesis: October 1820

As the historian Kenneth Caneva has argued in great detail, Ampère's enthusiastic reaction to Oersted's discovery was amplified by his lack of commitment to Laplacian concepts and institutions.[13] Rather than take a defensive stance and attempt to incorporate the new phenomenon into one of the disjoint Laplacian categories of magnetism or static electricity, Ampère was open to the possibility that a hitherto unrecognized force takes effect when electric fluid is in motion. He very quickly concluded that magnetic fluid simply does not exist; instead, he argued that magnetic effects are produced by electric circuits located within planes normal to the axes linking the magnetic poles of magnetized bodies.

L. Pearce Williams has devoted more attention than any other historian of science to the first few weeks of Ampère's investigations.[14] Williams has quite properly insisted that Ampère's lack of concern about chronology and less than trustworthy memory for dates make Ampère's own published comments about the sequence of his discoveries so dubious as sometimes to be more of a hindrance than an aid to the historian. In contrast to a methodical investigator such as Faraday, Ampère never had the discipline required to maintain a scientific diary. Experimental events and Ampère's thinking evolved together at such a rapid pace that, without a record of this type, he could not often recall accurately what the situation had been only a few days earlier. A reconstruction of his progress thus calls for judicious reliance on the chronological reports he did leave, combined with a sensitive attempt to resolve their inconsistencies. The major milestones are the weekly Monday meetings of the Académie; for several months, beginning on 18 September, Ampère made progress reports almost every week. With the constraints of time, his presentations were rushed and often had to be completed the following week. During the ensuing days, Ampère would often investigate far more than he could summarize accurately in his next short report; the Académie *Procès-verbaux* and the more detailed reports of Académie meetings that appeared in the *Bibliothèque universelle* at best provide highlights of what Ampère managed to report. As we attempt to unravel the first few months of his encounter with electrodynamics, it will be useful to have at hand a tentative chronology of the major events.

Major Events during Ampère's Early Research on his Force Law (1820)

4 September	Arago reports to the Académie on Oersted's discovery.
11 September	Arago repeats Oersted's demonstration for the Académie.
18 September	Ampère demonstrates the tangential orientation of a magnetic needle by an electric current when terrestrial magnetism is neutralized.
25 September	Ampère demonstrates for the Académie that conducting planar spirals attract and repel each other and respond to bar magnets in an analogy to magnetic poles. Arago follows Ampère's suggestion and magnetizes a needle by wrapping it with a conducting wire.
2 October	Ampère presents the Académie with the first part of his résumé article to be published in volume 15 of the *Annales de Chimie et de Physique*; by this time Ampère had discovered electrodynamic forces between linear conducting wires.
9 October	At the Académie, Ampère demonstrates electrodynamic forces between linear conducting wires (figure 4).

Late September or early October	Ampère uses a centrally suspended conducting helix wrapped around an axial current to replicate the rotational response of a suspended bar magnet to another bar magnet.
17 October	Ampère shows Biot and Gay-Lussac the new apparatus with which he discovered the action of terrestrial magnetism on a suspended circular conducting wire.
During the month after 25 September	Probably shortly after 17 October, Ampère gets anomalous results when he substitutes two helical conductors without compensating axial currents for the linear conductors in the apparatus he presented to the Académie on 9 October.
30 October	Ampère demonstrates the action of terrestrial magnetism on a suspended current loop. Biot and Savart make their first report to the Académie on their quantitative measurements of Oersted's effect.
6 November	At the Académie, Ampère presents his addition law for electrodynamic forces and uses it to interpret the action of helices. He uses a helix wrapped around an axial current to duplicate the action of a bar magnet on another bar magnet. He mentions that the motion of ether particles might be the basis for the addition law.
4 December	At the Académie, Ampère presents his symmetry principle and uses it together with the addition law to derive the angular factor in his electrodynamic force law. Biot gives a preliminary statement of the result of his new measurements with Savart.
11 December	At the Académie, Ampère reports on an experiment which he claims implies that his parameter k is negligibly small ($k = 0$); he explains why an experiment by Gay-Lussac and Thenard misleadingly seemed to indicate that k had an appreciable positive value. Assuming that $k = 0$, Ampère draws an analogy to heat radiation. He also reports on an inconclusive attempt to use quantitative

measurements to demonstrate the superiority of his theory to Biot's.

18 December   Biot and Savart report to the Académie on their second set of measurements of Oersted's effect.

26 December   At the Académie, Ampère describes an apparatus with a suspended magnet to test his addition law. He presents the second part of his résumé article for publication in volume 15 of the *Annales de Chimie et de Physique*.

Late December or early January 1821   Ampère designs an equilibrium apparatus with a suspended linear conducting wire in an effort to demonstrate his addition law.

Ampère's most detailed report on the events of September and October 1820 was published as a lengthy two-part memoir in the *Annales de Chimie et de Physique*.[15] Written hurriedly and in disjointed segments, it is a rich source of information in spite of its chronological errors and inaccuracies. The earliest account of the September and October meetings was published in the Belgian journal, the *Annales Générales des Sciences Physiques*.[16] Ampère sent a report which was printed as received. When combined with reports that quickly appeared in other journals, we thus have ample material at hand; the problem is to resolve glaring inconsistencies without imposing more order than was actually the case.[17]

As L. Pearce Williams has established, Ampère's initial discoveries came about in an effort to "complete" Oersted's discovery. By 18 September he could report that when the effect of terrestrial magnetism is eliminated, compass needles rotate fully so as to be normal to a line drawn to them from an adjacent linear current. Once in this position they experience either an attraction to the current-bearing wire or a repulsion if the orientation of the needle is reversed. He also leaped to the conclusion that the effect produced on a compass needle by terrestrial magnetism is due to electric currents that span the earth. Similarly, other magnets should contain hitherto undetected electric currents. This hypothesis implied that he should be able to duplicate interactions between magnetic poles by using conducting wires coiled into spirals. Ampère successfully did so at the Académie meeting of 25 September; it was probably during an investigation of this type that he observed attraction and repulsion between linear wires carrying electric currents.[18]

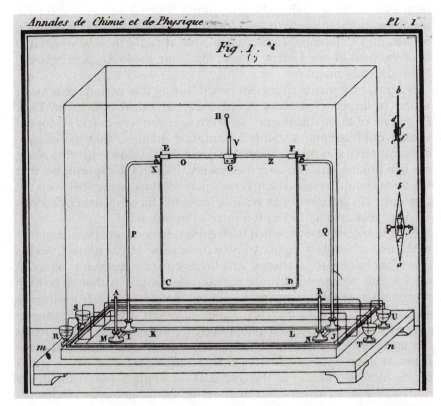

Annales de Chimie et de Physique Pl. 1.

Fig. 1.

4   Ampère's demonstration apparatus for electrodynamic forces between two parallel conductors; figure 1 from his "Mémoire . . . sur les effets des courans électriques," *Annales de Chimie et de Physique*, volume 15, 1820.

At the 9 October meeting of the Académie, Ampère used the apparatus shown in plate 4 to demonstrate an attraction between parallel linear conductors bearing currents in the same direction and a repulsion when the currents flow in opposite directions. AB is a stationary conductor, and the conductor CD is free to swing on pivots located at E and F. For example, when the current is arranged to flow in opposite directions through these two linear parts of the circuit, their mutual repulsion makes CD rotate away from AB.

In keeping with the reductionist mentality he shared with his Laplacian colleagues, Ampère saw his next task as the mathematical formulation of a microscopic phenomenological law expressing the electrodynamic force acting between any pair of electric circuit segments small enough to be considered "infinitesimal" in comparison to measurable magnitudes. Hopefully a formula of this type could be stated in differential form; in principle, it could then be integrated over the path of any electric circuit, including those Ampère believed to exist inside magnets.

Fragmentary manuscripts composed during this period are a major source of information about Ampère's activity during October, 1820. From one of these documents we learn that Ampère quickly adopted what would become a lasting assumption, namely, that the electrodynamic force should be an attractive or repulsive force directed along the line linking any two circuit elements.[19] Second, he speculated that the force should vary as the inverse square of the distance between the elements, "in conformity to what is observed for all genres of actions more or less analogous" to the force in question.[20]

For the special case in which both circuit elements are perpendicular to the line connecting them, Ampère drew some implications from his observations of forces between long linear circuit components he could manipulate experimentally into various orientations. Although he took no quantitative measurements, he noticed that the force was maximally attractive for parallel currents, vanished when they were mutually perpendicular, and became increasingly repulsive as they approached the anti-parallel orientation. Ampère concluded that the total force between these large conductors should include a factor which was a function of odd degree in the cosine of the angle between the directions of the two currents. His manuscript then continues as follows:[21]

> Furthermore, this function of the cosine of the angle between the directions of the two electric currents can have a simple form only when one considers infinitely small portions of these currents. It is probable that in that case it reduces to the first power of this cosine; at least this is the first supposition that it is suitable to test in the comparison of a hypothesis about the law of attractions and repulsions with the results of experience.

Ampère then speculated further that, in the more general case in which the two components are not perpendicular to the line joining them, the force should depend upon the angles α and β which they make with this line. Also, the cosine of the angle γ between two planes, each of which passes through the connecting line and one of the components, should replace the cosine of the angle between the direction of the two currents which had been argued for in the simpler case. The three angles α, β, and γ are illustrated in plate 5 with r representing the distance between the two circuit elements.

240

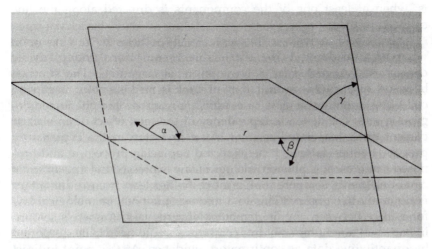

5 Ampère's angles for the mutual orientation of two current elements:
α, β, and γ.

Although Ampère left no record of the full mathematical expression he had in mind at this point, his early thoughts on the dependency of the force on the angles α and β were probably those registered by Babinet in the *Exposé des nouvelles découvertes sur le magnétisme et l'électricité*.[22] Ampère and Babinet did not publish this survey until 1822, but Babinet had written most of it by July 1821.[23] In sections 15 and 16 Babinet included an analysis of how the electrodynamic force should vary as a function of the angle α; his argument parallels the reasoning Ampère had applied to observations of how the forces between conductors of measurable length vary with the angle γ. Babinet concluded, apparently with Ampère's approval, that the force between two current elements vanishes whenever at least one of them is directed along the line connecting them.[24] Ampère again used his observations of attractions and repulsions between circuit components large enough to be manipulated experimentally to draw a conclusion about forces between infinitesimal current elements. Combining this reasoning with the manuscript record of the cos γ dependency, it is very likely that as early as mid-October 1820 Ampère had concluded that the force between current elements was proportional to the following expression:

$$r^{-2} \sin \alpha \sin \beta \cos \gamma$$

This would make the electrodynamic force between two infinitesimal circuit components vanish whenever either α or β becomes zero, that is, when at least one of the components is directed along the line connecting them.

Ampère had arrived at this tentative hypothesis by relying upon what he considered to be reasonable assumptions and plausible conjectures suggested by a few simple observations. The contrast between his procedure and that of Biot is striking. Biot began by collecting quantitative data to establish a macroscopic phenomenological law for the distance dependency of the total effect of an electric current on a suspended magnet. As a first step in the explanatory process, he then followed the expected Laplacian procedure and proposed a microscopic phenomenological law expressed in terms of small circuit elements. Ampère took a much bolder leap from rudimentary nonquantitative observations to a microscopic phenomenological law for a force between circuit elements. Nevertheless, Ampère's sojourn in Paris had taught him that this conjecture would have to be supported by quantitative data as confirmatory evidence. At this point indirect synthesis seemed to be the only applicable method of justification.

Ampère's Initial Research in Support of his Force Law: October, 1820

With a plausible hypothesis in hand, Ampère dedicated the second half of October to the design and construction of experimental apparatus. We know that as early as 25 September he had already started experimental investigations of the second major tenet of his electrodynamics, that is, his conviction that magnetic phenomena are caused by electric circuits in planes perpendicular to the axis linking magnetic poles. Expecting to be able to replicate the action of magnets using appropriately designed electric circuits, Ampère coiled conducting wires into spirals and helices. The planar spirals were intended to represent the circuitry at the pole of a bar magnet; helices represented the circuitry of the entire magnet.

Ampère's first helices were balanced on a pivot similar to that of a compass needle; perhaps as early as 25 September he found that this apparatus responded as expected to the presence of a bar magnet.[25] This was encouraging, but Ampère could not detect any corresponding response to terrestrial magnetism. Suspecting that friction in his pivot was preventing his helix from swiveling, Ampère switched to a more mobile suspension system by late in September. As shown in plate 6, the helix is now centrally suspended by a vertical extension of the same wire that makes up the helix.

242

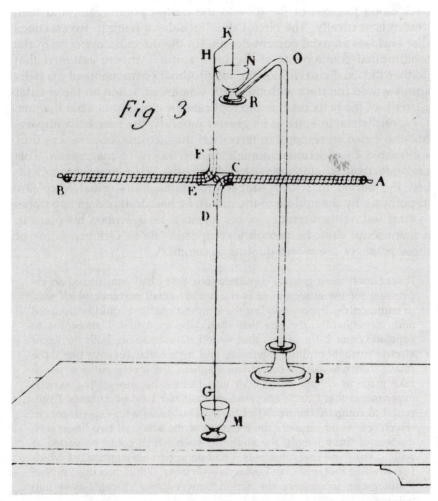

6   One of Ampère's axially compensated helices; figure 3 from his
"Suite du Mémoire sur l'action mutuelle entre deux courans électriques,"
Annales de Chimie et de Physique, volume 15, 1820.

This is accomplished by using a circuit in which the helix is assembled using two glass tubes. After descending vertically to the central
point of the suspension, the wire goes inside one of the tubes and
passes along its axis to the end of the tube. There it emerges and is
wrapped around this tube so as to return to the central point. The wire
then wraps around the second tube, enters the end of that tube, passes

back along the axis of the tube to the central point again, and then descends vertically. The circuit thus includes a helix in two sections that encloses an axial current flowing in the direction opposite to the longitudinal components of the helix spires. Ampère assumed that neither the axial current nor the longitudinal components of his helix spires would interfere with the earth's magnetic action on the circular currents of the helix intended to replicate the circuitry of a bar magnet.

Nevertheless, in spite of its greater mobility, the new helix apparatus also failed to respond to terrestrial magnetism; Ampère was thus confronted by a serious anomaly in his theory of magnetism. This uncomfortable situation persisted during the first two weeks of October, the period in which Ampère formulated his initial force law hypothesis. By the middle of the month he had designed an apparatus to start testing the accuracy of his formula; he described his plans in a manuscript draft he probably composed late in October. Some of these passages are worth quoting at length.[26]

> It was upon these general considerations that I had constructed an expression for the attraction of two infinitely small currents which was, in truth, only a hypothesis, but the simplest one that could be adopted and, consequently, the one that should be tried first. I attempted to conclude from it the effects that would have to result, both for linear electric currents of finite extension, and for circular currents like those I have shown to exist in cylindrical magnets, and for the currents which take place in copper wires bent into helices, because of the various experiments that I had performed on the latter kind of currents. I proposed to compare the results of these calculations with experiments in which one could measure the intensity of the action of two linear conductors of finite length, the angle between which could be varied at will . . . For these measurements, I had an apparatus constructed which I showed to Biot and Gay-Lussac last October 17th; I procured myself another one to observe the action between two currents bent into helices.
>
> The experiments I tried with these two instruments caused me to discover two new facts which complicated the results of them and consequently forced me to suspend the verifications of the results of my calculations that I had proposed to make with the aid of these apparatuses.

Both of Ampère's "two new facts" were of major importance for his subsequent research; in each case a discovery resulted from the unexpected behavior of new apparatus. One of these discoveries instigated Ampère's interest in equilibrium demonstrations. His other discovery should be discussed first, since this will clarify how Ampère initially

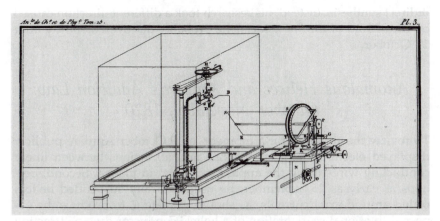

7 Ampère's apparatus for two conductors at variable orientations; figure 6
from his "Suite du Mémoire sur l'action mutuelle entre deux courans
électriques," *Annales de Chimie et de Physique*, volume 15, 1820.

intended to accumulate evidence for his force law hypothesis during
October. The relevant apparatus was the one Ampère claimed to have
shown to Biot and Gay-Lussac on 17 October. As shown in plate 7, it
was similar to the one in plate 4 except that the originally stationary
conductor AB was replaced by one that could be given a variable
orientation in a vertical plane.[27]

By varying the orientation of this conductor, Ampère could observe
effects on the mobile conductor CD and compare them with calculations based upon his force hypothesis. During his preliminary investigations of this instrument, he was surprised to see that the mobile
conductor was influenced by terrestrial magnetism. He now realized
that his helix devices had failed to respond in this way because of
their much smaller radial dimensions. What had been a serious anomaly
was thus resolved by a discovery using apparatus designed for an
entirely different purpose. Thereafter, in order to compensate for the
magnetic action of the earth on large suspended current loops, Ampère
simply used pairs of loops traversed by currents in the proper directions so that the torques produced on the two loops by terrestrial
magnetism would be in opposite directions. He used this method to
design a replacement for the initial uncompensated apparatus he had
shown Biot and Gay-Lussac on 17 October. Now he was in a position
to settle down to a laborious comparison of calculations and data
which hopefully would provide confirmatory evidence for his force
law hypothesis. There is no indication that Ampère ever initiated this

indirect synthesis. Instead, his research took a different direction caused by the second of the "new facts" he probably discovered shortly after 17 October.

Anomalous Helices and Ampère's Addition Law: October–November, 1820

To review the situation for a moment, on 9 October Ampère publicly displayed electrodynamic repulsions and attractions between linear conducting wires using the apparatus shown in plate 4. Second, perhaps as early as 25 September, he used centrally suspended helices coiled around axial currents, as shown in plate 6, to produce the expected rotational reorientation of a helical conductor due to attraction or repulsion by the pole of a bar magnet.

Now at some point during the hectic month after 25 September, Ampère also attempted to demonstrate how the electrodynamic interaction between two helical conductors duplicates that between two bar magnets. He left no records of his initial experiments; in his most detailed subsequent discussion of the subject he reported that he performed the test by using helices in place of the linear conductors in the apparatus shown in plate 4.[28] Although the apparatus with linear conductors was not presented to the Académie before 9 October, it is possible that Ampère constructed it and modified it for helices at some point during the preceding two weeks. A relatively early date is suggested by Ampère's claim that he made the resulting observation "long before knowing the cause of it."[29] On the other hand, in the long passage I quoted from what is probably the earliest relevant surviving manuscript report, Ampère implied that this observation gave him the second of the two "new facts" that delayed his plans to verify calculations based upon his hypothetical force law. Since Ampère explicitly said that he showed the apparatus responsible for one of these "new facts" to Biot and Gay-Lussac on 17 October, it appears that neither of these two discoveries took place until after that date. The available evidence simply does not allow any clear-cut determination of the date of Ampère's first experiment with two helical conductors; given what we know about the relevant apparatus, and in the light of Ampère's reference to 17 October, the second half of October is a plausible conjecture.[30]

Obviously Ampère did not give high priority to leaving a precise chronology of his research. This is particularly unfortunate in the case of his second "new fact," a discovery that posed a serious anomaly for his conviction that magnetic phenomena could be duplicated using spirals and helices. Much to his surprise, when he replaced the linear conductors in the apparatus shown in plate 4 by simple helices wound around glass tubes without internal axial currents, he found that the two parallel helices replicated the electrodynamic interactions of two parallel linear currents rather than two bar magnets, that is, they attracted each other when Ampère expected them to repel and vice versa. Ampère's reaction was typical of his general attitude toward experimental novelty. Considered simply as a "new fact," the unforeseen event would have held little interest for him if it had not violated his theoretical expectations. Faced with an apparently serious anomaly, Ampère did not publicize it until he could produce other phenomena conducive to an argument that could defuse the anomaly and transform it into confirmatory evidence.

This recovery process began when Ampère noticed how the misbehaving helices differed from those he had been using hitherto. Previously, Ampère had assumed that, when he coiled a wire into a helix to replicate a magnetic circuitry, the longitudinal dimension of each spire in the direction parallel to the axis of the helix was too small to have any noticeable effect. Probably during the last week of October or the first week in November, Ampère decided that this assumption was incorrect; he now argued that the unexpected behavior of his helical conductors was due to the fact that the cumulative effect of the longitudinal components of the spires was equivalent to a linear current flowing along the axis of each helix. Ironically, in apparatus such as that shown in plate 6, Ampère had unwittingly compensated for the longitudinal currents by running a current back down the axis of the helix in a direction opposite to that followed by the longitudinal components of the spires. He had done this simply to implement his new suspension method or, as he put it, "without foreseeing the advantages of it."[31] But when Ampère replaced the linear currents in the apparatus of plate 4 with helices, he had no reason to include compensating axial currents. In this situation, since the radii of his helices were relatively small, Ampère realized that the interaction between the longitudinal components of the spires of the two helices became predominant over the effect produced by the radial circular components and thus produced an effect equivalent to an interaction between two linear currents.

Ampère presented this argument to the Académie on 6 November, 1820.[32] He realized that his reasoning hinged upon the legitimacy of treating each of the "infinitely small portions" of an electric current as if it were made up of distinct components, followed by a determination of the force contributed by each individual component treated in isolation from the others. The legitimacy of adding together forces imagined to be produced by various components of the elements of an electric current became a permanent principle of crucial importance for the mathematical expression of Ampère's electrodynamics. He always referred to it as a "law," and I shall call it Ampère's addition law.

Unfortunately, Ampère's initial statements of this principle were far from clear. His ambiguous language was partially due to the fact that, by the time Ampère presented the law to the Académie on 6 November, his thoughts were already leaping ahead to possible applications. He realized that he had to present experimental evidence in support of the addition law; simultaneously, however, he was already trying to incorporate it into a derivation of the angular factor in his force law. From an experimental point of view, the situation called for a clear and concise statement of the law in question followed by compelling experimental evidence. But Ampère's impatience to use the law as a premise in mathematical derivations infected his early terminology with the ambiguities he was struggling with on the more abstract level of mathematical idealization. Further study of Ampère's addition law requires us to penetrate the smoke-screen of assertive rhetoric about "facts" and "laws" that protected his gradual creation of a new conceptual framework. This calls for a brief digression to clarify the issues at stake.

Throughout the month prior to 6 November, Ampère grappled with the central theoretical concept of his electrodynamics, namely the idea that a segment of an electric current that was small enough to be considered "infinitely small" in comparison to measurable magnitudes also could be assigned a direction in three-dimensional space. His explanation of anomalous helix behavior required an additional resolution of current elements into components; as might be expected, Ampère had some difficulty describing this operation. First of all, he assigned lengths to his supposedly "infinitely small" circuit segments in order to represent them geometrically. The idea Ampère tried to convey on 6 November was that the force in any given direction due to an infinitely small current element is equivalent to a sum of forces in that direction when these forces are imagined to be produced by the components into which the original element has been resolved. Geometrically speaking, the length of each component would represent the magnitude of the maximum force that component could exert on another current element at a given distance, that is, when the two current elements in question are parallel to each other.

248

Unfortunately, Ampère sometimes used language that could easily be interpreted to imply that the lengths of geometrically represented current elements correspond to the number of elementary molecular processes taking place at any instant within a given circuit segment. He thought of this microscopic activity as oscillations of the two electric fluids responsible for the electrodynamic forces attributed to the segment. He must have had this picture in mind when he wrote the following passage in an 1820 manuscript:[33]

"an infinitely small portion necessarily exerts an action proportional to its length, since in subdividing it into an arbitrary number of equal parts, its action is the sum of the actions of all these parts, which are necessarily equal to each other."

Ampère surely wanted his electrodynamic forces to be proportional to the "intensity" of electric currents. This quantity would represent the number of molecular processes taking place at any point in the circuit during a given unit of time. This intensity would presumably be the same in all components of a geometrically resolved current element; intensity thus is *not* represented by the lengths of Ampère's geometric depictions of current elements. The geometric analysis is intended solely to illuminate the angular dependency of the electrodynamic force. It does not appear that Ampère himself ever misunderstood these distinctions. His instinctive insight into the implications of his mathematical model seems to have been unimpaired by the ambiguous language that persisted even in the initial published version of the 4 December presentation in which he used the addition law to derive the angular dependency of the force law.

On the other hand, we can be sure that Ampère's early statements of his addition law so thoroughly confused current element "length" with current "intensity" that few of his readers could have reached any clear understanding. For example, in one of his earliest and most detailed published discussions of the law, Ampère wrote that in order to understand the law,[34]

one must imagine in space a line representing in magnitude and in direction the resultant of two forces which are similarly represented by two other lines, and suppose, in the directions of these three lines, three infinitely small portions of electric currents, the intensities of which are proportional to their lengths. The law at issue consists in the fact that the small portion of electric current directed along the resultant exerts an attractive or repulsive action equal to what would result, in the same direction, from the combination of the two portions of current directed along the components.

If the reader is not confused by the misleading preliminary remarks about "intensities," the statement of the addition law in the last sentence of this passage is fairly accurate. At the 6 November meeting of the Académie, Ampère tried to provide experimental evidence for the addition law using the helices that had suggested it to him. According to his later recollections of this presentation, he used a helix with a compensatory axial current to show that the earlier anomalous attractions and repulsions were now neutralized so that his apparatus remained in a stationary state of equilibrium. Ampère left no detailed record of how he carried out this demonstration. He may have used some combination of the apparatus shown in plate 4 and plate 6, or he may have wrapped a helix around a vertical glass tube.[35] Our ignorance on this point is unfortunate in that this may well have been Ampère's first encounter with an intentionally designed demonstration of a state of equilibrium caused by a balance of oppositely directed forces, that is, the type of demonstration he would incorporate into his equilibrium experiment technique. Considering how many issues were on Ampère's mind at this time, it is not surprising that he did not immediately attribute any special significance to this particular observation. He apparently devoted most of his 6 November lecture to an application of the apparatus shown in plate 6 to show how an axially compensated helix makes the expected interactions with bar magnets and compass needles.

At any rate, we can conclude that Ampère's initial efforts to justify the addition law were examples of the hypothetico-deductive mode of reasoning he called indirect synthesis; theoretical implications of the addition law and Ampère's theory of magnetism were confirmed by effects produced using compensated helices. More important, however, the state of equilibrium Ampère observed with one of these helices also set a precedent that he would build upon several weeks later. Meanwhile, Ampère realized that he also had to respond to the rival Laplacian interpretation of Oersted's discovery as Biot and Savart had presented it to the Académie on 30 October. As part of this response, Ampère used the addition law to give his first public derivation of the angular factor in his force law.

Ampère's First Public Discussion of his Force Law: 4 December, 1820

Although by 1826 Ampère had ample justification to claim that he could derive his force law from a well-chosen set of equilibrium demonstrations, this was certainly not the case during 1820. On 4 December he used symmetry arguments and his addition law to present a derivation of the angular factor in the force law. This derivation established the context for Ampère's subsequent attempts to improve the experimental basis of the addition law so as to justify his mathematical manipulations of idealized current elements. Ampère's "derivation" was actually an artful combination of idealization, mathematics, and approximation. His invention of the equilibrium experiment technique during the following 18 months was inspired by his effort to legitimate controversial stages in this argument.

Ampère's 4 December derivation concerned only the angular factor in his formula. At this time he continued to assume that the force should include a separable inverse square function of the distance between two current elements.[36] Second, he also assumed that the force would act along the line joining the two elements and that the force reverses its direction but retains its magnitude when the direction of the current in either of the two elements is reversed. He used the last of these assumptions to argue for a useful symmetry principle: the electrodynamic force between two current elements vanishes when one element is located in a plane that passes through the center of the other element and cuts it at a right angle.[37] Like the addition law, the symmetry principle applies to the idealized geometrical representation of infinitely small circuit elements depicted for mathematical purposes as tiny directed line segments.

Ampère began his 4 December derivation by taking two current elements with a mutual orientation determined by the three angles α, β, and γ as illustrated in plate 5. He then resolved each element into components along three conveniently chosen directions and considered the forces acting between all possible pairs of components when a component is selected from each of the two current elements. Geometrically, the components of a current element have no common point of intersection; they only combine mathematically to form the total element through a process that now is called vector addition. For the purposes of his derivation, however, Ampère imagined the components of a given element to be transported slightly so that all three of the element's components could be depicted as having their midpoints at the point originally chosen as the midpoint of the element itself. This was justified by the relatively small distances involved, in contrast to the distance between the two current elements.

With his components restructured in this way, Ampère could apply his symmetry principle to conclude that there are no forces between some pairs of components; he then invoked the addition law to add together the remaining forces. The resultant total force between the two current elements is given by the following formula where g and h are the two current intensities, k is an undetermined parameter, and r^{-n} is the dependency on the mutual separation r which Ampère assumed to be such that n = 2.

$$g \cdot h \cdot r^{-n} (\sin \alpha \sin \beta \cos \gamma + k \cos \alpha \cos \beta)$$

This expression is more complicated than the hypothesis Ampère had tentatively adopted in October; it includes an additional term weighted by the magnitude of the parameter k which was left undetermined by Ampère's 4 December derivation. In physical terms, k represents the relative strength of the force that would act between two collinear current elements in comparison to the force that would act if they remained separated by the same distance but were parallel instead of collinear.[38]

During the remainder of December Ampère argued on the basis of limited experimental evidence that k was small enough to be treated as insignificant in comparison to measurable quantities. Setting k = 0 reduced the new force law to the form of his initial hypothesis, a form Ampère continued to believe was correct until early in 1822. At that point another anomalous observation instigated a renewed scrutiny of the more general formula and resulted in the discovery that $k = -\frac{1}{2}$. But in December of 1820 these later developments were still far in the future. At this point it was imperative that Ampère devote more attention to experimental support for the addition law; this was the agenda that soon brought about the production of new equilibrium phenomena.

Ampère's First Equilibrium Apparatus: December 1820–January 1821

Ampère struggled with two closely related projects during the month following the 4 December meeting of the Académie. First, because Biot and Savart were energetically promoting their own Laplacian program, Ampère devoted much thought to what turned out to be unsuccessful attempts to design a decisive experimental test that would discredit their approach.[39] Second, the inconclusive results of these efforts forced Ampère to realize that he needed to call upon a new mode of argumentation if he was to generate acceptance of his conception of the *fait primitif* of electrodynamics. In particular, Ampère had to motivate acceptance of his addition law, a crucial premise in his derivation of the angular factor in the electrodynamic force law. In the light of Ampère's predilection for direct analysis, it is not surprising that he thought of improving the design of one of the experiments he had used during his initial presentation of the addition law on 6 November. On that date he had demonstrated that an axially compensated helical current does not produce the electrodynamic forces generated by a linear current. Although Ampère presumably performed or, at least, described this demonstration for the Académie on 6 November, the fact that he never published an accurate description of the original experiment is indicative of the disorderly state of affairs he sought to clarify during the following December.

Ampère's project was motivated by the mathematical techniques he had used in the derivation of 4 December. How could he provide an easily repeatable demonstration that could readily be interpreted as legitimation for his mathematical resolution of a force attributed to a current element into three forces assigned to the three geometrically depicted components of the original element? Clearly, there would always be a gap between physical circuitry and the idealized geometry of "infinitely small" current elements. Ampère's task was to make this gap relatively easy to cross by means of an experimentally stabilized stepping-stone. This was the problem that called equilibrium phenomena to Ampère's attention.

Although we do not know how Ampère presented his ideas verbally, his publications and notes leave a fairly clear record of his strategy. First, the seriously misleading confusion of current element length and current intensity was purged from subsequent statements of the addition law. For example, late in 1820, when Gillet de Laumont published a summary of Ampère's research, he relied heavily upon revised versions of passages originally written by Ampère for his own publications. In the description of the 6 November presentation of the addition law, Laumont replaced Ampère's original phrase "infinitely small portions of electric currents, the intensities of which are proportional to their lengths"[40] by a much more accurate reference to "electric currents of which the attractive or repulsive forces are proportional to their lengths."[41]

But in addition to this conceptual clarification, Ampère wanted to formulate an experimentally testable assertion that would readily suggest the desired transition from experience to his idealized mathematical model. In a passage published early in 1821, he claimed, somewhat misleadingly, that the addition law "reduces to" the following statement:[42]

If along the direction of an electric current, one specifies two points infinitely close together, and one substitutes for the little portion of electric current between these points another portion of this same current, along a line bent or contorted in any manner whatsoever, but terminating at the same points without deviating in any part to a finite distance, this substitution will not change in any manner the action exerted in any direction whatsoever by the little portion of current under consideration on another portion of electric current separated from the first by a finite quantity.

From Ampère's point of view, this assertion should be interpreted as a testable claim about wire segments large enough to be manipulated experimentally. He simply used the phrase "infinitely close together" to mean "very close together in comparison to other lengths or dimensions of the circuit." The addition law itself was actually applicable only to the idealized geometrical representation of current elements and thus could not be directly tested; Ampère hoped that by providing a striking demonstration of the experimentally testable analogue of the law, he could convince his colleagues to accept the law as a straightforward "deduction" from the experimental results.

Ampère publicized his project at the Académie meeting of 26 December. According to the surviving notes for his lecture, he stated the testable assertion as quoted above and then described his apparatus. Ampère's own sketch is shown in plate 8.

EG and FH represent two glass tubes equidistant from a central point N which acts as a point of suspension for either a mobile linear circuit segment or, as shown in plate 8, a small magnet. Either straight or twisted and contorted conductors can be inserted in grooves inside the glass tubes. Ampère's description of the experiments he planned to perform is worth quoting; this is the message he conveyed to the Académie on 26 December.[43]

The verification of the law of which I have just spoken consists in assuring oneself that the substitution of one of these conductors for the other in a given slot changes nothing in the action exerted, for example, on a small magnet suspended at N. On can perform a kind of counter-test (contre épreuve) of this experiment by placing two currents directed in opposite directions within the same slot and separated by a sheet of paper, using for this a single conducting wire, one extremity of which would issue from the tube EF, for example, and the other would return, while winding in some spires, to the mercury cup U located near the cup R and which replaces the other cup S. There would be no appreciable action.

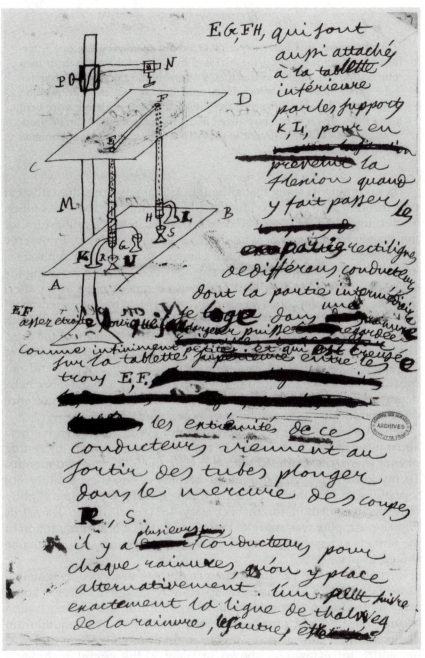

8 Ampère's sketch of the apparatus to be discussed at the 26 December, 1820, Académie meeting. (From the Académie's archives, *chemise* 164; photograph by Jean-Loup Charnet with the permission of the Académie des Sciences.)

Furthermore, this latter experiment would merely be the one I performed on a linear conductor enclosed in a glass tube and which then returned on the outside of this tube by surrounding it with some spires – the experiment from which I deduced the law which is in question here – with the sole difference that, instead of a conducting wire bent into a helix, one that can be twisted in an arbitrary manner can be used.

As soon as this instrument is ready, I will perform the experiments for which it is intended.

These comments show that although Ampère designed his circuit to pass through both of the conductors placed within the two glass tubes, he did not initially attempt to demonstrate a state of equilibrium for the suspended magnet or wire. Instead, expecting always to observe some rotational reaction, he only claimed to be able to replicate a linear conductor's contribution to any observed effect by replacing the enclosed linear wire by a bent and twisted one. Almost as an afterthought, he also described "a kind of counter-test of this experiment" using only one glass tube, but with a linear wire inside and with another arbitrarily twisted wire coiled around the outside of the tube. This arrangement was expected to leave the central suspended part of the apparatus unaffected in a state of equilibrium. Ampère did not consider this to be a major improvement over his 6 November helix demonstration and he did not as yet place any particular importance on the performance of equilibrium experiments.

Nevertheless, although Ampère had not even constructed his new apparatus prior to 26 December, he apparently noticed some interesting phenomena as he grappled with it during the following two months. To my knowledge, Ampère left no records of these exploratory observations and his initial reactions. By the time he wrote up a slightly retrospective memoir for the February 1821 issue of the *Journal de Physique*, Ampère was ready to launch into an elaborate description of the famous apparatus he later depicted in engravings such as the one shown in plate 9.[44]

In contrast to the variety of circuits Ampère described as tentative possibilities on 26 December, he specifically designed his new apparatus to produce a particular state of equilibrium; this demonstration became the first and prototypical member of the set of equilibrium demonstrations from which Ampère could eventually claim to derive his entire force law. As shown in plate 9, the circuit includes a vertical linear wire, GH, suspended equidistant from two longer conductors. One of these is a linear conductor, QP, and the other is an arbitrarily twisted conductor, SR, which follows a random, two-

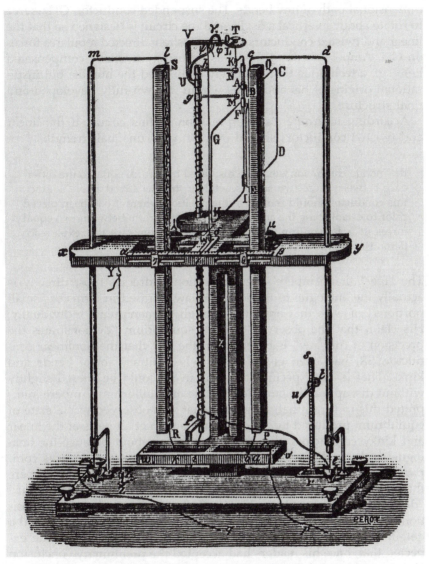

9   The final design for Ampère's first equilibrium apparatus, from his
Recueil d'observations électro-dynamiques, first published in 1823 though
dated 1822.

dimensional path of tiny bends. The suspended conductor, GH, is free to rotate about a vertical axis, FI, and the circuit is designed so that the linear and twisted conductors exert oppositely directed repulsive forces on GH. Ampère accomplished this by using an axially compensated helix, gf, a technique that creatively embedded the humble but inspirational origins of his apparatus within its own fully developed and final structure.

According to Ampère's interpretation of this circuit, if the linear and twisted conductors should ever act with unequal strengths:[45]

> the mobile conductor would be displaced by a force equal to the difference between these two actions, while if the law stated above is exact, this conductor should remain in the position where it had been placed prior to establishing the connections, in equilibrium between two equal forces. It is by confirming that this is in fact the case that experience demonstrates the precision of this law.

The "law" that Ampère's demonstration addresses most directly is actually the analogue to the addition law phrased in terms of "small portions" of wire that are large enough to manipulate individually. His claim that the observed state of equilibrium "demonstrates the precision of this law" is based upon the fact that the nonlinear conductor, SR, had been given an entirely arbitrary set of bends and kinks. That is, any specific little kink in SR could be altered slightly without disrupting the cumulative state of equilibrium. Ampère interpreted this to mean that, although what was observed was a state of equilibrium produced by the combined action of all parts of the linear and kinky conductors, this could only come about because the analogue to the addition law held for circuit segments of lengths comparable to the individual kinks in the circuit. Furthermore, Ampère thought that this preliminary conclusion sanctioned a straightforward "deduction" of the addition law itself, that is, the analogous proposition about "infinitesimal" circuit elements. Ampère never published a detailed discussion of this step in his reasoning; apparently he expected that, once his readers had accepted the preliminary conclusion about experimental circuit components, they would naturally extend it into a more abstract form by imagining the circuit components becoming increasingly small until they were mathematically "infinitesimal."

On the other hand, Ampère realized that some of his colleagues were reticent about following his lead in this manner. To ensure that they would at least recognize that his empirical discoveries were genuine, Ampère published a retrospective expository memoir early in 1821. While summarizing his 1820 research, Ampère drew a careful distinction between the experimental "fact" produced by his equilibrium apparatus and "the mathematical laws of attractions and repulsions between two metallic wires constituting part of a voltaic circuit."[46] He then included among these laws the following accurate statement of the addition law as he actually used it in practice:[47]

> if one considers one of these infinitely small portions as the diagonal of a parallelepiped, its action on the other is equal to the sum of the actions which would be exerted on the latter by three portions of the first conducting wire, directed along the three components (*arrètes*) which specify the three dimensions of the parallelepiped, and of the same length as these components.

One of the intriguing aspects of Ampère's research is the fact that it is very difficult to determine when he actually performed the equilibrium demonstrations he describes in such elaborate detail. He was not known for his manual dexterity, and some of his colleagues must have been dubious about "experiments" that generated no data and amounted to a single crucial observation. In the present case, Ampère left no clear record of when he carried out this equilibrium demonstration for the first time. In a draft for an 1823 publication, probably written early in 1822, Ampère included some interesting comments about the experimental basis that allowed him to derive the force formula:[48]

> First Fact: changing the direction; change of sign, same magnitude.
> 2°: replacement of the two components; experiments, but *a priori* if there are movements.

Ampère's "first fact" is the assumption required to derive the symmetry principle; no experimental evidence is cited. "2°" refers to the addition law and indicates that it might have to be asserted *a priori* if there are "movements." These undesirable movements must pertain to the mobile conductor in the equilibrium apparatus designed to support the addition law. If this passage was indeed not written until early in 1822, it is possible that Ampère still had not successfully performed his demonstration by that point.

At any rate, Ampère rather optimistically presumed that the addition law would be generally accepted as a straightforward implication of his equilibrium demonstration and his interpretation of that demonstration in terms of the small individual circuit segments of the kinky conductor in his apparatus. He maintained this expectation during the composition of the *Exposé* he and Babinet completed by July 1821.

Ampère's strategy was thoroughly in keeping with the reductionist mentality of the Laplacian milieu in which he had functioned for almost two decades. Biot had formulated his own conception of the electromagnetic "fundamental fact" in terms of an "infinitesimal" current element acting on a particle of magnetic fluid. In contrast to Ampère, however, Biot treated his current elements as if they were punctal; he did not incorporate current element orientation into his theory. Ampère's first equilibrium experiment thus provided important support for his alternative research program based on a phenomenological force law for *directed* current elements. The initial success of this program depended largely upon reductionistic explanations of electrodynamic phenomena. We have seen how Ampère deliberately produced some of these phenomena in response to what initially appeared to be serious experimental anomalies. Furthermore, by incorporating compensated helices into the circuitry of his first equilibrium apparatus, Ampère could display the engraving shown in plate 9 as an intriguing artifact of the conceptual and experimental route he had travelled.

In view of the importance that equilibrium demonstrations held for Ampère, it is useful to have a convenient way to refer to them. One method is to adopt the numbering system Ampère himself used in his famous 1826–7 *Théorie*.[49] In that scheme, Ampère numbered his demonstrations in the order he employed them in his full derivation of the force law; the apparatus depicted in plate 9 became Ampère's "second" equilibrium experiment. The "first" demonstration was one Ampère described in the 1823 edition of the *Recueil*; he used it to show that the magnitude exerted by a linear current preserves its magnitude when the direction of the current is reversed. As we will see, the 1822 invention of the "third" equilibrium experiment was particularly important for Ampère. Subsequent revisions of experiments "three" and "four" were frequent.

Ampère's invention of the equilibrium demonstration technique obviously did not come about in a single flash of inspiration. His longstanding preference for the direct analysis of simple theoretical conclusions from more complex experimental observations did give him a predisposition to pursue this procedure in the new domain of electrodynamics; this distinguished him from Biot and other more inductively minded Laplacian colleagues. Nevertheless, Ampère's first systematic study of equilibrium apparatus was a thoroughly unforeseen development as he refined his demonstrations with axially compensated helices during November, 1820. He realized that he could interpret the initially anomalous behavior of uncompensated helices as the cumulative effect of the small longitudinal components of the helix spires. Furthermore, the new equilibrium apparatus was used to provide evidence for Ampère's addition law, a crucial premise in his efforts to derive the force law between current elements. As we have seen, Ampère's argumentation for the addition law passed through an interesting sequence of both conceptual interpretations and the depictions that the historian David Gooding has called "construals," namely, "flexible, quasi-linguistic messengers between the perceptual and the conceptual."[50] For Ampère, the phenomenon demonstrated by the equilibrium apparatus shown in plate 9 was readily construed as a convincing illustration of the addition law. As we will see, he would be disappointed to find that few of his colleagues shared his point of view. Before turning to Ampère's strategy during 1821, we should note that at this point Ampère had no expectations that the equilibrium technique might provide a basis for a full derivation of his force law. Indirect synthesis still seemed to be a pragmatic necessity.

Indirect Synthesis: Ampère's initial determination of the parameters k and n

Together with his efforts to provide a derivation of the general angular function in his phenomenological force law, Ampère also tried to provide experimental evidence that would specify the values of the two parameters k and n. In most of his early memoirs, Ampère wrote the constant k as n/m (often misprinted as m/n) to indicate that it represents the relative strength of the force between two collinear current elements ($\alpha = \beta = 0$) as opposed to two elements that are mutually parallel and perpendicular to the line connecting them ($\alpha = \beta = \pi/2$; $\gamma = 0$). Aside from the argument Babinet included in his publication with Ampère in 1822, the main sources of information about Ampère's views on the subject are notes he prepared for Académie presentations during December 1820 and an article he published in the *Journal de Physique*.[51]

On 4 December Ampère informed the Académie that an observation by Thenard and Gay-Lussac of a "strong attraction between the two conducting wires of the large pile of the École Polytechnique" initially led him to believe that there might be an attraction between collinear current elements, and thus that k should have a finite positive value. Nevertheless, he quickly changed his mind and attributed their observation to "the vacuum produced in the atmospheric air by the electric fluid passing in a torrent from one conductor to another."[52] Ampère also reported that he had considered the possibility that k = 1. If this was true, the angular function in the force law would reduce to

$$\sin \alpha \sin \beta \cos \gamma + \cos \alpha \cos \beta$$

which is simply equal to the cosine of the angle between the directions of the two currents. Ampère noted that this would imply that two conductors perpendicular to each other should experience no mutual electrodynamic forces. At the 11 December meeting of the Académie Ampère claimed that he had tested this implication and he reported that "nothing indicated to me that it [k] had an appreciable value."[53] Ampère apparently used the apparatus illustrated in plate 7 with the two conductors orientated at right angles to each other. In 1822, following his discovery that k has a negative value, he recalled that during his early experiments,[54] "this value appeared to be all the smaller as the experiments I performed to determine it were increasingly exact. Since at that time I did not expect that this value was negative, I concluded only that it could be taken to be zero."

For the orientation Ampère was studying, the force between two current elements reduces to $(k - 1) \cos \alpha \cos \beta$. The force is attractive for all negative values of the product $\cos \alpha \cos \beta$ unless k is larger than unity. Ampère apparently observed only attractive forces and he considered their magnitude to be large enough for him to conclude not only that k was less than unity, but also that it was essentially zero. He overlooked the fact that any negative value of k would also have been compatible with his observations. Insofar as the logic of Ampère's argument is concerned, it is clear that he was still using the hypothetico-deductive reasoning of indirect synthesis; all his observations for the results of interactions between perpendicular conductors were compatible with the assumption that k = 0.

There is evidence that Ampère also based his belief that k = 0 on some observations with magnets. In a manuscript note he made the following remark:[55] "The small attraction between two magnets placed side by side proves that the term in $\cos \alpha \cos \beta$ is zero; it would give a strong attraction between them."

The observations referred to here must be those discussed by Babinet in sections 74–75 of the *Exposé* he co-authored with Ampère.[56] Two bar magnets are imagined to lie side by side in a horizontal plane with their poles oriented in the same direction. When similar poles are directly adjacent to each other, the two magnets repel each other. This repulsion is also observed when one of the magnets is displaced a small distance along the line of its axis. As this displacement becomes larger, the repulsion decreases and eventually becomes an attraction that becomes larger as the magnets approach the position where the pole of one of them is adjacent to the opposite pole of the other magnet. There is thus a "small" attraction when the pole of one magnet is near the center of the other magnet. In the *Exposé* these phenomena were attributed to forces acting between electric currents presumed to flow parallel to the contours of the two magnets. Assuming that collinear current elements exert no forces on each other, the "small" attraction noted is attributed to the fact that the only operative forces are those between the parallel currents flowing vertically within any of the directly adjacent vertical cross sections of the two magnets. If k had a positive value, the corresponding pairs of horizontal collinear elements on the tops and bottoms of the two magnets should also attract and presumably would result in a larger attraction than is observed. Ampère and Babinet did not consider the possibility that a negative value of k would make the forces between horizontal collinear elements repulsive and thus would merely tend to reduce the attractions produced by the vertical currents. The experiment by Thenard and Gay-Lussac may have been responsible for this oversight, since it seemed to indicate that k was positive.

Another of Ampère's observations during this phase is closely related and is worth mentioning as another example of an equilibrium demonstration that Ampère would later recognize as useful for a derivation of his force law. Gay-Lussac and Welter magnetized a steel ring and found that it had no effect on a compass needle until the ring was broken into disjoint pieces. Ampère's initial explanation of the lack of effect by the magnetized ring was that[57] "when two surfaces are covered with electric currents of the same intensity, and directed in opposite directions, their actions mutually cancel each other at a point with respect to which the two surfaces intercept equal portions of the surface of an infinite sphere." Ampère visualized the torus-shaped magnet as two semicircular helical conductors joined together at their ends. From this point of view, the combined helices might be expected to neutralize each other, at least in some circumstances. Ampère recognized that, using the form of the force law in which k is zero, the only component of a current element that makes a contribution to the electrodynamic force at a given point is the one that is normal to the line joining the current element to the point of interest. He thus tried to give his intuition about the ring magnet the status of an illustration of a more general principle, a possible analogy of electrodynamic forces to thermal radiation from a surface. A detailed treatment would have required consideration of the distance dependency of the two surfaces in question. Ampère made only some vague comments on why the surface analysis gave different results for the torus magnet and a single straight helical conductor.[58] A full quantitative analysis of the Welther and Gay-Lussac discovery would not be accomplished until two years later, when Félix Savary made one under Ampère's direction.[59] As we will see, the experiment then took on a considerably elevated role as one of Ampère's equilibrium demonstrations.

The fact that Ampère did not yet have any special expectations about equilibrium demonstrations is also illustrated by his early plans to justify his assumption that his phenomenological force law should include an inverse square distance dependency. According to recollections Ampère recorded early in 1822, he initially planned to compare quantitative measurements with hypothetical calculations based upon the general form of the force law and his conception of the electric circuitry of a magnet.[60] He recalled that this project was on his mind during the time he designed the circuit for the equilibrium demonstration illustrated in plate 9. The lower right-hand corner of that illustration includes an independent apparatus from which Ampère planned to suspend a magnetized needle at a variable height above a horizontal current. By displacing the needle from equilibrium he intended to use it as a "magnetic pendulum" to measure how the electrodynamic force acting on it varied with distance; as in the Biot-Savart experiments, the frequency of the needle's oscillations could be related to the strength of the force.

In retrospect, Ampère gave two reasons for abandoning this project. First, he preferred "to do these experiments directly on a mobile portion of the same voltaic current, and not on a magnet."[61] Purely electrical experimentation would avoid what in this case would be a distracting reliance on Ampère's theory of magnetism. Second, the measurements were not carried out, thanks to the second set of data reported by Biot and Savart. Ampère was perfectly willing to let others gather data, provided that he was allowed to provide the proper interpretation. As he soon learned, this was a task that called for aggressive rhetoric and a clear agenda.

8

Ampère's Electrodynamics (1821–1822)

The Guiding Assumptions of Ampère's Research Program

We have seen that Ampère's initial response to the new electromagnetic and electrodynamic phenomena was an enthusiastic attempt to trace them to an appropriate phenomenological force law. This project was part of a research program which we should understand in general terms before considering how Ampère developed it during 1821 and 1822. First, Ampère advocated a revolutionary realignment of the domains of French physics. Although he did not directly challenge the well established two-fluid theory of electricity, he did call for an outright rejection of the magnetic fluids assumed by the corresponding program in magnetism. The central postulate of Ampère's program was that all electrodynamic, electromagnetic, and magnetic phenomena are due to interactions between electric fluid particles as a result of their motions within electric currents. The explanatory task of electrodynamics should be to give an empirically adequate mathematical formulation of these interactions compatible with Newtonian dynamics. It should be noted that Ampère did not conclusively stipulate whether the electrodynamic interactions between electric fluid particles are analogous to action-at-a-distance electrostatic forces, or whether they have a restricted range so that a particle can only affect those particles directly adjacent to it. Although Ampère worried about this issue a great deal, he ultimately left it unresolved.

Second, Ampère interpreted the restraint imposed by Newtonian mechanics to mean that any mathematical formulation of a force acting between a pair of electric fluid particles must not only conform to the general statement of Newton's third law of motion but must also be a function solely of the distance between the particles and must be directed along the line joining them. This requirement did not rule out the possibility that, due to the specific directions of electric fluid particle oscillations within electric currents, more complicated force expressions might be needed to represent composite interactions between "infinitesimal" segments of an electric current. We have already noted that Ampère's famous phenomenological force law was a complicated function of the mutual orientations of two current elements, elements that actually contain multitudes of oscillating fluid particles and which are only "infinitesimal" in comparison to measurable magnitudes.

A full execution of the explanatory task set by the assumptions of Ampère's program would have been a detailed stipulation of the electric fluid motions that give rise to the phenomena attributed to electric currents. Due to his commitment to an all-pervasive luminiferous ether, Ampère hoped that the conjunction of his physics and his metaphysics might for the first time produce more than superficial speculation. He continued to believe that space itself was a noncorporeal substance made up of ether particles which, in their undisturbed state, individually consist of superpositions of equal quantities of the two oppositely charged electric fluids. Furthermore, he held that the molecules of each chemical element have a characteristic electric charge that is ordinarily neutralized by a surrounding ethereal atmosphere of an appropriately opposite net charge. Early in 1821 he argued that the alleged "flow" of an electric current through a conductor actually results from a periodic creation of a "chain" of nearly simultaneous separations of the two electric fluids within each molecular atmosphere, followed almost immediately by their collective return to the initial equilibrium condition; he thus denied that there is any actual "flow" of either of the electric fluids through the circuit.[1] Nevertheless, although Ampère continued to develop his thoughts about electric current throughout the 1820s and early 1830s, he could not make them sufficiently precise to stipulate a fundamental law for interactions between moving electric fluid particles. A similar fate awaited his speculations that the attractive and repulsive interactions between conductors of electric current might be due to transverse propagation of a state of polarization within the electric fluids that constitute the particles of the surrounding ether. Ampère would only have accomplished the full explanatory task set by the guiding assumptions of his program if he had been able to provide fundamental laws applicable to the conditions responsible for the entire range of electrodynamic phenomena.

Although this goal eluded Ampère, this does not imply that it was detrimental to or, at best, only incidental to the more modest success he did achieve. Psychologically, Ampère was driven by his conviction that the ultimate goal of science is the reduction of all phenomena to the real properties of a fundamental ontology, a bedrock of objective truth. Maintaining his idiosyncratic but systematic misuse of Kantian terminology, he continued to refer to this foundation as the "noumenal world" responsible for the world of phenomena. His conviction that electrodynamics could provide deep new insights into the "noumenal world" was a major incentive for the energy he expended, at the relatively advanced age of 45, in his first extended foray into theoretical physics. While all too many experimenters wandered aimlessly in a labyrinth of empirical details, Ampère's research remained snugly harnessed to theory construction and revision.

Lacking any quantitative ether mechanics for the electric fluids, Ampère shouldered the preliminary task of showing that all electro-dynamic phenomena could be attributed to a phenomenological force law applied either to directly observable electric circuits or their microscopic counterparts within magnets. In this pragmatic context, the microstructure of an electric current was not specified except to note that it consisted of moving electric fluids; the "direction" of a current was simply defined operationally in terms of the experimental apparatus. Successful application of Ampère's strategy would thus preserve the plausibility of the central assumption of his program while at the same time generating a sequence of increasingly accurate phenomenological theories, stated in terms of electric currents rather than particles of electric fluid.

We have already seen how Ampère's program prescribed a heuristic procedure for theory revision. On the basis of an experimental study of electric currents, he proposed a tentative phenomenological force law for pairs of circuit elements; he assumed that this law was ap-plicable to both large-scale circuits and the microscopic circuitries of magnets. Second, in an effort to demonstrate that all magnetic phe-nomena are consequences of the electrodynamic force law, Ampère tried to stipulate the geometry of magnetic circuitries. In this respect, the application of the law required an idealized model of the circuit within any given magnet. With such a model in hand, the cumulative effect of the electrodynamic forces between all relevant pairs of circuit elements could be calculated mathematically and compared to experimental observations. Small discrepancies between calculations and observations were attributed to inadequacies or oversimplifications in the model, a situation that could be rectified most readily by introducing additional complications. For example, as will be described in detail in due course, Ampère argued that experimental observations by Faraday and Gaspard de La Rive could best be explained by assuming that molecular electric currents within bar magnets are tilted with respect to the axis of the magnet at angles that increase with the distance of the molecules from the axis. Because Ampère had originally supposed that all magnetic circuits are perpendicular to the axis, his new stipulation of magnetic circuitries represented an important theory revision conforming to the guiding directives of his program.

As might be expected, some of Ampère's colleagues objected that revisions of this type were illegitimately *ad hoc* and were introduced simply as desperate attempts to preserve a flawed theory from refutation. Ampère was thus forced to address this issue with all the rhetorical skill at his command.

Indirect Synthesis and Ampère's Rivalry with Biot

Preliminary Skirmishes: 1820

By the end of October, 1820, Ampère had used the addition law and the symmetry principle to derive the angular component of the force law; he later claimed to have mentioned this to Arago on 8 November.[2] He also thought about making some quantitative measurements to determine how the force between a suspended magnet and a linear current varies with distance. He dropped this project when Biot and Savart reported their first series of measurements on 30 October. As we have seen, they cited their data as support for their own phenomenological law that the force between a linear current and a magnetic pole is inversely proportional to the distance between them. During November Ampère was too preoccupied with his own work to devote any serious attention to Biot's claims. Tentatively adopting the value zero for the parameter k, and assuming an inverse square distance

268

dependency, Ampère tried to calculate the cumulative forces between various combinations of currents and magnets. These calculations turned out to be far more difficult than expected; initially he had trouble even dealing with the simplest case of two linear currents.

According to Ampère's subsequent recollections, it was at the 4 December meeting of the Académie that he learned of the second set of observations by Biot and Savart, those from which Biot concluded that the force exerted by an "infinitesimal" segment of conducting wire on a particle of magnetic fluid varies with the sine of the angle between the direction of the current and the line joining the fluid particle to the wire segment.[3] At this point Ampère realized that the historical sequence of experimental discoveries was not unfolding in his favor. From Ampère's perspective, the Biot-Savart measurements misdirected attention to the details of Oersted's discovery at the expense of Ampère's discovery of forces between electric currents. Furthermore, of course, Biot had reported the "results" of his first set of measurements in terms of forces exerted on magnetic fluid particles that Ampère was convinced did not exist.

Ampère responded by devoting parts of his 4 and 11 December Académie lectures to a comparison of the two points of view. He reported that he had used his force law to show that the cumulative force between two linear currents, one very long and the other relatively short, should be inversely proportional to the distance between them. Second, using the same initial conditions, he gave an argument analogous to the one Laplace had provided as a supplement to Biot's 30 October report; that is, Ampère showed that, regardless of what the angular component is in his microscopic phenomenological force law for individual current elements, an inverse square assumption for the distance dependency yields a cumulative force that is inversely proportional to distance. From a mathematical point of view, Ampère's and Laplace's derivations had the same form; when each current element is assumed to exert a force proportional to the inverse square of distance, the total effect of a long wire turns out to be inversely proportional to distance, regardless of how the individual circuit element forces might vary with direction.

On the other hand, this mathematical argument was embedded in two thoroughly incompatible physical interpretations. Biot and Ampère disagreed about the entities upon which forces are presumed to act and about the direction in which these forces act. According to Biot, each current element exerts a force on magnetic fluid particles distributed throughout a magnet; these forces act in a direction perpendicular to both the wire and the line connecting the current element and the magnetic particle. Ampère insisted that electrodynamic forces act between current elements of the wire and elements of other electric currents within the magnet; furthermore, these forces are "central" in that they act along the line connecting any two interacting elements.

Each man introduced coy references to his opponent's research to support his own interpretation. For example, Ampère cited Biot's "precise measurements" as verification for the inverse square distance dependency in Ampère's force law, a citation that Biot hardly found flattering. He responded in kind during the early months of 1821. Meanwhile, Ampère used the December meetings of the Académie to expound some nonquantitative conclusions he had drawn from his assumption that the electric currents within cross-sections of a bar magnet take the shape of rectangles corresponding to the magnet's contour. He showed that the orientation taken by the magnet when it is acted on by a conducting wire is the one that is expected when the action of the wire is analyzed in terms of attractions between parallel currents and repulsions between anti-parallel currents. But admitting that Biot's theory was equally amenable to these phenomena, Ampère could only point out that his own theory did not have to postulate magnetic fluid particles and relied solely upon central forces that act along the line connecting the relevant circuit elements.

On 10 December Ampère carried out what was for him an extremely rare set of quantitative measurements. At least we know that on the following day he appeared before the Académie to claim that he had applied Biot's method of oscillations "with the aid of a chronometer that M. Breguet was kind enough to lend me yesterday."[4] Ampère reported that he had once again suspended a small bar magnet near a vertical current. With the magnet displaced from equilibrium by vertically rotating it through a small angle b, Ampère calculated the force between an element of the vertical current and one of the currents assumed to traverse the side of the magnet adjacent to the current. He found that this force should include an additional factor of cos b in contrast to the case when the magnet is horizontal and cos b becomes unity. Although Ampère initially thought the agreement between his calculations and measurements proved the superiority of his own theory, he soon realized that Biot's theory once again made the same predictions. Unfortunately he did not save his data for our perusal; it is easy to imagine him destroying his records in disgust.

Piqued by this reversal, Ampère tried a more ambitious "crucial experiment" during January of 1821. Although once again he would be disappointed, this experiment would have a remarkable subsequent history.

The "Crucial Test" of January, 1821

The Biot-Savart measurements weighed heavily on Ampère's mind throughout December of 1820. Two different arrangements of wires were used to study the oscillations of a suspended magnet free to rotate in a horizontal plane:

1   A long vertical wire was suspended at a variable horizontal distance from the point of suspension of the magnet.
2   A long and symmetric V-shaped wire was suspended in a vertical plane at a variable horizontal distance from the point of suspension of the magnet.

In both cases the suspended wires were in a vertical plane that passed through the center of the magnet. When the magnet is displaced from its equilibrium position, it oscillates with a frequency that depends upon the torque or "moment" exerted on it by forces emanating from the wire. Since rotation takes place only in the horizontal plane, the relevant moment will be determined by the horizontal components of these forces. From Biot's point of view, each current element was expected to exert a force on each particle of magnetic fluid within the magnet. Insofar as observable effects are concerned, the magnetic fluids could be treated as if they were concentrated at the two poles of the magnet with a density responsible for the strength of the magnet.

In their first set of measurements, Biot and Savart found that the total horizontal force exerted by the wire is inversely proportional to the perpendicular distance between the wire and the center of the magnet. Biot attributed this macroscopic phenomenological law to inverse square forces exerted by each current element and acting in a direction perpendicular to both the wire and the line connecting the current element to the magnetic particle. With this direction specified, the forces exerted on magnetic particles by the elements of any wire in a vertical plane should always be in a horizontal direction and should fully contribute to the moment that makes the magnet oscillate when displaced from equilibrium. Thus there should not be any variation in the observed oscillation frequency for various inclinations of the linear wire with respect to the horizontal, as long as, in a vertical plane passing through the center of the magnet, the perpendicular distance between the magnet and the wire is kept constant.

Ampère decided to compare this implication of Biot's theory with the corresponding calculations based upon his own force law. For Ampère the action of the wire on the magnet is due to forces exerted

271

by each current element of the wire on each element of the currents within the magnet. He initially assumed that the magnetic circuits follow concentric circles in planes normal to the axis of the cylindrical magnet. Using the version of the force law in which k = 0, Ampère introduced some simplifying approximations and carried out the required integrations of the force expression over the length of the wire and the volume of the magnet. His result for the moment was an expression proportional to $(1 + \sin^2 b)/c$, where b is the angle of inclination of the wire with respect to the horizontal, and c is the perpendicular distance between wire and magnet. Thus, contrary to the prediction of Biot's theory, Ampère's calculation implied that the relevant moment should vary with b, the angle of inclination; in fact, the moment should double in magnitude when b changes from zero to 90 degrees. Ampère concluded that if he let a vertical and horizontal wire act on the magnet simultaneously, he would have[5] "a simple and easy means to decide by experiment which one it is of these two manners of conceiving the phenomena that is in agreement with the results of observation." More specifically, Ampère designed his test using a small magnet suspended in equilibrium in the magnetic meridian; the electrodynamic forces are produced by currents in a circuit that takes a right-angled bend and satisfies three additional conditions.

1 One branch of the circuit is horizontal and the other is vertical.
2 Either the currents in the two branches both flow toward the apex of the right-angled bend, or they both flow away from it.
3 The magnet is equidistant from each of the two branches, and this distance is large enough for each of the branches to be considered to act as an indefinitely long wire with respect to the magnet.

Under these conditions the two branches of the current should tend to rotate the magnet in opposite directions. According to Biot's theory, the two moments should be of equal magnitude and the magnet should remain stationary. According to Ampère's calculation, the horizontal current should exert a moment with twice the magnitude of that produced by the vertical current, and the magnet should rotate out of its initial state of equilibrium. Furthermore, since Ampère's calculation makes the same prediction when the magnet is replaced by a single circular current suspended in the plane of the magnetic meridian, this circuit too should rotate away from equilibrium, due to the combined action of horizontal and vertical currents.

Ampère made these predictions on 15 January, 1821. He performed

272

the experiment with Despretz five days later on his forty-sixth birthday. Regardless of whatever celebrations may have been planned, the outcome was a resounding confirmation of Biot's theory and a disappointing refutation of Ampère's prediction. Ampère's fanfare stopped abruptly. The experiment was allowed to pass unmentioned for over two years. Furthermore, when the outcome was finally noted in Ampère's *Recueil* and in a memoir by Savary early in 1823, it was cited as a confirmation of Ampère's theory. By that time Savary had shown that the observed phenomena were indeed implied by an application of the new version of Ampère's force law in which the constant k takes on the value of $-\frac{1}{2}$.

This episode thoroughly demonstrates the relative importance Ampère assigned to theoretical generalities and individual empirical observations. Faced with an unfavorable outcome to the experiment he himself had heralded as a decisive test, he conveniently ignored the unfortunate result and continued to develop his fledgling research program. His behavior calls to mind Einstein's similar response to apparent "refutations" of special relativity theory by various versions of the Michelson-Morley experiment. Confident that the theory was essentially correct, Einstein remained calm and patiently awaited the inevitable reports of flaws in the instrumentation or inaccurately specified initial conditions. Ampère, though far less exposed to public scrutiny, seems to have taken a similar attitude.

Second, the January 1821 experiment also indicates that at that point Ampère was not as yet attributing any special significance to equilibrium demonstrations. Suppose for the moment that Ampère was already searching for a set of these demonstrations as a basis for a direct analysis of his force law. Would not the equilibrium condition he and Depretz observed have struck him as precisely the kind of demonstration he needed? In calculating the moment acting on the suspended magnet, Ampère had used the truncated version of his force law in which k = 0. But on 4 December he had shown that the most general angular function allowed by the addition law was

$$\sin \alpha \sin \beta \cos \gamma + k \cos \alpha \cos \beta$$

Now if he had been willing to place his confidence in the outcome of the experiment with Despretz, it would have been natural for him to recalculate the moment using the general force expression and then compare the result with the observed equilibrium to determine the value of k. Indeed, if this calculation is carried out, the moment is proportional to

$$(1 - k) + (1 + 2k) \sin^2 b$$

Thus, in order to conform to the experimental observation that the moment is the same for $b = 0$ and $b = \pi/2$, k must be $-\frac{1}{2}$. More generally, k must be $-\frac{1}{2}$ in order for the moment to be independent of the angle b, as was found to be the case by observations taken at other angles of inclination. Had Ampère been carrying out a direct analysis of equilibrium demonstrations in 1821, he might well have discovered the correct value of the parameter k over a year earlier than he did.

Ampère versus Biot: 1821

During the first half of 1821 ill health prevented Ampère from continuing the calculations and experimental observations that had climaxed in the disappointing experiment of 20 January. He had never enjoyed a particularly robust constitution; the feverish work pace he had set for four months contributed to an onslaught of severe chest pains early in 1821. Writing to Bredin the following August, Ampère complained that[6]

one of my sharpest regrets is to have spent eight months without calculating the consequences of my formula on the action of electric currents so as to compare them to the facts. Since I have been forbidden my pipe because of my chest, I have become almost incapable of working.

Ampère reported that he spent much of this time in his garden, "as if I was already beginning to relapse into childhood." April was a particularly bad month; Ampère related his condition to Bredin.[7]

Ah, if only you knew how, through thinking and writing, I became like an idiot for a while, and then was afflicted with pains in the chest and difficulty in breathing which are forcing me to follow a regime of leeches, ass's milk, etc.... This morning, before rising, I applied 15 leeches to my chest and then had recourse to a séance with the medium who cured my sore throat two years ago; although this was the first session, I fell asleep rather promptly.

What were Ampère's thoughts as he dozed off into some form of hypnotic or mesmerized condition? Perhaps memories of Dr Petetin's reduction of mesmerism to the action of electric fluid crossed his mind and struck a harmonious note in keeping with his present agenda as a physicist. If so, he left no records. He had learned to compartmentalize his professional life and he did not jeopardize his credibility as a physicist by speculating about mesmerism in the wrong circles. His philosophical activity already made him suspect to men like Biot and Poisson. In addition to his physical trials, Ampère was afflicted by an "inexprimable douleur morale,"[8] an inexpressible moral sadness or depression, which by June had him begging Bredin to write and "force me to leave these external thoughts which have entirely dominated me since last summer."[9]

After fulfilling his academic obligations, what little energy Ampère could muster for electrodynamics was devoted to correspondence with Arago, Erman, de La Rive, Van Beek, Maurice, Van der Eyk, and old friends in Lyon and Geneva. Ampère's letters reveal his dismay in the face of the less than enthusiastic reception his program met among conservative physicists. To some extent he attributed this to short-sighted opposition by supporters of the *status quo*. Writing to Roux-Bordier in February, he could even find the situation amusing, up to a point.[10]

> You are quite right in saying that it is unbelievable that for twenty years the action of the voltaic pile on a magnet had not been tested. Nevertheless, I think we can assign a reason for it; it lies in the hypothesis of Coulomb on the nature of magnetic action. This hypothesis has been believed like a fact; it absolutely dismissed any idea of an action between electricity and the so-called magnetized wires. The prejudice against it had reached the point that, when Arago spoke of these new phenomena to the Institut, they were dismissed just like the stones that fell out of the sky had been when Pictet read a memoir on these stones to the Institut. They had all decided that it was impossible. It is the very same prejudice that at present prevents the admission of the identity of the electric and magnetic fluids, and the existence of electric currents within the terrestrial globe and within magnets, just as for years it prevented the admission that chlorine was a simple substance. One resists as much as possible changing the ideas to which one has become accustomed. It is amusing to observe the efforts that certain minds make to try to reconcile the new facts with the gratuitous hypothesis of two magnetic fluids distinct from the electrical fluids merely because they have become accustomed to that idea.

Nevertheless, Ampère remained confident that his own program would prevail as soon as his experimental work became more widely known and properly understood. He knew that the 1820 survey he had written for the *Annales de Chimie et de Physique* was poorly organized; as he admitted in a letter to Maurice, writing was not his best mode of communication.[11] "I can make myself understood only by talking. I imagine, in spite of myself, that when someone reads what I have written, he mentally rephrases what I intend to say just as I rephrase what I read." Ampère's enthusiasm often made his verbal presentations even more in need of "rephrasing" than his publications. Nevertheless, he found it exasperating that physicists had difficulty distinguishing between his theoretical construals or interpretations and the experimental demonstrations he wanted them to repeat for themselves. For example, in a March 1821 letter to Maurice, Ampère explained his terminology for terrestrial currents and vented some frustration.[12]

> However long a letter I might write to you, it would be no more clear than my memoir. No one understands it? I am certain that what they are trying to understand is the theory, with which I do not yet want people to concern themselves; besides, it will be adopted as soon as the facts are well known. It is the facts, *the facts*, which it is a matter of becoming fully acquainted with. Can't the Lyon physicists distinguish between new facts and theories in my memoir, in spite of the lack of order which rules there due to my writing it in response to new experiments?

Ampère was convinced that when physicists became familiar with his demonstrations of forces between electric currents they would recognize the correct "fundamental fact" as the key to understanding more complicated phenomena. This optimistic attitude was responsible for the experimentally oriented memoirs Ampère published in 1821. For example, his papers for the *Annales de Chimie et de Physique* had relatively little theoretical content and straightforwardly described apparatus suitable for replication of all the demonstrations he had performed through March of 1821. That is, he described all successful demonstrations and deftly omitted any mention of the disappointing January experiment with Despretz. Included in this collection was the equilibrium demonstration he had designed the previous December to promote belief in the addition law. Ampère attributed to it no special methodological significance and even remarked that "the same experiment can be done more simply" using the compensated helices he had used for his 6 November 1820 demonstrations.[13] From a theoretical point of view, these two demonstrations were not "the same experiment" at all. The helix experiment merely illustrated an application of the addition law; the more general experiment was intended to be construed as an experimental proof of the law. But this memoir was not the place to call attention to this distinction; Ampère was intent on demonstrating "the facts." In the memoir he sent to the *Bibliothèque universelle* in March,

Ampère did cite his equilibrium demonstration as the one chiefly responsible for "verifying" the addition law, and Delambre passed along Ampère's detailed description for his annual Académie report.[14] By March 1821 this equilibrium demonstration was no longer the simple "counter-test" it had been the previous December. On the other hand, there is no indication that Ampère expected to apply the equilibrium technique elsewhere.

Meanwhile, some relatively minor experimental discoveries were reported to Ampère by Gaspard de La Rive, Paul Erman, Van der Eyk, Van Beek, and Fresnel. In general, these were nonquantitative variations of the experiments Ampère, Biot, and Arago had already performed to illustrate forces between currents and magnets or magnetization of metallic objects using currents. Ampère welcomed these new facts as evidence for his own theory. He no longer insisted, as he had during his early years in Paris, that the prediction of previously unobserved phenomena was a necessary condition for the confirmation or "verification" of a theory. He now considered it sufficient that all the new discoveries were simply compatible with his theory in the sense that it "could have been used to predict them in advance."[15] His unsuccessful prediction during January may have moderated his zeal for prediction as a necessary accomplishment for fledgling research programs.

Throughout 1821 and 1822 Ampère used the term "prediction" in a general sense that often included any agreement between theoretical implications and experimental observations, regardless of how they were related in time. Writing to Paul Erman in April 1821, Ampère declared that his electric theory of magnetism was established

as solidly as physical theory can be . . . since, in only admitting it at first as a hypothesis, it serves to predict and make known in advance all the magnetic phenomena formerly known, those which M. Oersted has discovered, and the new properties whose existence in voltaic conductors I have made known . . . When one finds such an agreement between the facts and the hypothesis from which one started, can one recognize it merely as a simple hypothesis? Is it not, on the contrary, a truth founded on incontestable proofs?

In the same letter Ampère calmly harvested Erman's experimental discoveries as further confirmatory evidence.[16]

The observations described in the memoir which you have been so good as to send me are all the more new proofs of it. For, if I am not mistaken, they could all be predicted according to the theory in which magnets are considered to be assemblages of what I call electric currents.

Unfortunately for Ampère, most members of the elite Laplacian group that was still influential in the Académie des Sciences were not willing to interpret the new phenomena as "proofs" of Ampère's theory. During the early months of 1821 Biot promoted his own program to reduce what Ampère called "electrodynamic" phenomena to a magnetic basis. He emphasized the fact that he could use a single microscopic phenomenological law to account for all of the quantitative data he had collected with Savart. On the other hand, he realized that he was on a par with Ampère in his inability to provide a foundation of truly fundamental laws, laws which in Biot's case would specify interactions between individual particles of magnetic fluid. The Biot-Ampère debate thus entered a phase in which each man attempted to garner support for his own phenomenological law.

We have already noted how determined Biot was to wedge Oersted's discovery into one of the conceptual pigeonholes of Laplacian physics. Pending a full reduction to magnetic interactions, he proposed a phenomenological law for the action of a current element on a particle of magnetic fluid, that is, a force proportional to $r^{-2} \sin V$, where V is the angle the current direction makes with the line drawn between current element and magnetic particle. By April of 1821 he was ready to extend the scope of his theory to include Ampère's discoveries of forces between electric currents.

Before considering how Ampère and Biot clashed at the public meeting of the Académie on 2 April, an important preliminary point should be clarified. Biot's theory was not one of the "transverse magnetism" theories expounded by Wollaston, Berzelius, Althaus, Erman, and Prechtl during 1820 and 1821. Biot merely made schematic use of imaginary tangent magnets to specify the *direction* of the force expressed by his phenomenological law and thus to offer some clues about the fundamental force he hoped to eventually deploy. On the other hand, transverse magnetism theorists believed that they could specify an actual distribution of magnetic poles on the surface of a conducting wire. They considered the phenomena Biot and Ampère attributed to electric currents actually to be the result of forces acting between combinations of induced magnetic poles. These transverse magnetism theories were not well received in France. They were based upon crudely executed and entirely nonquantitative demonstrations of a few effects that were claimed to be of particularly revealing significance. No attempt was made to show that electrodynamic phenomena were calculable consequences of a mathematically expressed force law.

J. J. Prechtl was responsible for the most carefully worked out transverse magnetism theory.[17] Prechtl believed that when a metallic wire connects the poles of a voltaic cell, an alternating sequence of positive and negative (austral and boreal) magnetic poles is produced around the circumference of each cross section of the wire, just as would be produced by tiny bar magnets either emerging from or passing through the axis of the wire like the spokes of a wheel. The magnetic effects produced by these surface poles would thus be equivalent to a closed chain of tiny bar magnets, each of which would be "transverse" to the axis of the wire, that is, tangent to the wire's contour. Prechtl tried to explain all Ampère's "electrodynamic" phenomena by using this claim that a wire becomes a very complicated magnet when it completes a voltaic circuit. For example, the forces between parallel conductors were attributed to interactions between magnetic poles directly adjacent to each other on the two wires.

On 12 March, 1821, the Académie received an unsolicited memoir on transverse magnetism from Prechtl. Biot, Arago, and Ampère were assigned the duty of reporting on it, and although they apparently never did so, this should not be interpreted as a sign of indifference on Ampère's part. There is manuscript evidence that some of his earliest thinking about electrodynamics involved a considered rejection of transverse magnetism. For example, in an undated draft for a letter to an unspecified correspondent, Ampère recalled that his decision was based on the fact that his own electric theory of magnetism provided a ready explanation for why a bar magnet remains magnetized when broken into two pieces.[18] Furthermore, he was convinced that all transverse magnetism theories were conclusively refuted by the discovery Gay-Lussac and Welter had made in 1820. Their magnetized ring produced no magnetic effects until it was broken into pieces; if each cross section of a conducting wire was presumed to be a tiny magnetized ring, how could this wire ever produce the effects that Oersted had discovered?

In sharp contrast to the transverse magnetism theories, Biot originally only pointed out how the *direction* of the force exerted by a linear conductor on a magnetic pole could be stipulated by imagining that the particles of magnetic fluid located at the pole are acted on by each cross section of the wire in the same direction as would be the case if a tiny bar magnet was located on the circumference of each cross section. As a visual aid or a mnemonic device for a speedy recall of the results of Oersted's experiment, Ampère might have tolerated this ploy simply as a misleading but harmless expression of Biot's point of view. However, perhaps following Prechtl's lead, Biot concluded his public lecture for the 2 April *séance* of the Académie by extending his technique to the analysis of the directions of the forces between linear currents. Casting Ampère in the unlikely role of deft experimenter, Biot claimed that his rival's discoveries provided useful "analogies" for what would eventually be a fully magnetic theory of these phenomena. According to Biot, Ampère's discovery of attractions and repulsions between linear conductors was important as an indication of the operative magnetic orientations.[19]

> One should conclude two things: first, wires exert actions on each other perfectly analogous to those which they exert on magnetized needles; second, the distribution of these forces in each of their particles is analogous in its direction to that in magnetized needles themselves.

Of course, for Ampère the importance of his discovery was quite different. Forces between conductors illustrated what Ampère considered to be the "fundamental fact" responsible for all magnetic and electrodynamic phenomena, the force between infinitesimal current elements.

Engaged in his first scientific debate at this level of public scrutiny, Ampère considered Biot's manipulation of the open session of the Académie to be a personal affront. It was only on Friday, 30 March, that he learned of Biot's plan to provide his historical *aperçu* the following Monday. Fearing that a distorted presentation was imminent, Ampère feverishly drafted a short lecture of his own as an alternative *exposé méthodique* of the historical sequence, or as he put it, "a short indication of the facts which seem the most worthy of attention."[20] For Ampère, these facts included his demonstration of the directing action of terrestrial magnetism on current loops and, most prominently, his replication of the action of bar magnets using helices, a demonstration that provided "direct and multiple proofs of the identity of electricity and magnetism."[21] Ironically, Biot did not object to this statement; he simply disagreed with Ampère about which of the two domains was the causal basis of the other. Grumbling continued at subsequent meetings of the Société Philomathique, with Biot complaining that Ampère had been a bit "indelicate" and Ampère responding that he had spoken with good reason.[22]

Ampère's concern was well-founded. Although Biot never successfully derived his phenomenological law from Laplacian first principles, in April 1821 there was still reason to think that he might. Biot himself was too busy trying to defend his particle theory of optics from Arago's promotion of Fresnel's wave theory to make any progress during the remainder of 1821. When he brought out the second edition of his *Précis* in June, 1821, the section on electrodynamics was simply an edited reprint of his April 1821 lecture to the Académie. In January, 1823, still smarting from Arago's polemics and soundly defeated by Fourier in the 1822 election for the Académie *secrétaire*, Biot withdrew to his country estate to lick his wounds and write the third edition of the *Précis*. This 1824 edition included the most detailed exposition of his views on "the magnetization impressed on metals by electricity in motion."[23] Nevertheless, Biot never gave a clear statement of how to attribute the magnetization of a material molecule to the action of an electric current without violating the Laplacian taboo on interactions between electric and magnetic fluids. His 1824 manifesto accentuated this problem more than it resolved it. Resolutely attributing all of what Ampère called "electrodynamic" phenomena to the action of a "truly molecular magnetism impressed on the particles of the metallic bodies by the voltaic currents which traverse them,"[24] Biot claimed that the central issue to be resolved was "how each infinitely small molecule of the connecting wire contributes to the total action of the section to which it belongs" and thereby brings about the "molecular magnetism" of the wire.[25] Furthermore, as he remarked in his April 1821 lecture, "this effort in some sort of *divination* is the goal of almost all physical research."[26]

From Biot's point of view, explanation of all the new phenomena ultimately had to be traced to a specific distribution of magnetic fluid particles. In this respect, the phenomenological laws he had discovered with Savart were only useful tools for summarizing some of the measurable data. They allegedly owed their accuracy to the fact that, in some unknown manner, they expressed the composite action of the fundamental forces operative within the magnetic fluid distribution. Biot's initial use of the term "divination" was intended to apply to the hypothetical assumption of a fundamental force and the distribution to which it should be applied. However, since neither Ampère nor Biot could derive their phenomenological laws from fundamental forces, they saw no alternative but to pose these phenomenological laws as equally hypothetical "divinations". Due to the "infinitesimal" current elements cited in their microscopic phenomenological laws, there was no reason to expect that they could be "directly" derived from experiments performed with circuits of finite dimensions. Ampère was characteristically more candid than Biot in admitting that the hypothetical method of indirect synthesis guided his early attempts to stipulate his force law. In an 1822 memoir he wrote that his original procedure was "perfectly described" by Biot's term "divination."[27] On the other hand, Biot rather deviously claimed that he had arrived at the directional component of his law through "calculation" and that Laplace had "deduced" the inverse square distance dependency from the experimental data.

By mid-1821 both men were ill at ease with the situation. Ampère's only consolation during months of illness was to point out that newly discovered phenomena "could have been predicted in advance" by perspicacious applications of his theory; his satisfaction was tempered by regret that, in point of fact, he had not been healthy and quick-witted enough to do so. But he was also pondering the nature of the electric currents he believed to be responsible for magnetism. This issue suggested an experiment that brought Ampère tantalizingly close to the discovery of electromagnetic induction.

Molecular Magnetism and Ampère's First Induction Experiment: July, 1821

It was not until August, 1831, that Michael Faraday discovered how to produce a temporary electric current in a coiled wire by either starting or stopping another current in a nearby coil. Characteristically patient and thorough, Faraday specified the circumstances responsible for the induced current and was properly credited with the discovery of electromagnetic induction. Twice during the preceding ten years Ampère came very close to making this discovery himself. The circumstances that distracted him from making a full and accurate investigation illustrate the profound contrast between his style of physics and that of Faraday, who was largely unconcerned about mathematical formalism and reductionistic explanations in terms of the microscopic structure of materials. He delighted in the observation of new phenomena and he resolutely investigated the conditions that produce them. Ampère was a mathematical physicist with a vested interest in explanatory goals specified by an elaborate research program. Unlike Faraday, he was not intrigued by novelty for its own sake; empirical information primarily interested him when he recognized that it might either provide evidence for the correct form of his force law or help determine the electric circuitries of magnets. It was the second of these two topics that inspired Ampère's investigations of induction in July, 1821, and again in September, 1822.

Ampère himself did not use the term "induction" until Faraday began doing so following his 1831 discovery. Before that time Ampère typically discussed the issue in terms of whether electric current could be "excited" or "produced" by the "influence" of another current. This issue already arose during September and October, 1820, the first few weeks of Ampère's research in electrodynamics. In one of his earliest publications on the subject he claimed that whenever the two electric fluids are decomposed or polarized at a point on a metallic surface, the two fluids move in opposite directions along the surface and reunite on the other side of the object. They thereby complete a circuit of minimum length which affects the adjoining parts of the metallic object:[28] "This current determines in the molecules of the steel the disposition analogous to that of elements of a voltaic pile which tends to continue it, and pull along (*entraîne*) currents, and consequently, similar dispositions, in the remainder of the body." Ampère thus at least tentatively believed that one current within a metallic body could "pull along" (*entraîne*) another current in an adjacent region. Similarly, Ampère and Arago applied this idea in their description of the magnetization of a needle or wire when it is placed inside a helical circuit wrapped around an insulating tube of glass:[29] "The current of each spire pulls along (*entraîne*) one similar and directed in the same direction on the surface of the steel, and consequently others in its interior." In both of these early passages, the claim is that one steady current can *entraîne* another steady current; no effects are attributed to sudden changes in current magnitude. In November, 1820, Ampère and Fresnel attempted to reverse the Arago magnetization procedure to produce a current in a wire wrapped around a bar magnet. After rectifying an erroneous interpretation of preliminary observations, they agreed that they could not produce currents using stationary magnets; they did not investigate the possible results of putting the magnet in motion within the surrounding coil of wire.[30]

Until the very end of 1820 Ampère assumed that currents within bar magnets were centered on the axis of the magnet and contained within planes normal to the axis. By January, 1821, Fresnel had convinced him that molecular currents in these planes would be a more reasonable hypothesis. The existence of large-scale currents implied effects that were not observed experimentally, such as the production of heat. The molecular hypothesis in turn allowed for two alternative amplifications. Magnetization might be brought about either by the "production" of molecular currents or by achieving a uniform orientation of previously existing randomly oriented molecular currents. During 1820 Ampère apparently preferred the "production" option for macroscopic currents; the new molecular hypothesis called for a re-examination of the entire question.

Fresnel was probably the single most influential factor motivating Ampère's next attempt to detect induced current. Fresnel continued to espouse the molecular hypothesis, and he drafted two short memoirs that offered an impressive list of favorable experimental evidence. Although never published by Fresnel, they were preserved in Ampère's papers and must have been the result of their conversations. One of them is dated 5 June, 1821, so they probably represent a summary of arguments developed over the preceding few months. In particular, Fresnel claimed that the molecular and macroscopic hypotheses had different implications, due to the fact that a steel cylinder can be magnetized by surrounding it with a helical current. Assuming that the magnetization comes about because current in the spires of the helix produces either molecular or coaxial currents on the surface of the cylinder and parallel to the spires, it might be expected that a straight current should also produce either molecular currents or a macroscopic current in an adjacent linear conductor. Furthermore, if molecular currents are produced, Fresnel claimed that these currents would be located "in the plane passing through the two wires," and therefore "there would be no longitudinal current."[31] Fresnel thus suggested that a failure to produce electrodynamic effects in a conductor after exposing it to a parallel current would be evidence in support of the molecular hypothesis.

This was the experiment Ampère performed for the first time in July, 1821; to amplify the action of his primary current, he wrapped it into a vertical coil and then suspended a circular conductor directly inside it. Several familiar pieces of equipment contributed to Ampère's experimental design. During his initial demonstrations of the influence of terrestrial magnetism on current loops he had often suspended a loop using thread or wire. He found that the loop did not quite rotate fully so as to be normal to the earth's magnetic meridian. He attributed this to a resisting torsion in his suspension, and he rectified it by switching to a mercury-lubricated pivot suspension.[32] However, in both the 1821 and 1822 induction experiments Ampère returned to wire or thread suspensions; his assessment of the torsion factor was an important part of his interpretation of the 1822 observations. Second, an interesting experiment by Paul Erman caught Ampère's eye when it was reported in the *Bibliothèque Universelle* early in 1821. In a published open letter to Erman dated 20 April, 1821, Ampère pointed out that from the electric theory of magnetism, it is obvious that[33] the two branches of a horseshoe magnet should jointly attract or repel a given vertical voltaic conductor . . . It is only necessary to recall that attraction takes place when the currents of the magnet, in the part nearest the conductor, are in the same direction as that of the connecting wire, and repulsion when they are in contrary directions.

Erman had used a bar magnet detector in his own experiment; Ampère would use bar magnets in 1821 and a more powerful horseshoe magnet in 1822. Similarly, in March, 1821, Gaspard de La Rive had used a bar magnet to demonstrate attractions and repulsions of a vertical current loop.[34] Consequently, by early in 1821 Ampère was thoroughly familiar with apparatus in which steady currents in vertical current loops were rotated using magnets. Fresnel's suggestive arguments about molecular currents in magnets then served as a catalyst for Ampère's invention of his July 1821 induction apparatus.

As shown in plate 10, Ampère suspended a circular copper loop, HIG, just within the circumference of a stationary circular coil of copper wire, DEB. A small gap separates the suspended loop from the stationary coil; when a current is sent through the coil DEB, it becomes the primary circuit for Ampère's attempt to induce current within the suspended loop. Writing to Albert van Beek early in 1822, Ampère recalled that this arrangement "seemed to me the most convenient to excite electric currents by influence in this circle, if this was possible; however, upon presenting to it the action of a strong magnet, I did not perceive that it took any movement, in spite of the great mobility of this type of suspension."[35] Unfortunately, Ampère left no clues about how he manipulated his primary current and what variables he considered relevant. No doubt he expected to produce steady current, if any, and he expected this to be revealed by a rotation of the copper loop due to a repulsion or attraction by the bar magnets he had placed in fixed positions on the platforms ng and kp.

Ampère's letter to van Beek included a summary of Ampère's initial interpretations of what he had observed. After bemoaning the poor health that had prevented him from establishing the molecular hypothesis by conclusive experiments, Ampère gave the following assessment.[36]

Nevertheless, I performed one, in the month of July, 1821, which completely fixed my opinion in this respect, although it proved only in an indirect manner that the electric currents of the magnet take place about each molecule. What this experiment proves directly is that the proximity of an electric current does not excite any current at all, by influence,

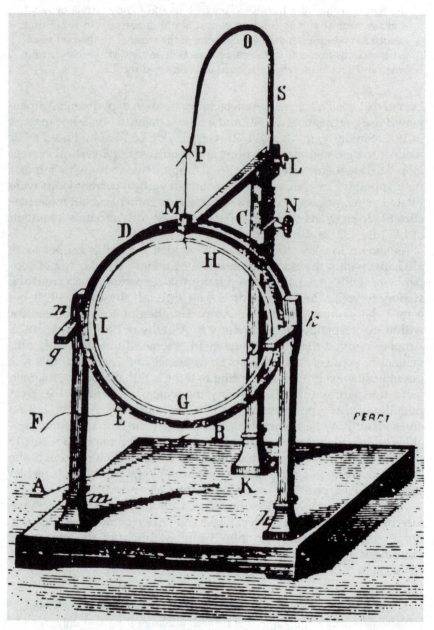

10 Ampère's induction apparatus, from his reply to a letter by Van Beek, printed in the *Journal de Physique*, volume 93, 1822.

in a metallic circuit of copper, even under circumstances most favorable to this influence . . .

It was from this experiment that I concluded, at the time when I performed it, that the electric currents, the existence of which I had already admitted around each particle of magnets, existed just as well before the magnetization in iron, nickel, and cobalt; however, due to the fact that they were oriented in all sorts of directions, no external action could result from them, some tending to attract that which the others repel.

Ampère's condensed account calls for some interpretation. First, he concluded that electric currents cannot produce other currents *of any kind whatsoever* in any other conductor, even under optimal conditions. That is, Ampère felt that he had shown that it is impossible to produce either macroscopic or molecular currents through the influence of an existing current. No macroscopic currents were produced, since the loop failed to respond to the magnets as it should have if a current was flowing. On the other hand, according to Ampère's theory of magnetism, the production of molecular currents should temporarily have magnetized the copper loop; if this had happened, once again the loop should have responded to the bar magnet detectors by an appropriately attractive or repulsive rotation. Ampère then claimed that from this "direct" implication of the experiment, it followed that when electric currents *do* magnetize metallic bodies, as in the method he and Arago had designed using helices, they must do so by giving a uniform orientation to pre-existing currents rather than by inducing these currents. To reach this conclusion, Ampère assumed that any randomly oriented molecular currents in his copper loop were too rigidly positioned to be given a uniform orientation by his coil.[37] Failure to magnetize the loop thus implied a failure to induce currents. Ampère realized that this was a dubious inference, since it presumed that he had employed experimental conditions optimal for both magnetization and detection. He also realized that subsequent use of a stronger magnet might detect induced current too weak to be revealed by his original apparatus.

Second, Ampère claimed that "in an indirect manner" his experiment proved that the pre-existing currents must be of molecular dimensions. He reasoned that, since the experiment ruled out the possibility of induced macroscopic currents as the cause of magnetism, it also ruled out magnetization through the uniform orientation of these currents. Ampère seems to have relied upon his prior knowledge that no effects of pre-existing macroscopic currents had ever been detected in materials prior to magnetization. Nevertheless, it is possible that Ampère also felt that if a random distribution of macroscopic currents had initially been present within his suspended loop, the surrounding coil should have given them a uniform orientation so that their cumulative effect should have been detected by his bar magnets. This would particularly have been the case if Ampère's "loop" was a cylindrical sheet of copper, as was implied by Auguste de La Rive when he described the later 1822 apparatus as using "a sheet of copper" (*une lame de cuivre*).[38]

In short, the immediate upshot of Ampère's unsuccessful 1821 attempt to detect induced electric currents was a renewed commitment to the molecular theory of magnetism he had tentatively adopted in January of that year. From Ampère's perspective, the experiment was useful only insofar as it had a "negative" result, that is, if induced currents are not detected, this is evidence in support of his molecular hypothesis; magnetism must be produced by the orientation of pre-existing currents which surely are not macroscopic. On the other hand, detection of an induced current would not help to resolve this issue; the induced current might be either macroscopic or molecular.

Although Ampère realized that a more powerful magnet might detect induced current at some future date, his trust in the molecular hypothesis was bolstered by the explanations it provided for a wide variety of experimental discoveries announced in France by October, 1821. For example, in an enthusiastic article dated 7 October, 1821, Albert van Beek claimed that he had produced decisive evidence in favor of the molecular hypothesis.[39] While using electric discharges to alter the polarity of magnetized steel plates, van Beek discovered that this change was localized to within a limited distance from the discharge. Second, he found that a magnetized cylindrical solid manifested no magnetic effects until it was split into two pieces longitudinally. Both of these discoveries were to be expected according to the molecular hypothesis, but they were puzzling for the alternative account. This evidence gave Ampère confirmation of the molecular hypothesis independent from the outcome of his induction experiment. He thus needed to rely upon that experiment only to discriminate in favor of the uniform alignment version of the molecular theory; he did not feel compelled to go out of his way to repeat it.

Furthermore, by the middle of October, 1821, Ampère was thoroughly engaged in an assessment of a collection of "rotary effect" discoveries reported by Faraday. This inaugurated a period of research for Ampère of equal if not greater intensity than the one with which he had responded to Oersted's discovery of electromagnetism. Before considering that episode in detail, we should emphasize that Ampère's reaction was directly linked with his worries about the molecular hypothesis. In addition to his experimental discoveries, Faraday included a criticism of what he understood to be Ampère's latest theory of magnetism.[40] Faraday pointed out that the poles of a bar magnet are slightly distant from the ends of the magnet and are not precisely at the ends, as they are when a helical conductor is used to duplicate the alleged circuitry of the magnet. Ampère's response was to reassert his molecular hypothesis and point out how its implications differed from those of the macroscopic theory Faraday had attacked.

Ampère's new explanation of the slightly displaced poles of a bar magnet was perfectly in keeping with the heuristic guideline for theory revision that was central to his research program. Prior to his adoption of the molecular hypothesis, Ampère had followed Fresnel's lead and attributed the action of a bar magnet to the relatively large intensity of the currents located near its center.[41] But this claim about a variation in current intensity within the magnet was an *ad hoc* response to experimental observation; it was not motivated by any theoretical expectations. By the time Félix Savary wrote up some notes as an appendix to the French translation of Faraday's memoir for the *Annales de Chimie et de Physique*, Ampère had thought of another explanation that did not require this assumption. The same effect might be produced simply as a result of the mutual attraction of the current loops within the magnet, an attraction which would tend to make them draw together toward its center.[42] This explanation at least had a theoretical basis; nevertheless, it would seem to require a corresponding variation in the density of the material molecules bearing the magnetic currents. Ampère soon realized that a more reasonable hypothesis was to keep the density uniform with fixed molecular positions and notice that, owing to the collective interactions of the current loops, the planes of all the molecular currents would not remain normal to the magnetic axis. Instead, they should become increasingly tilted with respect to the normal, depending upon how far the molecular current is from the center of the magnet.

The option to vary current loop orientations as a function of distance from the magnetic axis was not available for the old macroscopic model in which all currents were centered on the axis. Furthermore, Ampère had met Faraday's criticism through a nonquantitative but direct application of the electrodynamic force law; he did not have to introduce additional hypotheses, such as Fresnel's variation in current intensities. Ampère thus claimed that he could have made this revision of the molecular hypothesis even without the incentive provided by Faraday's critique. He applied his usual methodological rhetoric to claim that his program "could have predicted" Faraday's discoveries had not experimental detection outrun his calculations.[43] Ampère initially hoped that analysis of Faraday's rotary effects would provide conclusive confirmation of the molecular hypothesis. He was disappointed in this effort, but his attempt generated additional interesting and important theoretical revisions. Before considering these developments in detail, another aspect of the 1821 induction experiment should be noted.

The illusive but ultimate goal of Ampère's research program was an ether mechanics that would explain the accuracy of his phenomenological force law through a derivation from fundamental laws for moving electric fluid particles. During 1821 Ampère incorporated the results of the 1821 induction experiment into an early and very sketchy contribution to this project. Some of his thoughts were recorded in the notes Félix Savary appended to the French translation of Faraday's memoir. After describing the negative result of the induction experiment, Savary included the following remarks.[44]

> However, it is not the case that M. Ampère does not grant that there may be compositions and decompositions of electricity produced in a conducting body by the influence of those of a neighboring conductor in communication with the two extremities of the pile; however, since they would then be precisely the same as in an equivalent space where there would not be any ponderable body, there cannot result from this any effects analogous to those of an electric current due to the electromotive action of a voltaic element or of a particle of a magnet. All attractions or repulsions produced between two bodies by the electric currents which flow through them evidently require that the currents of each of these bodies be produced by a cause which resides within them.

Ampère approved of these comments because he was considering the possibility that his force law might be derivable from the motions of ether particles. He speculated that the motions of electric fluid within currents produce motions in the surrounding ether that propagate isotropically. From this point of view an electric current in a conductor would experience electrodynamic forces only when the motion of the ether differs from one side of the conductor to the other. This would require the existence of two independent currents or the interaction of at least two parts of one circuit. The fact that this condition was not satisfied in the induction experiment might explain its negative outcome. This issue played a major role when Ampère repeated his experiment with a more powerful magnet in September, 1822. In the meantime, Ampère's attention shifted to Faraday's discovery of rotary effects.

Direct Analysis and Ampère's Adoption of the Equilibrium Demonstration Technique

Rotary Phenomena and Molecular Magnetism

Faraday completed his memoir on the new rotary effects on 11 September, 1821, and he sent the news to Ampère on 18 October. Although the chest pains that had plagued Ampère all year had subsided somewhat, within a month he was once again pushing his health to its limits. He wrote to Bredin at the end of November.[45]

> I am writing you from the meeting of the Institut while waiting for it to begin. Never have I had such exorbitant work, truly beyond my strength. I've come down with severe stomach aches by writing day and night . . . I fell asleep while writing to you as someone was reading a memoir – impossible to keep awake.

Ampère's exertions were provoked by Faraday's report of new effects encountered during his patient investigation of Oersted's discovery. Faraday had become intrigued by how linear conductors tended to follow circular paths around the poles of a bar magnet, insofar as they were permitted to do so by the constraints of his apparatus. Eventually he produced an uninterrupted rotation, either of a magnetic pole around a wire or of a wire around a magnetic pole. In the first case, the current entered a basin of mercury through a fixed vertical wire and the free upper end of a vertical bar magnet rotated around the end of the wire. For the converse effect, the current entered the mercury through a suspended wire, and the free end of the wire rotated around the upper pole of a fixed bar magnet.

French physicists immediately recognized the importance of these phenomena. Faraday's memoir was translated for the *Annales de Chimie et de Physique* by Riffault, and Savary added some "explanatory" notes approved by Ampère. For the most part, Savary's notes interpreted the new effects in terms of Ampère's research program; for example, Savary's discussion of the rotating wire began with the rather glib comment that "these circular motions are easily explained according to the theory of M. Ampère."[46] To achieve this "easy" explanation, Ampère assumed that the operative forces were those acting between the current elements of the vertical wire and the elements of molecular currents flowing in tiny circular loops in the horizontal plane of the upper extremity of the magnet. The next step was to apply the force law to this set of interacting currents. But rather than attempt to carry out an integration over a molecular circuit, Ampère used a symmetry

291

argument to show that the operative forces should combine to produce a horizontal force propelling the wire through a circular orbit in the observed direction. Even though this analysis was based upon the truncated version of the force law in which k = 0, the explanation of the *direction* of the rotation was successful, due to the special geometry of the situation. Although the explanation was stated in terms of molecular currents, an exactly analogous explanation could have been provided using large-scale magnetic circuits centered on the axis of the magnet. These particular rotary phenomena thus could not decide the issue between Ampère's two alternative theories of magnetism.

Ampère had already begun a search for related effects when he reported to the Académie on Faraday's discoveries on 19 November. By 3 December he had invented an improved apparatus to repeat Faraday's demonstrations, and on 10 December he used this apparatus to demonstrate the rotation of a wire, using as a source either terrestrial magnetism or a circular circuit. Similarly, on 7 January, 1822, he used an electric current to make a bar magnet spin about its longitudinal axis, a phenomenon that Faraday had not been able to produce. Upon the shifting sands of these discoveries, Ampère now tried to erect a beachhead in support of the molecular hypothesis. He initially thought that he had the demonstration he needed when he found that an electric current could make a bar magnet spin about its own axis. The effect was produced by sending a current into the end of a vertically suspended magnet, using a vertical wire directly above the magnetic axis; the lower end of the magnet was placed below the surface of a mercury bath where the circuit was completed. On a preliminary analysis, the two hypotheses gave different results; the molecular hypothesis predicted rotation while the macroscopic alternative did not.

However, it was just at this juncture that Savary, Davy, and Faraday all called attention to effects produced by currents within liquid conductors. Ampère now realized that both the rotation of a magnet around a conductor and his spinning magnet effect are primarily due to currents within the liquid conductor of his apparatus; the forces exerted by the rigid conductor that brings the current into the liquid make a relatively minor contribution. Furthermore, the new explanation of the spinning magnet in terms of currents within the mercury turned out to be equally amenable to either the molecular hypothesis or its alternative, and Ampère was stymied once again in his search for a decisive experiment.

Now, in all these explanations, Ampère had only accounted for the *direction* of the observed motion; to do so he devised directional rules for circuit element interactions which were valid consequences of his force formula regardless of what value was given to the parameter k, as long as this value was not greater than zero. As a result, since Ampère attempted no calculation of force *magnitude*, his belief that k = 0 was not placed in jeopardy by these tests. Ampère, of course, felt that he had experimentally established the correct value of k as early as December, 1820. His investigation of rotary effects was not pursued with that goal in mind. On the other hand, he naturally hoped that he might be able to use rotary effects as decisive evidence against Biot and Prechtl. With this thought in mind, Ampère invented an apparatus which brought him to a methodological turning point, the recognition that equilibrium demonstrations might provide a derivation of his force law by direct analysis.

Ampère's "Third" Equilibrium Apparatus

On 27 March, 1822, Ampère concluded an open letter to Albert van Beek by mentioning that he had just performed some rotation experiments with a new and more efficient apparatus.[47] As shown in plate 11, Ampère's original equipment used as the mobile conductor a horseshoe-shaped copper strip soldered to a copper hoop at its base; the two vertical copper strips, OL and OM, played the same role as Faraday's single vertical wire. The efficiency of the apparatus was limited by the fact that the current in the mobile conductor was a relatively weak one produced by the voltaic action between its copper base and the zinc wall of the surrounding basin. Although Ampère reported a fairly brisk rotation when a cluster of vertically oriented bar magnets was anchored beneath the center of the apparatus, the movement was much slower when the magnets were replaced by a horizontal circular circuit or the magnetic action of the earth.[48]

According to his letter to van Beek, Ampère put his new apparatus into operation on 27 March; it is shown in plate 12. The major improvement was that now a single circuit included both the mobile conductor and the circular coil that caused the rotation. The current in this circuit was produced by an independent voltaic cell and could be made strong enough to readily demonstrate the rotation of the mobile conductor. The technical problem that complicated and delayed this innovation arose from Ampère's discovery that, in order to produce a rotary effect, the mobile part of a circuit could not be allowed to connect two points on the axis of rotation. This restraint ruled out the employment of a mobile conductor formed using a closed circuit with the axis of rotation as a diameter.

Prior to March, 1822, Ampère had made extensive use of circuit components that did include the rotation axis. A brief summary of

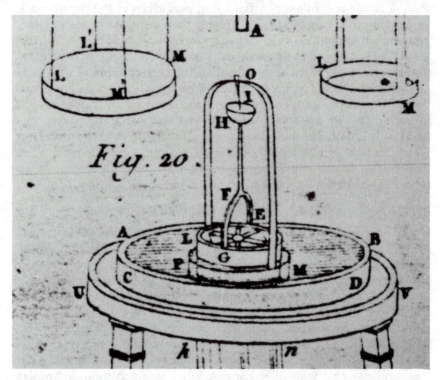

11 Ampère's initial rotation apparatus, from his "Note sur un appareil à l'aide duquel on peut vérifier toutes les propriétés des conducteurs de l'électricité voltaïque," *Annales de Chimie et de Physique*, volume 18, 1821.

these investigations will help us understand the innovations that came shortly thereafter. Ampère had of course used circular mobile conductors in his 1820 demonstrations of the electrodynamic effects produced by terrestrial magnetism. We have also noted his plans to determine the distance dependency of his force law by observing the oscillations of a magnetic needle suspended at a variable height above a horizontal current. Although he abandoned this particular project following the second Biot-Savart report, Ampère realized that further quantitative measurements were still in order, preferably using direct interactions between electric currents without the intervention of magnets. He encountered severe mathematical difficulties when he attempted the integrations required to predict measurable effects; he expected to be more successful with relatively easy cases of currents confined to circular or semicircular circuits. To make the required measurements,

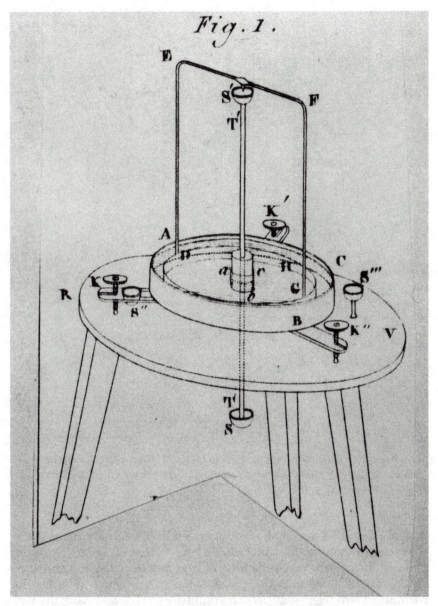

12 Ampère's improved rotation apparatus (March, 1822), published in
    Annales de Chimie et de Physique, volume 20.

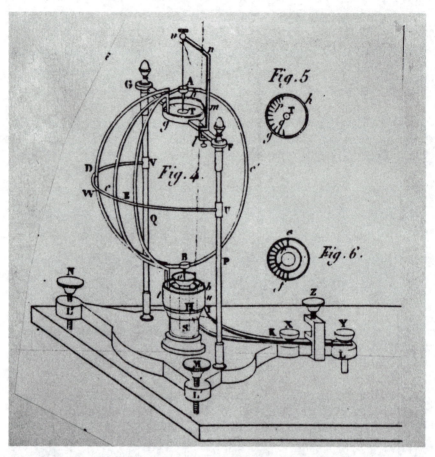

13 Ampère's oscillatory apparatus (1822–1823), from plate 10, figure 4 of his *Recueil d'observations électro-dynamiques*, first published in 1823 though dated 1822.

Ampère designed the apparatus illustrated in plate 13. Contemplating a procedure similar to that followed by Biot and Savart, Ampère planned to count the oscillations of the mobile circular conductor ACBC' about its equilibrium position when it is acted upon by the currents in two fixed semicircular conductors, ADB and AEB; the two fixed conductors could be set at any desired azimuth angle with respect to the central axis of suspension, AB. He also planned to use a modification of the same instrument to measure the action of a magnet on the circular component by counting the oscillations of the mobile loop when it is displaced from the plane in which it comes to equilibrium with respect to the magnet.

Ampère probably designed this instrument early in 1822; he may never have entirely assembled it although he did claim to "present" it to the Académie in June of 1823.[49] The mobile circular conductor resembled the one Ampère used in his rotation apparatus in that the current descends in both halves of the loop and thus makes it immune to terrestrial magnetism. Nevertheless, it was not suitable for the production of a rotary effect because each half of the loop connects two points on the axis of rotation. Ampère had used analogously constructed rectangular circuit components ever since his October, 1820, discovery of the action of terrestrial magnetism on closed circuits. The oppositely directed moments produced by the earth's magnetic action on the two halves of the rectangle left it in equilibrium with respect to terrestrial magnetism. Plate 14 shows a typical example and is in fact the mobile conductor Ampère eventually used in what he called his "first", "second", and "third" equilibrium demonstrations. During some of his initial investigations of the rotary phenomena reported by Faraday, Ampère learned that conductors of this type could not be put into continuous rotation. He finally solved the problem by keeping the lower end of the mobile conductor off the rotation axis and completing the circuit by having the current flow into a mercury bath. This was the insight required to invent the new and improved rotation apparatus of March, 1822.

With this instrument at his disposal, Ampère could be confident that when a potentially mobile conductor remained stationary, this situation was due to the equilibrium of the forces acting on it and not simply to the fact that the current in the mobile conductor was too weak to overcome the frictional resistance of the apparatus. But to understand why Ampère decided to use part of this apparatus to make the observation that became his "third" equilibrium demonstration, we need to reconsider the arguments he was directing at his rivals, Biot and Prechtl.

Speaking at the public session of the Académie on 8 April 1822, Ampère insisted that Faraday's discovery of rotary effects[50]

> proves that the action which emanates from voltaic conductors cannot be due to a specific distribution of certain fluids at rest within these conductors, as is the case for ordinary electric attractions and repulsions. This action can only be attributed to fluids in motion within the conductors.

Within the constraints of a short lecture Ampère did not pause to distinguish transverse magnetism theories from Biot's "molecular

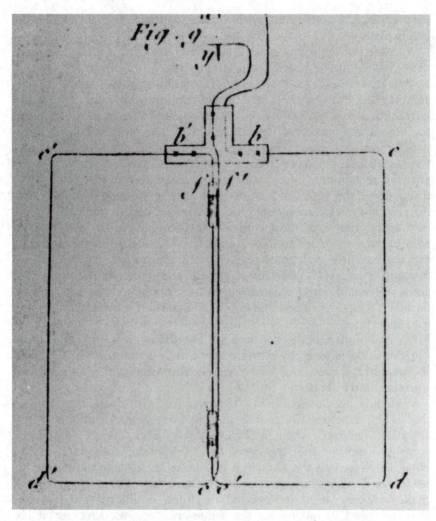

14 Ampère's mobile rectangular conductor, from his "Note sur un appareil à l'aide duquel on peut vérifier toutes les propriétés des conducteurs de l'éléctricité voltaïque," *Annales de Chimie et de Physique*, volume 18, 1821.

magnetism." His point was that the continual production of *vis viva* required to overcome the frictional resistance of the rotary apparatus could not arise from a static distribution of force centers, molecular or otherwise, without violating Newton's third law of motion. Neverthe-less, transverse magnetism theorists claimed that a conducting wire owed its electrodynamic effects to a specific distribution of magnetic poles on its surface. Ampère commented as follows:[51]

> I had examined this hypothesis before opting for the one which I have adopted, and I rejected it more in response to the general order of the facts than by basing myself upon direct proofs. These proofs today result from the new phenomena I have just reviewed because they are peculiar to the mobile portions of voltaic conductors which do not form nearly closed circuits, and which thus cannot be duplicated with magnets. On the other hand, as I showed long ago, all the phenomena which magnets present can be duplicated by bending wires so as to form nearly closed circuits.

Ampère's reference to conductors that "do not form nearly closed circuits" was based on his knowledge that no one had been able to produce a continuous rotation of a rigid conductor that comprised a complete circuit. A rotation could only be produced in a "portion" of the circuit, and this component obviously was "not closed" in the sense that it made up only part of the full circuit.

Now according to transverse magnetism theories, every portion of a conducting wire owes what Ampère called its "electrodynamic" properties to a static distribution of magnetic poles on its surface. Consequently, Ampère challenged them to replace the rotating mobile conductors in the rotary demonstrations with some combination of magnets. In the same spirit, though with less justification, he also claimed that they should be able to produce a rotation of a magnet using the same sources he had used to rotate a wire segment: magnets, coiled conducting wires, and terrestrial magnetism.

Although Ampère's effort to shift the burden of proof to his opponents was clever strategy, the situation was far from clear-cut. Taking nothing for granted, he tried to use magnets to produce a continuous rotation of a wire with both of its extremities on the axis of rotation. He needed to be aware of all the effects that magnets could produce because, according to his own theory, he would have to replicate all such phenomena using electric circuits in place of the magnets. As shown in plate 13, the apparatus Ampère had designed to study the distance parameter in his force law included a circular mobile conductor such that both halves of the loop connected points on the axis of rotation. During November and December, 1821, Ampère had assured himself that no combination of magnets could produce rotation of a mobile conductor of this type. As shown in plates 11 and 12, his own rotary apparatus relied upon a U-shaped mobile conductor with its branches parallel to the axis of rotation. By 27 March, 1822, he had perfected this demonstration by including within a single circuit both the mobile conductor and the coil of wire causing it to rotate.

But the invention of the 27 March apparatus shown in plate 12 made it possible for Ampère to test another inference he had noted at some point during the preceding few months. Fortunately, his short two-page note has survived as a record of his thinking. Using his symmetry principle, Ampère argued for the third of what he labelled "3 things to be noted," namely, that the force exerted on a current element by a long straight wire will have a component parallel to the wire that is equal to the force exerted in that direction on the component of the current element in the direction perpendicular to the wire. As was so often the case during the first 18 months of Ampère's adventure with electrodynamics, this fact is a consequence of his force law regardless of what value is taken by the parameter **k**. Ampère then drew additional conclusions and referred to the crude drawing shown in plate 15.[52]

> A closed circuit can never be transported parallel to a wire. However, it seems to me that it can be made to rotate indefinitely in the same direction in this apparatus:
> This is because the sum of the moments of the parallel forces is not zero, even though their sum is zero. In the circuit ABCD, the forces exerted on the four sides will sum to zero, but their moments will not; this is because those on AB will vanish, and these moments will not be able to cancel those of the corresponding forces on CD, since the moments of the forces on AD and AB have respective lever arms shorter than those of the forces on CD.

Although the diagram is crudely drawn, it undoubtedly represents a rectangular conductor of the type shown in plate 14 suspended above a horizontal coil of wire. Ampère's note includes no calculations in support of his assertion that the relative magnitudes of the moments exerted on various parts of the mobile rectangle make possible a steady rotation; nevertheless, this conclusion does follow from an application of the force law with k = 0. Ampère thus felt he had discovered the route to an indubitable refutation of transverse magnetism theories; if he could produce the rotation he had predicted, he knew from extensive experience that it could not be replicated using magnets in place of the conducting coil. Furthermore, he drew another implication to test the molecular hypothesis for magnetic circuits.[53]

> It follows from this that in the hypothesis where the currents are around the axis, two magnets could be arranged so as to make one of them

300

15 Ampère's early drawing of the "third" equilibrium demonstration, from the archives of the Académie des Sciences (*chemise* 206bis). (Photograph by Jean-Loup Charnet with the permission of the Académie.)

continually rotate around the other one, and this would be impossible according to the hypothesis of molecular currents. This is thus an experiment which will decide the question. I do not doubt that it will decide in favor of the latter.

Ampère reasoned that if magnetic circuitries are indeed macroscopic, the conducting coil in his hypothetical apparatus could be replaced by the end of one magnet and the mobile conductor could be replaced by the end of another magnet oriented perpendicular to the first. If no rotation occurred using these magnets, the molecular hypothesis would have to be the correct alternative.

But of course this test was based on the assumption that a circular coiled wire could indeed produce the rotation of a closed circuit that included the axis of rotation. By the end of March Ampère had the apparatus he needed to make the required observation. Using his new rotation apparatus shown in plate 12, he replaced the U-shaped conductor he had used to produce rotary effects with the rectangular component that would not rotate under the influence of magnets. Expecting to see a rotation conforming to his prediction, he found that the apparatus remained stubbornly stationary. Under the circumstances, his initial reaction was quite probably one of bitter disappointment. With a single dramatic demonstration he had hoped to not only give the *coup de grâce* to transverse magnetism, but also to establish at last his molecular hypothesis for magnetism. Now neither argument was viable and he was left staring at a maddeningly stationary piece of apparatus.

Nevertheless, although Ampère had suffered setbacks of this type all too often, he also had a knack for learning from them. During the following two months he applied some extensive mathematical analysis to the phenomenon he had observed, a phenomenon which, after all, involved a rather interesting state of equilibrium. By 10 June, 1822, he was prepared to make a report to the Académie; in the meantime he had once again transformed an anomaly into an equilibrium demonstration unexpectedly amenable to direct analysis.

The Corrected Force Law: $k = -\frac{1}{2}$

Ampère left little information about his initial observations with his new apparatus. According to one manuscript note, he first used it in "the month of March";[54] since he also reported that part of his apparatus was not available until 27 March, this probably represents the earliest possible date. Similarly, he left no record of how he came to recognize the significance of his observations. One year later, he simply wrote that the experiment was motivated by "the same considerations" that had prompted his recognition of the necessary conditions for a rotary effect; to carry out the new experiment, "I merely had to replace the magnets, which I was having act on my mobile conductor, with the spiral conductor."[55] Perhaps Ampère was embarrassed by the fact that analysis of this experiment revealed that for 18 months he had been in error about the value of the parameter k in his force law. A long-standing error in the statement of the central phenomenological law of his program was not something he wanted to publicize unduly.

Although Ampère's belief that $k = 0$ had probably never been stronger than it was in the fall of 1821 when he began his investigation of rotary effects, the actual experimental evidence was meager. When the Thenard and Gay-Lussac experiment had seemed to indicate that there was an attractive force between collinear current elements, Ampère attributed their observation to a vacuum effect. Similarly, when one of Prechtl's articles was translated for Ampère in Lyon during October, 1821, he read that Prechtl had observed that the pieces of a fractured conductor are held together by a current, an effect that also suggested the possibility of a small positive value for k.[56] Ampère's own observations had convinced him that if k had a positive value, it was negligibly small. He considered this to be sufficient evidence to conclude that k was zero; the possibility that k had a negative value required there to be a mutual repulsion between collinear current elements, a situation Ampère considered too counter-intuitive to merit further consideration.

At any rate, between March and June of 1822 Ampère decided to carry out a mathematical analysis of the anomalous new state of equilibrium he had observed. He reported to the Académie on 10 June and used the illustration shown in plate 16 in subsequent publications.

Using the general form of the force law in which the force is proportional to

$$r^{-n}(\sin \alpha \sin \beta \cos \gamma + k \cos \alpha \cos \beta)$$

Ampère argued that the observed equilibrium implied that

$$k = \tfrac{1}{2}(1 - n).$$

When combined with the long-standing assumption that $n = 2$, this relation meant that $k = -\tfrac{1}{2}$.

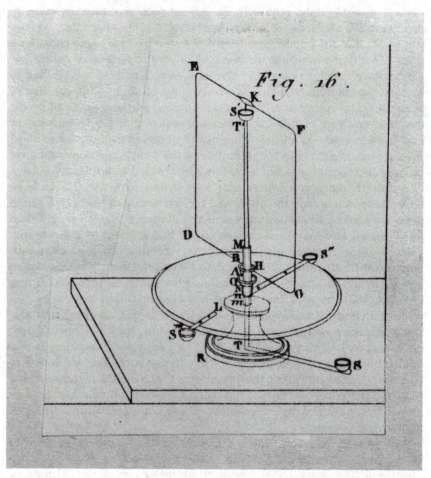

16   Ampère's 1822 "third" equilibrium demonstration, from his
"Expériences relatives à de nouveaux phénomènes électro-dynamiques,"
Annales de Chimie et de Physique, volume 20, 1822.

Ampère calmly presented his derivation as a *fait accompli* and he
provided no explanatory comments about its origins. He began by
transforming the force law into a second-order differential expression
for the change in the separation distance, r, due to changes in the
positions of either of the two current elements involved. It may have
been in connection with this transformation that Ampère once again
became interested in the general force expression in which the parameter k is not set equal to zero. Using the symbol d to represent the

change in a function due to an infinitesimal displacement of one current element, and the symbol d' for the corresponding change due to a displacement of another element, Ampère rewrote the force between two current elements as an expression proportional to[57]

$$r^{1-n-k}d(r^k d'r)$$

The mathematical application of the force law to the experimental conditions required an interesting preliminary set of approximations, idealizations, and interpretations. First, although Ampère was initially disappointed to find that his apparatus did not produce a continuous rotation, he did observe some small oscillations of his mobile conductor. He attributed these slight departures from equilibrium to the fact that his horizontal loop was not perfectly circular. He idealized the situation by claiming that a perfectly circular coil would produce perfect equilibrium. He also assumed that, regardless of the shape of the mobile conductor, it would be held in equilibrium as long as it linked two points on the rotation axis. More controversially, Ampère interpreted the equilibrium state to mean that no rotational moment was exerted on the mobile conductor by any element of the loop; he did not consider it feasible that equilibrium might be the result of a mutual cancellation of moments produced by different elements of the loop. He soon acknowledged that this was a dubious assumption and he began searching for another demonstration that did not require it. For the time being, however, using all these assumptions and idealizations, Ampère wrote the force between an element of the circular loop and an element of the mobile conductor and imposed the requirement that the integration of this expression over the length of the mobile conductor, that is, between any two points on the axis, must vanish regardless of the shape of the conductor. This is a problem in the calculus of variations, and Ampère's skill with those techniques now stood him in good stead.[58] Invariance with respect to the shape of the conductor gave the result:

$$k = \tfrac{1}{2}(1 - n)$$

It is a bit surprising that after laboring for 18 months under the illusion that k = 0, Ampère concluded his derivation with the matter-of-fact comment that "such is the relation that experiment demonstrates to hold between k and n. When n = 2, one has $k = -\tfrac{1}{2}$."[59] No doubt Ampère was disinclined to dwell upon his error and preferred to emphasize the progressive aspect of his achievement. At any rate, he realized that his discovery of the correct value for k involved a major methodological innovation. He had remained in error as long as he had restricted himself to the hypothetico-deductive method of indirect synthesis championed by Biot as "divination". He had only discovered the correct value through a direct analysis of an appropriately chosen experimental demonstration. In his memoir for the 10 June meeting of the Académie, Ampère recalled his earlier assumption that only the hypothetico-deductive mode of reasoning was applicable for a microscopic phenomenological law that involved infinitesimal current elements too small to be individually manipulated experimentally. He then summarized his new method.[60]

> It consists in substantiating by experiment that the mobile parts of conductors, in certain cases, are exactly in equilibrium between equal forces or equal moments of rotation, whatever may be the shape of the mobile part, and then to try to directly determine, with the aid of calculation, what the value of the mutual action between two indefinitely small portions must be in order that the equilibrium be in fact independent of the shape of the mobile part.

In addition to the fact that Ampère discovered that $k = -\frac{1}{2}$ through analysis of an equilibrium demonstration, his willingness to adopt the technique of direct analysis reflects the preference he had expressed during his lectures at the École Normale five years earlier. The hypothetico-deductive reasoning of indirect synthesis provided only a limited degree of confirmation; it was always possible that future observations might not agree with calculated predictions. From Ampère's perspective, direct analysis offered a more secure route to the "certitude" he sought; to question the validity of conclusions reached by this method required either an objection to the idealizations that permitted the mathematical analysis or a proof that all the operative forces had not been acknowledged. Although Ampère realized that these objections could be raised, the formal elegance of the technique held a powerful charm for him.

Nevertheless, Ampère was still not unduly optimistic that direct analysis would generate further discoveries. For example, we should recall that when Ampère derived the general form of the force law in December, 1820, he relied upon the symmetry principle which in turn depended upon the assumed equality of the magnitudes of the forces exerted by a current before and after its direction is reversed. In his June, 1822, memoir Ampère did not justify this assumption by means of an equilibrium demonstration. He eventually did so with what he called his "first" equilibrium demonstration, but his argument only reached print in the edited version of the June memoir when it was reprinted in the 1823 edition of the *Recueil*.[61]

Furthermore, Ampère was dubious that direct analysis would provide a derivation of the value of the distance parameter n. Realizing that his reliance upon the Biot-Savart experiments required a prior acceptance of his electrical theory of magnetism, he had designed the entirely electrical apparatus shown in plate 13. The apparatus was not "presented" to the Académie until June of 1823 and Ampère probably never actually put it into operation. It is hard to imagine Ampère patiently counting the oscillations of the mobile part of his instrument and correlating this variable with distance. This was Biot's style rather than Ampère's. The painstaking accumulation of quantitative data was a last recourse he always managed to postpone.

Ampère's hesitant state of mind about the methodological stance he should take is evident in the manuscript notes he compiled during the preparation of the 1823 edition of the *Recueil*. In March, 1823, he began the revision of the first "note" to the 8 April, 1822, lecture he had published in the 1822 edition. Hitherto he had not described the instrument shown in plate 13, and in a March 1823 manuscript draft we find the following comment.[62]

> Long ago I proposed to have this instrument constructed and to consign to this note the results of the experiments for which it was to serve. This is why I have postponed its publication for almost a year, the preceding part having been printed in the month of April 1822. ~~Today I have completely abandoned this means of determining, by~~ experiments performed directly on the mutual action of two voltaic conductors, the law according to which this action varies between two infinitely small portions insofar as their distance increases or decreases.

Ampère's deletion in the manuscript is a symbolic indication of his hesitation at the thought of "completely abandoning" all recourse to evidence provided by indirect synthesis. He had been temporarily encouraged by the fact that in February Savary had analyzed an equilibrium demonstration to show that n = 2. But Ampère's resolve was short-lived and he eventually rejected the entire passage and replaced it with a full description of his apparatus for measuring the oscillations of a circular conductor. Two months later when he "presented" this apparatus to the Académie, he wrote a descriptive note for the *Procès-verbaux* and mentioned that his new instrument would provide[63]

307

a new means for verifying the precision of the formula . . . a formula which has been established by another means of measuring the forces – the observation of situations in which a mobile conductor remains in equilibrium between equal forces.

Ampère's concern to provide this supplementary evidence was not superfluous. His direct analysis of equilibrium demonstrations was a bold methodological innovation in a domain of confusing new phenomena. Few physicists were as yet willing to concede that Ampère had correctly construed his elaborately constructed states of equilibrium.

Ampère would have to argue extensively for the legitimacy of his technique during the following years; he repeatedly revised his set of equilibrium demonstrations so as to make his derivations more convincing. This effort was fueled by demonstrations of a multitude of new electrodynamic phenomena, many of which he devised himself. The fall of 1822 was a particularly exciting period, a period that once again brought Ampère tantalizingly close to the discovery of electromagnetic induction.

9

Defense and Elaboration of the Theory

Within two years of Oersted's 1820 discovery of electromagnetism, Ampère had created the research program he would develop and defend during the subsequent decade. Convinced that he understood the electrodynamic basis of both magnetism and electromagnetism, he continued to revise his equilibrium demonstrations to improve the empirical basis of his phenomenological force law. He also repeated his induction experiment, this time with "positive" results that frustrated his hope that it might provide a decisive demonstration of his molecular theory of magnetism. By 1823 Ampère's experimental investigations were subordinated to mathematical considerations. He became fascinated by the mathematical formalism of his program and the rival theories of Poisson and Biot. Following Poisson's 1824 publication of an imponderable fluid theory of magnetism, Ampère demonstrated the equivalence of the forces produced by a surface layer of magnetic elements and an electric circuit around the surface contour, a relationship referred to as "Ampère's theorem" throughout the nineteenth century. More generally, he argued that his own theory was preferable to those of Biot and Poisson because it could account for all magnetic, electromagnetic, and electrodynamic phenomena without postulating the existence of the magnetic fluids. These arguments culminated in Ampère's most influential publication, his *Théorie des*

Phénomènes électro-dynamiques, published in two versions in 1826 and 1827.

While a detailed discussion of all these issues is impossible here, I have selected some representative episodes to illustrate Ampère's experimental and mathematical ingenuity. We will also see how his conviction that electrodynamic effects are produced by the motion of electric fluid particles sets him apart from subsequent field theorists such as James Clerk Maxwell and William Thomson.

The Context for the 1822 Geneva Experiments

We have noted how Ampère's invention of his "third" equilibrium apparatus marked a significant turning point for his research program. Although delighted by the direct analysis of this demonstration, Ampère realized that his argument included dubious steps that might be eliminated by a more cleverly designed apparatus. He also realized that the new negative value for the parameter k implied that there should be a repulsive force between any pair of current elements oriented in the same direction along the line joining them. The combined effect of a series of these forces should be a repulsion between any two collinear segments of a linear circuit. A successful demonstration of this effect thus would be a dramatic confirmation of the new version of his force law. Ampère also wanted to repeat his induction experiment with a more powerful magnet. His investigation of rotary effects had been disappointing in that it generated no decisive evidence in favor of his molecular theory of magnetism; this issue had been the incentive for his initial exploration of induction in 1821.

Ampère explored all these topics through an interesting set of three experiments carried out in Geneva with Auguste de La Rive in September 1822. As was his wont, Ampère attributed importance to his results in direct proportion to their relevance to the progressive development of his research program. While in retrospect we might expect that he would primarily call attention to the induction experiment, Ampère had his own agenda. From his point of view the "successful" outcome of the induction experiment meant that it initially appeared to have little significance.

The Second Induction Experiment

We have already noted that it was not until August, 1831, that Michael Faraday discovered that he could produce a temporary electric current in a coiled wire by either starting or stopping another current in a nearby coil. Although Ampère did detect an induction effect in 1822, he did not understand how it was related to intensity changes in his primary current. His belief that some form of induction persisted as long as he maintained a current in his primary circuit thoroughly disqualifies him from any claim to Faraday's discovery. Nevertheless, his failure to accurately investigate induction was not simply an unreflective oversight. Lacking Faraday's exploratory mentality, Ampère lost interest in the phenomenon as soon as he thought it to be irrelevant as a deciding factor for or against the molecular version of his electric theory of magnetism.

Ampère's 1822 investigation of induction was carried out using the same experimental arrangement he had designed in July, 1821. His only revision was to replace the bar magnet detectors by a more powerful horseshoe magnet. As shown in plate 10, Ampère once again suspended a circular copper loop, HIG, just within the circumference of a fixed surrounding circular coil, DEB. The horseshoe magnet was situated in a horizontal plane with its arms perpendicular to the plane of the loop. By placing the tip of one arm of the magnet at the center of the loop and the tip of the other arm outside the loop, the two poles of the magnet would be on opposite sides of a relatively short and nearly linear segment of the loop. The effective range of the magnet did not extend beyond this region between and adjacent to its two magnetic poles. The sole function of the magnet was to detect any induced current produced by the primary current in the coil, that is, if any appreciable macroscopic current or collection of microscopic currents is produced in the suspended copper loop, the magnet should make the loop rotate out of its original position in the plane of the stationary primary coil. In September 1822 Ampère and Auguste de La Rive observed a rotation of this type as soon as they sent a current through the primary circuit of the coil. Furthermore, the suspended copper loop remained at or near its new orientation until the primary current was terminated; it then returned to its original position within the plane of the coil.

Although it is not easy to replicate Ampère's equipment, the historian Eric Mendoza has verified the accuracy of Ampère's observations.[1] To prevent depletion of his power supply, Ampère probably sent a current through his coil for a maximum period of approximately 30 seconds. In modern terminology, the prolonged deflection of the loop away from its initial position occurs because it functions as an over-damped ballistic galvanometer in the presence of the strong magnet, that is, the return to the initial position is so delayed by the damping effect produced by the magnet that there is no significant return from maximum deflection during the relatively short time period in which Ampère maintained his primary current. In Mendoza's replication of Ampère's experiment, he observed a maximum angular deflection of 13 degrees; after 30 seconds this value had only dropped to 10 degrees. Termination of the primary current 4 seconds later resulted in an instantaneous rotation of the loop back to a position just beyond its initial plane of suspension at the beginning of the experiment; the loop then slowly rotated through the small angle required to place it in that plane.

Although this modern description agrees with Ampère's own report, it is not obvious how he interpreted what he observed; due to the minor importance he attributed to the effect, he and La Rive only provided incomplete descriptions couched in highly ambiguous language. As soon as Ampère returned to Paris on the morning of 15 September, 1822, he hurriedly composed a brief summary of all three of his Geneva experiments. His notes, which survive in manuscript, apparently became the basis for the oral Académie report he began the following day and completed the next week.[2] The surviving document includes a description of the induction apparatus lifted almost verbatim from the published version of Ampère's letter to Albert van Beek about the 1821 experiment. Ampère simply added a paragraph to report the new result. The passage is worth quoting as an indication of Ampère's assessment of his discovery.[3]

> The closed circuit placed under the influence of the coiled electric current, but not in any communication with it, was alternately attracted and repelled by the magnet; consequently, this experiment should leave no doubt concerning the production of electric currents by influence provided there is no reason to suspect the presence of a little iron in the copper from which the mobile circuit was formed . . . This fact concerning the production of electric currents by influence, very interesting in itself, is, moreover, independent of the general theory of electrodynamic action.

Ampère's audience at the Académie could not have found this superficial gloss very informative unless he embellished it with extensive explanatory comments. It is not clear from this passage whether Ampère claimed to have produced a single current passing through the entire copper loop or whether he meant a collection of molecular currents when he referred to "the production of electric currents by influence."

312

Although Ampère was equally ambiguous in other reports, there is sufficient evidence elsewhere to conclude that he believed that he had temporarily magnetized his copper loop by inducing steady molecular currents that persisted as long as another steady current was flowing in the adjacent primary circuit.[4] In this respect the new outcome of the experiment came as a disappointment to Ampère. In 1821 he had argued that, since an electric current could not be induced by another current, the molecular currents responsible for magnetism must exist with random orientations prior to any magnetization process. In 1822 he continued to assume, as he had in 1821, that molecular currents ordinarily exist in metals such as copper but are not subject to reorientation; he thus interpreted his 1822 experiment as an induction of molecular currents rather than as an alignment of pre-existing currents.

Second, because Ampère assumed that molecular currents exist in other metals prior to magnetization, and because he thought that his new experiment had shown that molecular currents could be produced by induction, he could no longer attribute magnetism entirely to the alignment of pre-existing currents. It appeared that he no longer had any means to decide whether induction or alignment was an exclusive cause of magnetism or whether they both contributed. His disappointment over this issue contributed to his lack of interest in further exploration of induction phenomena. Without placing undue trust in Ampère's notoriously inaccurate memory, the following passage from an 1833 letter to Faraday does at least confirm this assessment.[5]

> I had only one goal in performing these experiments at that time . . . it was to resolve this question: do the electric currents to which magnetic attractions and repulsions are due pre-exist around the molecules before the magnetization . . . or are they indeed produced at the instant of the magnetization by the influence of neighboring currents?

In the same letter Ampère recalled that, following his detection of induced current in 1822,

> I thought that the great question of the pre-existence or non-pre-existence of molecular currents in metals susceptible to magnetization could no longer be determined in this manner, and that it would have to remain undecided until it could be resolved by other means; I no longer placed much importance on these experiments, and I erred in not studying them in depth.

Ampère's hasty and incomplete investigation of induction is typical of his general aversion to exploratory experimentation that had as its primary goal the discovery of novel phenomena. Ampère turned his attention elsewhere as soon as it appeared that he could not invoke what he knew about induction to resolve the questions that interested him. I have already noted Eric Mendoza's explanation of why Ampère observed a steady deflection of his copper loop throughout the period of activity of his primary current. Ampère himself did not discover that induced current is produced only during changes in the primary current. On the basis of his prior experience with current loops suspended in the magnetic field of the earth, he attributed the relatively small deflection he observed when he turned on his primary current to a prolonged state of equilibrium between electrodynamic forces and a counteracting torsion in the loop suspension; that is, he believed that the torsion of the suspension prevented the loop from continuing its rotation so as to take an orientation normal to a line joining the poles of the horseshoe magnet. Similarly, when he terminated his primary current, the resulting oppositely directed rotation of the loop back to its initial position was attributed to the torsional force of the suspension acting without any balancing electrodynamic force. Furthermore, Ampère merely noted that the direction of rotation of the loop varied with the current direction in the primary circuit. Since he thought that induction is a prolonged and continuous effect of the primary current itself, he did not discover that the direction of the induced current depends upon whether the strength of the primary current is increased or decreased.[6]

None of these essential characteristics of induction was discovered prior to Faraday's more systematic research nine years later. There was only one aspect of induction, as Ampère understood it, that he at least temporarily considered relevant to the goals of his research program. Ultimately, he hoped to show that the propagation of electrodynamic forces could be understood in terms of an ether mechanics based upon the two electric fluids. As we will see, during 1825 he demonstrated a mathematical equivalence between the forces exerted by Poisson's magnetic elements and closed electric circuits. On the basis of his incomplete knowledge of induction, he speculated that the dynamic effects of electric currents might be due to the induction of tiny currents or equivalent magnetic dipoles "in the surrounding space."

While from a judgmental point of view Ampère's hasty study of induction might lend itself to a denigration of his addiction to theory in contrast to Faraday's more humble empiricism, it is also true that Ampère avoided the indiscriminate pursuit of novelty that beguiled all too many of his contemporaries. French physics journals continued to publish a seemingly interminable sequence of "new electrodynamic effects," many of which were blown out of all proper proportion by empirical explorers who claimed to have discerned a revolutionary "primitive fact" for the domain. Ampère was convinced that his own discovery of electrodynamic forces had properly filled that niche two years earlier; even if this attitude prevented what might have been an influential discovery in the case of induction, it also shielded Ampère from the random bombardment of facts that might have destroyed his sensitivity to the structural unity of electrodynamics.

The other Geneva Experiments and the Exposé méthodique

In sharp contrast to the "disappointing" discovery of induction effects, Ampère's two other Geneva experiments thoroughly confirmed his expectations. One of these provided a new basis for his derivation of the important relation $2k = 1 - n$. The equilibrium demonstration Ampère had invented for this purpose early in 1822 used a circular coil that was observed to produce no rotation or deflection of the potentially mobile part of the apparatus. Ampère realized that his analysis of this situation could be questioned because his derivation was based upon the dubious assumption that each infinitesimal component of the coil individually contributed no rotational moment to the mobile conductor. He then noticed that he could also derive the relation between k and n using a semicircular coil as his source. In the notes Ampère prepared for his presentation to the Académie, he claimed that this was a more trustworthy starting point because:[7]

> in this experiment, the immobility of the conductor, suspended in such a manner as to turn freely around the vertical axis, thus cannot be attributed to the compensation of two equal and opposite actions produced by two portions of the fixed semicircular conductor, because it takes place in all orientations given to the mobile conductor.

While this bit of bluster may have sufficed for a short verbal presentation, in reality Ampère was simply bluffing. He must have realized that his new demonstration could be criticized on essentially the same grounds as its predecessor. As Ampère later admitted in 1826, what he had actually observed in both cases was only a state of equilibrium produced by his entire circuit; in the second demonstration he simply replaced a circular set of current elements with a less symmetrical set. Nevertheless, for lack of any better starting point, Ampère continued to cite the Geneva demonstration during the following three years. Only in the autumn of 1825 did he design a completely different apparatus to replace it.

Although the Geneva equilibrium demonstration was of distinctly greater importance to Ampère than the induction discovery, he always regarded it as a stopgap measure in lieu of future developments. The third of the Geneva experiments stood out as far more important. The immediate motivation for this experiment was Ampère's June, 1822, prediction that there should be a repulsion between any two collinear portions of an electric circuit. We have already noted Ampère's predilection to cite successful predictions as particularly strong confirmatory evidence. Furthermore, there was good reason for Ampère to be concerned about his credibility. Late in August, 1822, his colleague Jean-Baptiste Demonferrand wrote from Paris to warn him that during Ampère's absence Claude Pouillet had addressed the Académie in support of Biot's claim that the key to understanding electrodynamics was a phenomenological force law for the transverse interaction between an electric current segment and a magnetic fluid particle.[8] Demonferrand further advised Ampère that Pouillet had made it clear that it was "according to this theoretical route that he is going to continue his research on the mutual action of currents." In the face of this rival program, an impressive counterattack was in order.

Ampère also realized that a successful confirmation of a repulsive force between collinear current elements would resolve a nagging obstacle to his efforts to link electrodynamic forces to ethereal vibrations. An experimental test for the existence of this force thus assumed importance for a multitude of reasons; not surprisingly, it was triumphantly highlighted during 1822 and 1823 when Ampère published several versions of his *Exposé méthodique*, an experimentally based semi-axiomatic presentation of electrodynamics.[9]

Fortunately for Ampère, the experiment itself was considerably easier to describe than his induction demonstration. To test for a repulsion between collinear current elements, he designed a circuit in which the action of the forces in question would be expected to produce a motion in a thin conducting wire floating on the surface of a shallow dish of mercury. As shown in plate 17, the conductor srqpnm consists of

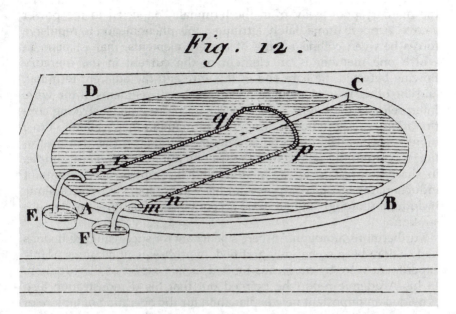

17 Ampère's floating wire experiment; figure 12 from A. de La Rive's "Mémoire sur l'action qu'exerce le globe terrestre sur une portion mobile de circuit voltaïque," *Annales de Chimie et de Physique*, volume 21, 1822.

two parallel straight sections srq and mnp connected at one pair of their adjacent ends by a third wire segment pq of unspecified shape. The dish of mercury is divided into two independent halves by a nonconducting partition AC. The two linear portions of the wire float on the mercury surface on opposite sides of the partition. The entire wire is wrapped with silk insulation except at the bare ends, s and m.

To carry out the demonstration, an electric current is sent into the mercury from a terminal E, for instance, which is directly aligned with one of the linear segments of the floating wire. The current then passes through the mercury into the bare end of the wire and completes its circuit by flowing successively through the first linear segment, the connecting section that passes over the partition, the second linear segment on the other side of the partition, and finally through the mercury to a second terminal where it returns to its source.

Ampère and Auguste de La Rive reported that as soon as a current was sent through the circuit, and regardless of the direction of this current, the originally stationary floating wire was propelled across the mercury pool away from the terminals connected to the power source. Ampère immediately attributed this phenomenon to repulsive forces between collinear pairs of current elements, that is, pairs in which one member is an element of the current in the mercury flowing between the bare end of the wire and the adjacent terminal, and the other is an element of one of the linear segments of the wire. Interpreted in these terms, the experiment represented a striking confirmation of the prediction Ampère had made to the Académie three months earlier. The importance Ampère ascribed to this demonstration was promptly reflected in the way he publicized it. For example, in sharp contrast to his ambiguous and incomplete descriptions of induction, the text he composed for his verbal report to the Académie includes a thorough and accurate account of the floating-wire demonstration.

Furthermore, although Ampère's penchant for successful predictions had seldom been so happily gratified, his subsequent citations of this experiment were by no means limited to this aspect of the situation. In his Académie report he pointed out that his demonstration also provided an important new explanation for the operation of a piece of apparatus typically referred to as the *moulinet électrique*, an explanation Ampère had briefly alluded to the previous June. The *moulinet* is easily constructed, using a short piece of brass or iron wire. Both tips of the wire are honed to sharp points and are then bent perpendicular to its length; the wire is mounted on a pivot like the needle of a compass. When the originally stationary wire is given an intense electric charge through the pivot base, it rotates on the pivot like a turnstile; hence the name *moulinet électrique* or "electric mill." The rotation had customarily been explained by presuming that electric fluid spreads into the surrounding atmosphere from the sharpened tips of the wire, where the fluid concentration is greatest. The rotation was thus accounted for by ascribing to the escaping fluid a momentum of equal magnitude and opposite direction to that of the rotating wire of the *moulinet*.

While seemingly innocuous, this attribution of a small but measurable momentum to moving electric fluid had implications that worried Ampère. The most general guiding assumption of his research program was that all electrodynamic phenomena are ultimately due to fundamental forces acting between pairs of moving electric fluid particles. Since these forces were postulated to be irreducible to any more elementary constituents, Ampère insisted that they must conform to Newton's third law of motion, implying that they would have to be directed along the line joining the two fluid particles. On the other hand, there was no corresponding constraint for the force stipulated by his phenomenological law for "infinitesimal" current elements; that is, it was at least possible that this force was not "central." This possibility was due to the fact that Ampère pictured each current element containing a multitude of oscillating fluid particles, and the direction of these oscillations could be expected to make a contribution to the force law. Ampère cited this as the cause of the angular factors in the force formula. In addition, however, if the electric fluid particles within an electric current moved with an appreciable momentum, they could impart a momentum of the same order of magnitude to the electric particles Ampère believed to constitute the surrounding ether. Since the phenomenological force law for current elements does not take into consideration momentum transmitted to the ether, the existence of an appreciable momentum of that type would make it possible that an empirically adequate force law for current elements might not act along the line connecting the current elements.

Now throughout the first two years of his research, Ampère had assumed that his phenomenological force was central; he even argued that this property made his force law more trustworthy than the noncentral Biot-Savart force promoted by Claude Pouillet during the summer of 1822. As long as he did not have to assign an appreciable momentum to moving electric fluid, he could continue to cite the centrality of his force law as a consequence of Newton's third law without abandoning his assumption that electrodynamic forces are ultimately caused by forces between moving electric fluid particles. This was an important stratagem for Ampère in the context of the French penchant for central forces.

In spite of the abstruse nature of the theoretical issue at stake, Ampère felt compelled to address it. He recognized that the floating wire experiment suggested a new explanation of the *moulinet électrique*. By ascribing the rotation of the *moulinet* to repulsive electrodynamic forces between the collinear parts of the currents linking each tip of the *moulinet* wire to the adjacent air molecules, Ampère did not have to assign a finite momentum to the particles of escaping electric fluid. Thus he could continue to highlight the centrality of his phenomenological force law while also posing an electric ether of negligible mass as the mode of propagation of this force. Inspired by his successful trip to Geneva, Ampère quickly wrote, published, and reprinted a crisp expository memoir. Using a format designed to revive memories of the development of electrostatics by Coulomb and Poisson, Ampère

319

proclaimed that a historical presentation of his topic was no longer appropriate; sufficient data were now at hand to allow him "to present them in the order that naturally results from their mutual dependence." The floating-wire experiment was brought to the very forefront of this "methodical exposition." All the observations Ampère subsequently presented as the experimental basis for his phenomenological force law were described as more complicated variations of the *premier fait* Ampère had demonstrated with his floating wire.

The contrast between the induction demonstration and the floating-wire experiment is symbolic of Ampère's priorities as a theoretical and experimental physicist. It nicely illustrates his characteristic subordination of experiment to specific confirmatory and didactic tasks set by a rigorous research program. The floating-wire experiment was originally performed to justify Ampère's revision of the value of the parameter k in his force law. It was then lifted out of this context and reassigned a more general didactic function as the fundamental experimental fact in a "methodical exposition" of his program. Ampère had no inclination to vary his experiment in the exploratory frame of mind so typical of Faraday. For example, in spite of numerous repetitions of the induction experiment during subsequent years of teaching experimental physics at the Collège de France, neither Ampère nor any of his associates discovered the transient nature of induction and the associated direction of the induced current. The experiment became a set piece of demonstrative apparatus, and it is easy to imagine how the "correct" result of the experiment was sternly conveyed to Ampère's students. Any temporary failure to duplicate Ampère's observation was no doubt attributed to a blunder on the part of the experimenter and quickly rectified. Similarly, the floating-wire experiment also involves a transient effect brought about by the initial surge of current through the apparatus. The importance Ampère assigned to his steady-state interpretation of the experiment made it virtually inevitable that he would never tinker with his original apparatus to reach a more accurate understanding of the phenomenon. Later nineteenth-century textbooks often cited the experiment as a blunder on Ampère's part.[10]

Another way to pose the contrast between Ampère and Faraday is in terms of Ampère's readiness to transform his experimental observations into idealized mathematical representations. He endorsed the view typical of his French colleagues by holding that legitimate physical concepts must be "of a nature to be submitted to the procedures of the integral calculus, procedures to which we believe all questions of this type should be reduced when one wishes to form precise ideas of them."[11]

Theoretical Exposition and Mathematical Formalism

Savary's 1823 Derivations

Ampère's electrodynamics went through another significant transition early in 1823. With his most creative experimentation behind him, he now directed most of his energy to the composition of synoptic presentations of his program in a mathematically elegant form. These efforts would culminate in Ampère's best-known publication, his 1826 *Théorie des Phénomènes électro-dynamiques*. During this period, for the first time, he had unusually talented collaborators in Félix Savary and Joseph Liouville, two of his former students at the École Polytechnique. With Ampère's approval, Savary had composed the "Notes" to the 1821 French translation of Faraday's memoir on rotary effects; beginning with the memoir he completed early in 1823, Savary now made much more creative contributions.[12] But more than his creativity, it was Savary's discipline and ability to concentrate at length on specific problems that proved especially valuable to Ampère. There is room to speculate that, without Savary's aid, Ampère might never have found time to complete the detailed calculations required to apply his force law to magnetic phenomena. Burdened by financial problems, his candidacy for a position at the Collège de France, and repeated demands that he publish an edition of his analysis lectures for the École Polytechnique, Ampère was hard pressed during the early part of 1823 just to find time to proof-read Savary's work and make minor additions.

With Savary's talents at his disposal, Ampère could soon take pride in notable signs of progress. In the spring of 1823, following a hectic period of characteristically slapdash editing, Ampère published the revised edition of his *Recueil*. It included edited versions of many of his prior publications as well as Savary's memoir; it represented the most complete single summary of his accomplishments to that date. Savary's contribution was well publicized by Ampère. He wrote several complimentary reviews for influential journals and wrote to La Rive that Savary's presentation of his work to the Académie marked "a kind of epoch in the history of dynamic electricity."[13]

Ampère had good reason to be pleased. Savary had carried out the first thorough application of Ampère's force law to all three categories of magnetic interactions, namely, those between bar magnets and straight wires, between pairs of bar magnets, and between bar magnets and terrestrial magnetism. In order to apply the necessary mathematics, Savary represented a bar magnet by an "electrodynamic cylinder," that is, as a cylinder made up of a series of circular current loops all with a fixed radius very small in comparison to the total length of the resulting cylinder. In accordance with Ampère's program, the basic force in operation is the electrodynamic force acting between two current elements. When the relevant expression for this force was integrated over the circuits of the magnets or wires in question, Savary found expressions equivalent to Coulomb's inverse square law for magnets and the Biot-Savart law for magnet-wire interactions. From Ampère's point of view, these phenomenological laws are simply convenient ways to approximate the cumulative effects produced by electrodynamic forces. Similarly, the term "magnetic pole" was simply a convenient way to refer to an experimentally useful point in magnetic circuitry.

Savary's memoir included two other noteworthy accomplishments. Noting that Ampère had simply assumed that the distance parameter n in the phenomenological force law has the value 2 "by analogy" to the gravitational force, Savary presented a new derivation of a relation between n and k. To do so he carried out an equilibrium analysis of an old 1820 observation by Welter and Gay-Lussac. They had observed that a magnetized torus produces no magnetic effects unless the torus is broken into pieces. Representing the magnetized torus as a closed electrodynamic cylinder, Savary showed that the requirement that the torus has no electrodynamic effect on an arbitrarily oriented current element results in the expression $nk = -1$. When combined with Ampère's 1822 experimentation that showed that k is negative, and the result of Ampère's "third" equilibrium experiment, $2k = 1 - n$, Savary's conclusion produced the values $n = 2$ and $k = -\frac{1}{2}$. This analysis by Savary thus provided Ampère with a "fourth" equilibrium demonstration and allowed him for the first time to present a derivation of both parameters in the force law. Even before Savary published his memoir, Ampère suggested that the magnetic torus be replaced by a toroidal helix with a compensating axial current; as usual, Ampère wanted his equilibrium demonstrations to be entirely electrical and thus as independent as possible from his theory of magnetism.

Second, Ampère was also vindicated by one of Savary's less prominent calculations. We have noted how, in January of 1821, Ampère incorrectly predicted the outcome of the embarrassing experiment he had performed with Despretz. Ampère had tried to predict the outcome of the experiment by relying upon the incorrect value of zero for the parameter k; when his prediction failed, Ampère conveniently maintained silence during the ensuing two years. Using the new correct value of k, Savary could now cite the experiment as yet another piece of confirmatory evidence for the corrected force law.

Although Ampère publicized Savary's work by writing favorable reviews, during most of 1823 he had no time to make any significant contributions himself. This situation changed toward the end of the year. Along with the completion of his editing chores for the publication of his analysis lectures, Ampère began work on several important electrodynamics papers that would be incorporated almost verbatim into the 1826–7 *Théorie*.[14] They were composed during a period of alarming creativity by Ampère's rivals: Biot, Pouillet, and Poisson.

Biot and Pouillet

Biot had withdrawn from the public scene in Paris following his unsuccessful campaign against Fourier for the secretariat of the Académie in 1822. Laplacian physics was in decline and Biot was particularly disgruntled by Arago's successful promotion of Fresnel's optics at the expense of Biot's corpuscular theory. To maintain the survival of his research program for magnetism, Biot extended his patronage to Claude Pouillet, who carried out the required derivations and thus played a role analogous to that of Savary for Ampère. The analogy between the two went further in that, like Savary in 1823, Pouillet presented a detailed mathematical application of his patron's theory at an August meeting of the Académie in 1822. But it was a sign of the times that only a short résumé of Pouillet's presentation was published while Savary's work was rushed into print and received prominent reviews. Temporarily disinclined to participate in public debate, Biot vented his spleen in a third and final 1824 edition of his *Précis élémentaire de physique expérimentale*. He greatly expanded the section on electromagnetic phenomena and responded to earlier criticisms by Savary concerning the experimental basis of the Biot-Savart law. With Pouillet's help, he showed how this phenomenological law could account for a wide range of phenomena, particularly rotary effects. He also remained adamant that future research should be directed toward the reduction of these effects to purely magnetic forces. Nevertheless, this research avenue came to an abrupt dead end. Although the Biot-Savart law retained its value as a phenomenological law, it was never reduced to fundamental laws of magnetism. Ampère could thus respond to Biot by citing the broader scope of his own electrodynamic law, even though it too was phenomenological rather than fundamental.

On the other hand, Biot's Laplacian colleague Poisson took an even more conservative approach. Rather than try to account for the new electromagnetic and electrodynamic phenomena, he developed a purely magnetic theory in strict accordance with Laplacian standards.

Poisson's Theory of Magnetism

In parallel with his two electricity memoirs of a decade earlier, during 1824 Poisson developed an analogous imponderable fluid theory of magnetism in two lengthy memoirs he eventually published in 1826.[15] Unimpressed by developments inspired by Oersted's discovery in 1820, Poisson staunchly preserved the Laplacian tradition by treating magnetism as a domain entirely distinct from that of electricity. With a single introductory remark, he cavalierly dismissed the issue that so preoccupied Ampère.[16]

> The identity of the magnetic fluid and the electric fluid does not necessarily result from the important facts that have been discovered recently. Fortunately, the solution of this question is not at all relevant to the object of this memoir; our analysis is independent of the particular nature of the boreal and austral fluids; our goal is simply to determine the resultants of their attractions and repulsions, and, if it is possible, how they are distributed within magnetized bodies.

With this narrow agenda firmly in mind, Poisson followed the same carefully delineated three-stage procedure that had guided the composition of his electricity memoirs: the adoption of Coulomb's physical principles, the precise mathematical formulation of these principles, and the mathematical derivation of observable consequences to be compared with experimental measurements. His most fundamental physical assumption was that magnetic effects are produced by two magnetic fluids that act on each other through fundamental forces that vary as the inverse square of the distance between their particles. Poisson also assumed that the austral and boreal fluids are confined by some unknown force to the interior of small "magnetic elements." At any point within the body, Poisson called the excess of either of the two fluids with respect to the other the "free fluid" at that point. In the non-magnetized state, the value of the free fluid vanishes everywhere, but when a body is exposed to a magnet the two magnetic fluids become polarized within each magnetic element and form a thin layer of free fluid on the surface of each element. Poisson's analysis was based on the fact that any particle of magnetic fluid within a magnetic element must be in equilibrium with respect to magnetic forces. Once Poisson determined the equations that express this state of equilibrium, he could use series expansions similar to those he had used in the electrostatics case to calculate distributions of magnetic fluid and the resulting external forces.

One early stage in Poisson's lengthy discussion deserves extended attention. This is his preliminary calculation of the force exerted by a magnetic element on a single particle of magnetic fluid located outside the element. Ampère showed that Poisson's magnetic force is analogous to the force between a small circular electric current and a single circuit element located outside the circular circuit. This demonstration then contributed to Ampère's argument that, although Poisson's derivations might produce results that agree with experimental measurement, the data were equally supportive of Ampère's theory.

Poisson began his calculation by locating the external fluid particle at a point M with rectangular coordinates (x,y,z); he let C be an arbitrary point inside the magnetic element with coordinates (x',y',x'). A point M' on the surface of the magnetic element was stipulated by coordinates $(h\chi, h\xi, h\zeta)$ with respect to C, where h^3 represented the volume of a cube of volume equal to that of the magnetic element. The following definitions completed Poisson's terminology.

ρ the distance MC between an external fluid particle and a point within the magnetic element

ρ_1 the distance MM' between an external fluid particle and a point on the surface of the magnetic element

ε magnetic fluid thickness normal to magnetic element surface (a function of position on the surface)

μ free fluid density (a function of position)

$h^2 dS$ an infinitesimal surface area on the magnetic element surface

Using these definitions and Poisson's sign conventions, the force exerted on a single austral particle located at M by the free fluid $\mu h^2 \varepsilon dS$ located at M' is $\mu h^2 \varepsilon dS / \rho_1^2$. The three rectangular coordinates of this force at M will be partial derivatives with respect to x, y, and z of the quantity $\mu h^2 \varepsilon dS / \rho_1$. As a preliminary transformation, Poisson used the fact that h is much smaller than the other distances so that he could apply one of his favorite approximation techniques and expand $1/\rho_1$ as a Taylor's series and drop all higher terms in h. Integration over the entire surface of the magnetic element produces the quantity q:

$$q = \alpha' \frac{d\left(\frac{1}{\rho}\right)}{dx'} + \beta' \frac{d\left(\frac{1}{\rho}\right)}{dy'} + \gamma' \frac{d\left(\frac{1}{\rho}\right)}{dz'}$$

where:

$$\alpha' = \int \chi \mu \varepsilon dS \qquad \beta' = \int \xi \mu \varepsilon dS \qquad \gamma' = \int \xi \mu \varepsilon dS$$

Using these definitions, Poisson wrote the three components of the force at M due to all the free fluid located on the surface of the magnetic element:

$$\lambda = h^3 \, dq/dx \qquad \lambda' = h^3 \, dq/dy \qquad \lambda'' = h^3 \, dq/dz$$

Although as usual Poisson did not pause to comment at length on the physical significance of these quantities, he did call the set (α',β',γ') the body's "distribution of magnetism." Later theorists of magnetism would refer to it as the magnetic moment per unit volume; this is because each of the three components is equal to the magnetic density of the magnetic element multiplied by the corresponding component of the distance between the centers of mass of the two magnetic fluids. Poisson's "magnetic distribution" (α',β',γ') can thus be thought of as a stipulation of a small dipole representative of the magnetic effect produced by the magnetic element in question. With this interpretation in mind, Poisson defined a new quantity δ which is proportional to the distance between the centers of mass of the two fluids:

$$\delta^2 = \alpha'^2 + \beta'^2 + \gamma'^2$$

Poisson noted that a new coordinate system can be chosen so that in that system the transformed values of β' and γ' will vanish and δ will simply equal the new value of α'. This new coordinate system has its x axis along what Poisson called the "axis of δ," that is, the line joining the centers of mass of the two fluids. Using his expression for q, Poisson then calculated the three force components $(\lambda,\lambda',\lambda'')$ in terms of δ, the angles a, b, and c which the δ axis makes with the original coordinate system, the angle i between the δ axis and the direction CM, and the angles l, l', l'' which are the direction angles for CM with respect to the original coordinate system.

$$\begin{aligned}
\lambda &= -(h^3\delta/\rho^3)\,(3 \cos i \cos l - \cos a) \\
\lambda' &= -(h^3\delta/\rho^3)\,(3 \cos i \cos l' - \cos b) \\
\lambda'' &= -(h^3\delta/\rho^3)\,(3 \cos i \cos l'' - \cos c)
\end{aligned}$$

Finally, he calculated the resultant of these three components:

$$(h^3\delta/\rho^3)\,(3 \cos^2 i + 1)^{\frac{1}{2}}$$

For the staunch Laplacian Poisson, the magnetic state of a body is thus best characterized by stipulating *lines of magnetization* determined by the orientations of imaginary little magnetic needles or dipoles directed along the axis of each magnetic element.

Ampère's Preliminary Response

Although Laplacian physics was losing its appeal for young physicists, Poisson's model still posed a sophisticated alternative to what Ampère considered to be the correct electrodynamic basis of magnetism. Ampère was convinced that each of Poisson's magnetic elements or equivalent dipoles was in reality a tiny circular circuit in a plane normal to the line of magnetization. Nevertheless, it was not obvious how Ampère should proceed in arguing his case. Should he be content with a demonstration that his own theory could explain all the relevant phenomenological laws, including the Biot-Savart law and Coulomb's law for magnets? He could take this approach without any detailed concern about Poisson's rival model. On the other hand, Ampère could also try to explain why Poisson's approach was as successful as it was; that is, he could try to show that the structure assumed in Poisson's model, and the forces he applied to it, generated accurate results simply because this was an alternative way of describing the action of circuits and electrodynamic forces. This approach would require detailed analysis of Poisson's model and a demonstration of how it could be translated into an equivalent description of the magnetic state in terms of microscopic electric circuits.

Ampère never made a clear-cut decision in favor of one or the other of these two strategies. He did give more attention to the first approach, the explanation of phenomenological laws using his own model and force law. Nevertheless, he did so in a manner that emphasized the relationship between his electrodynamic model and the magnetic model of his rivals. For example, some of his most mathematically sophisticated derivations of the Biot-Savart law were based upon analysis of magnetic fluid particle distributions equivalent to the electric circuits Ampère felt were actually operative. On the other hand, Ampère did not extensively explore the equivalence of Poisson's model and his own at the microscopic level of individual magnetic elements and "infinitesimal" electric circuits. This is not surprising in view of the fact that Poisson based his analysis on fundamental forces between individual particles of magnetic fluid; there is no counterpart to an individual magnetic particle in Ampère's model. Furthermore, for lack of ether dynamics, Ampère was forced to base his analysis on a phenomenological law between current elements rather than a fundamental law between particles of electric fluid.

The 1824 Précis

Beginning in 1823, Ampère's work on these problems passed through several stages. First, by December of that year he had made a discovery which thereafter he repeatedly referred to as one of his "great things." At some point shortly before that date he found an interesting and suggestive new way to represent the electrodynamic force between two circuit elements, ds and ds', that is, he showed that the force exerted on ds' is mathematically equivalent to the area of a triangle formed by the midpoint of ds' and the projection of the other segment ds onto a plane normal to ds'. This theorem would allow him to translate Poisson's surface integrals over distributions of magnetic elements into one-dimensional integrals around the contours of these surfaces. This became the basis for Ampère's argument that Poisson's theory was simply an alternative representation of what were actually electrodynamic forces. Ampère could then argue that his program was preferable to Poisson's because there was no need to postulate the existence of two magnetic fluids in addition to the motion of electric fluids within electric circuits.

Ampère included his first area theorem in the long excerpts of a memoir he presented to the Académie in several stages beginning in December of 1823. He published it in two forms during 1824; I will refer to this memoir simply as Ampère's *Précis*, the key title word of the monograph version.[17] He began with a diagram for the interaction between two current elements as indicated in plate 18. Element ds is represented by BMm, element ds' is represented by aAb, and they are separated by a distance r. The segment AG is normal to ds' and is in the plane formed by the element ds' and the line connecting the two elements. Ampère resolved the force between the two elements into components parallel and normal to ds'. As he had shown in 1822, the parallel component vanishes when integrated over a complete circuit. Ampère thus concentrated on the normal component; he defined AN as the projection of AM on the bAG plane and let ϕ be the angle bAN. He then considered the force exerted by the circuit in the normal direction AG and used some trigonometric relations and integration by parts to show that this force is:

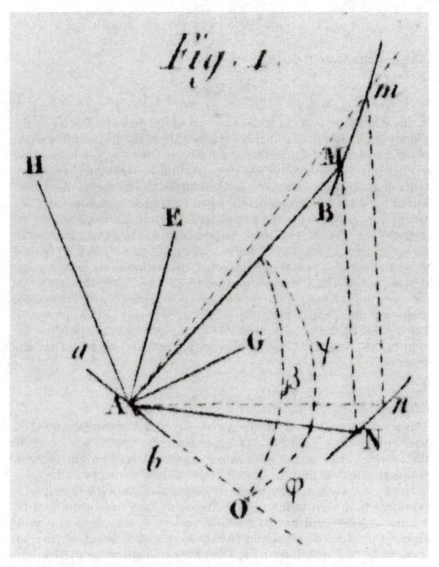

18   Figure 1 from Ampère's 1824 memoir on electrodynamic phenomena
in *Annales de Chimie et de Physique*, volume 26.

$$-\frac{1}{2}\,ii'ds'\int\frac{\overline{AN}^2}{r^{n+1}}\,d\phi$$

Ampère pointed out that the expression

$$\frac{1}{2}\overline{AN}^2d\phi$$

is the area of the triangle formed by A and the projection of ds on the plane bAG. After noting that the magnitude of the force in this plane is independent of the direction of the element ds', Ampère introduced further notation to explore the area relationship. He let U represent twice the integration over a closed circuit of the projected areas produced by the circuit elements ds when each area is divided by r^{n+1}. Using this terminology, the force exerted by the entire circuit on ds' in the plane bAG is $-\frac{1}{2}ii'Uds'$. Second, Ampère let (ξ,η,ζ) be a set of direction angles for a normal to the plane bAG, and let (A,B,C) be new projections of the previously projected area onto the three new coordinate planes. The three magnitudes (A,B,C) could be thought of as components of a magnitude D with direction angles (ξ',η',ζ') where the cosines of these angles are given by A/D, B/D and C/D respectively. Letting AK be the projection of AD on the bAG plane and ψ' be the angle DAK, Ampère showed that an alternative expression for the force is:

$$-\frac{1}{2}ii'Uds' = -\frac{1}{2}Dii'ds\,\sin\psi'$$

The force component in a given plane could thus be calculated in terms of the magnitude D. In later publications Ampère would call D the *directrice* of the circuit with respect to the external current element. D is a directed magnitude located at the current element ds'; the force exerted by the closed circuit on ds' is perpendicular to D and proportional to the sin of the angle between ds' and the direction of D.

Since Ampère did not pause to comment on the physical significance of the *directrice*, it would be rash to speculate about whether he thought of it as anything more than a mathematical entity. He and Savary typically interpreted any cumulative effect of simple electrodynamic forces between current elements as simply a computational result without intrinsic physical significance. At any rate, Ampère immediately applied his new expressions to carry out an alternative derivation of the results Savary had reached earlier in 1823; that is, he showed that the interaction between two solenoids lies along the line that connects their poles and is an inverse square function of distance;

it is thus equivalent to Coulomb's phenomenological law for the interaction between bar magnets. Similarly, he showed that calculation of the interaction between a solenoid and a current element reproduces a form of the Biot-Savart law. By showing that the phenomenological laws of Coulomb and Biot could be traced to the cumulative effect of simple electrodynamic forces, he could conclude that there was no need to postulate the existence of magnetic fluid. At this point he did not attempt the more reductionist strategy of trying to provide a direct translation of Poisson's magnetic elements into electric circuits.

Académie Presentations: 1825–1826

Ampère reconsidered the relation between his program and Poisson's in several memoirs drafted during 1825 and early 1826. In November, 1825, he presented a paper to the Académie which included results that eventually would be incorporated into the *Théorie*; in 1826 he published an abstract that included some interesting comments.[18] Most important, he claimed that he had calculated the force between two "infinitely small" electric circuits and had found that only the proportionality constant differed from Poisson's expression for the force between two magnetic elements. In the case of the electric circuits the constant was one half the product of current strengths and circuit areas; in the magnetic element derivation the constant was the product of magnetic "intensity" and the lengths of the axes of the magnetic elements. From Ampère's point of view, "this result establishes completely the identity of everything that can be deduced from the two species of considerations by which electrodynamic phenomena have been explicated."[19]

It is not clear what calculations Ampère performed to justify his claim. Although in his *Théorie* Ampère included calculations of the force between two small circuits, he did not provide analogous calculations for two magnetic elements. As we will see, his argument in the *Théorie* was less direct than the straightforward comparison he claimed to have carried out in 1825. The manuscript version of the November, 1825, memoir shows that Ampère noted that the force between two tiny circular circuits oriented in the same plane varies as the inverse fourth power of the distance between them. This calculation was presented in detail in the *Théorie*.[20] He then pointed out that this is the same dependency on distance that pertains to two very small parallel bar magnets when their poles are assumed to attract or repel according to inverse square forces. It may be that in 1825 Ampère was temporarily

331

satisfied with this analogy and did not attempt the complicated calculations called for by Poisson's program. He simply pointed out that an electric circuit should have the same effect as an assemblage of small electric circuits covering the surface spanned by the actual circuit, and each of these small circuits was equivalent to a small magnetic dipole normal to the surface.

In his published abstract Ampère reported that he had calculated the force exerted by a surface layer of magnetic elements acting on a magnetic pole, according to the inverse square forces postulated by Poisson's magnetic fluid theory.[21] He allegedly found that the three components of the force were proportional to the areas projected on the coordinate planes by tiny arcs of the surface contour. In other words, the force components were proportional to the set of quantities (A,B,C) he had found in his 1824 *Précis* for the components of the *directrice* D. Ampère commented in considerable detail about the implications of this identity.[22]

Whether one supposes that each magnetic element owes the action it exerts to two molecules, one of austral fluid and the other of boreal fluid, or that this action results from electric current forming, in a plane perpendicular to the axis of the element, an infinitely small current, the purely mathematical result that I have just announced subsists, and one can scarcely avoid concluding from it, in the two manners of explicating phenomena presented by voltaic conductors, that the action exerted by these conductors is produced by the formation in the surrounding space of the magnetic elements of which I have just spoken, above all if one recalls an experiment I performed in Geneva in 1822 and those through which Mr. Becquerel has generalized and completed the result I obtained, namely, that the electric current impresses in effect in all bodies the type of magnetization in question, a magnetization which disappears as soon as the current is interrupted.

Ampère thus retained his earlier interpretation of the Geneva induction experiment and applied it to "the surrounding space" of conductors. Furthermore, he went on to attribute the apparent violation of conservation of *vis viva* by rotary effects to the formation of "new magnetic elements in the surrounding space" when the mobile conductor changes position. Rather than rely upon alleged distributions of magnetic fluids, Ampère's injunction was that:[23]

the electric current of the conducting wire must impress in the particles of the neutral fluid spread in space a movement of rotation which is propagated gradually (*de proche en proche*) over the surfaces of which

332

I have just spoken and from which there results as many infinitely small electric currents as there are of these particles, since they are each composed of molecules of positive electricity and molecules of negative electricity.

Ampère's inaccurate interpretation of his own induction experiment thus temporarily lent support to an ether theory as a foundation for the reduction of magnetic effects to the action of electric currents. From Ampère's point of view, Poisson's magnetic element calculations could be applied when convenient as long as it was realized that these magnetic elements were actually small electric circuits induced in the space surrounding conducting wires. However, early in 1826 Ampère had not yet provided calculations to back up his assertions about the identity of magnetic elements and electric circuits.

The Brussels Memoir

With this lacuna in mind, Ampère once again tackled the problem of magnetic surfaces and their equivalence to electric circuits in a long memoir he later claimed to have completed for the most part early in 1826; the published version appeared in the *Mémoires* of the Belgian Académie Royale in 1827.[24] His primary result was a mathematical theorem that became his new foundation for the transformation of surface integrals into contour integrals. Using the elaborate diagram shown in plate 19, Ampère considered an arbitrary surface GHK with a point M exterior to it. He drew the line MP parallel to the x axis and let two planes through this line cut the surface with the orientations of the planes differing by a small angle $d\phi$. These planes cut out a slice of the surface GghH which Ampère designated $d\sigma$. The projection of the two planes on the yz plane forms a triangular region PLl. Within the area $d\sigma$ Ampère constructed an infinitesimal surface element Nnn'N' which he called $d^2\phi$ and which has a projection on the yz plane of Qqq'Q'. He let NO be a normal to the surface at N with direction angles a, b, and c and where NO makes an angle i with MN. Finally, he let ρ represent the distance MN and u the projection of this distance on the yz plane.

With this terminology in place, Ampère noted that the area PQq is equal to $\frac{1}{2}u^2d\phi$. Differentiating this quantity, Ampère derived the following theorem:

$$d\frac{u^2d\phi}{\rho^3} = \frac{[3(x-x')\cos i - \rho\cos a]\,d^2\sigma}{\rho^4}$$

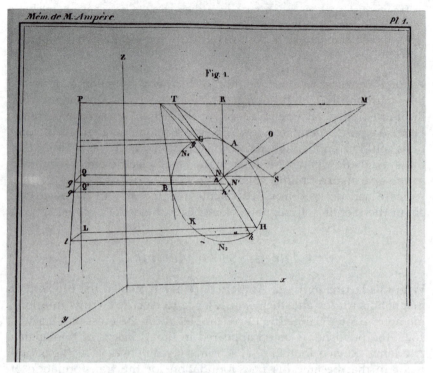

19   Figure 1 from Ampère's Brussels memoir, printed in *Nouveaux Mémoires de l'Académie Royale des Sciences et Belles-Lettres de Bruxelles,* volume 4.

where (x,y,z) are coordinates for M and the primed coordinates refer to N. The form in which the theorem is stated was carefully chosen for application to Poisson's magnetic theory; the results of Poisson's force calculations may well have suggested the theorem to Ampère in the first place. The theorem allowed him to transform surface integrations into integrations with respect to a single angular variable ϕ. Noting the striking resemblance of the second half of his expression to Poisson's force components (λ, λ', λ''), Ampère immediately applied his theorem to the geometry of Poisson's model to calculate the magnetic force between a magnetic surface and an exterior magnetic molecule. He could once again rely upon the terminology of plate 19 where the surface GHK now represented a magnetic surface of uniformly spaced magnetic elements each of volume h^3 and of separation distance k between elements. Letting δ be the ratio h/k, and δ the "intensity" of the action of each magnetic element, Ampère used Poisson's expression for the x component of the force due to a single magnetic element to write the force in that direction between an area $d^2\sigma$ on the magnetic surface and an exterior magnetic molecule at the point M:

$$-\frac{m^2 h\delta \left[3\left(x-x'\right)\cos i - \rho\cos a\right] d^2\sigma}{\rho^4}$$

When this expression is integrated over the surface to calculate the x component of the total force exerted, Ampère applied his preliminary theorem to rewrite this surface integral as a one-dimensional integral around the contour of the surface:

$$X = m^2 h\delta \int \frac{u^2 d\phi}{\rho^3}$$

After writing similar expressions for the y and z components, using angles χ and ψ analogous to ϕ, Ampère noted the physical interpretation implied by these results.[25]

One sees that the total action exerted by the assemblage of magnetic elements of the surface σ on the molecule M is precisely the same as in the case where each element ds of the contour s would exert on the molecule M an action represented by a force of which the three components parallel to the axes would be

$$m^2 h\delta \frac{u^2 d\phi}{\rho^3}, \; m^2 h\delta \frac{v^2 d\chi}{\rho^3}, \; m^2 h\delta \frac{\omega^2 d\psi}{\rho^3}$$

From Ampère's point of view, this intriguing relationship was still only an incomplete indication of the actual basis of magnetic phenomena. Ampère did not believe in the existence of "magnetic molecules" and he intended to eliminate all references to them from electrodynamic theory. Nevertheless, in the present context he retained the transformation he had discovered in order to derive the Biot-Savart law for the action of an electric current on a magnetic molecule or magnetic pole. Using the expressions noted in the above passage for the three components of the action exerted by a current element ds, Ampère pointed out that the area $u^2 d\phi$, for example, is equal to the projection on one of the coordinate planes of the triangular area formed by the magnetic molecule and the circuit segment ds. Letting θ be the angle between the direction of ds and the line joining it to the magnetic molecule, Ampère showed that the resultant of the three force components has the value:

$$m^2 h \delta \frac{ds \sin \theta}{\rho^2}$$

where this force is in a direction normal to the plane containing ds and the line joining it to the magnetic molecule. This expression is of course simply a version of the Biot-Savart phenomenological law, and Ampère concluded that he had discovered a general identity between magnetic surfaces and electric circuits,[26]

> from which it follows that the action on a magnetic molecule by the surface in question covered by magnetic elements is identical to that which a conducting wire would exert on the same molecule if it was substituted for the closed contour which circumscribes this surface. It is in this way that one accounts for this law in the first hypothesis in which everything should be reduced to the mutual action of austral and boreal magnetic molecules.

Ampère's own view was of course that electromagnetic interactions are all due to more basic interactions between electric currents. Nevertheless, by showing that the Biot-Savart force could be reduced to interactions between magnetic elements, Ampère could argue that there was no need to follow Biot and postulate a transverse force between current elements and magnetic molecules; he then could use his demonstration of the equivalence of magnetic elements to tiny electric circuits to complete the reduction of the Biot-Savart force to his own electrodynamic forces. This argument reached its most influential form when the first version of Ampère's *Théorie* went to press in November of 1826.

Théorie des Phénomènes électro-dynamiques, uniquement déduite de l'expérience

In keeping with his propensity for revision, Ampère published his best-known essay in two different formats. Although the initial version was written for the Académie *Mémoires*, prior to that publication Ampère added a table of contents, revised his long appendices, abbreviated the title, and published the essay as a monograph in November of 1826.[27] The Académie version did not appear until the following year.[28] The table of contents in the monograph edition was a belated but thoughtful addition to the 214 pages of text Ampère delivered to the Académie *Mémoires* without section divisions or topic headings;

this text must have confronted most readers as an impenetrable thicket. Although his introductory methodological remarks are an interesting statement of Ampère's position with respect to the changing scene in French physics, they provide little insight into the convoluted structure of the essay itself.

Nevertheless, with due allowance for Ampère's rambling style, the general structure of the *Théorie* can be summarized fairly succinctly. Following his methodological introduction, Ampère provided a new derivation of his force law from a revised set of equilibrium experiments. In the midst of this derivation he introduced his new concepts of *directrice* and *plan directeur* for a closed circuit with respect to a current element. Following this long digression, he concluded the derivation of the parameters in the force law and then applied the law to solenoids and straight wires to derive results analogous to the phenomenological laws of Biot-Savart and Coulomb. He also applied the *directrice* formalism to argue that his force law generates results equivalent to the forces derived by Poisson, using a magnetic fluid theory of magnetism. Finally, he assessed the three competing explanatory hypotheses and argued that his own was preferable to both Biot's and Poisson's. In this argument he emphasized his reliance upon central forces and his program's broad explanatory scope, based upon a minimal number of assumptions about hypothetical entities such as electric and magnetic fluid particles. The demonstration of "Ampère's theorem," the equivalent effect of electric circuits and Poisson's magnetic layers, is a major accomplishment and deserves extensive discussion. First we should examine Ampère's famous preliminary statement of his "Newtonian" methodology.

Direct Analysis and "Newtonian" Methodology

The methodological preamble to Ampère's *Théorie* contains the most widely cited passages he ever wrote. In this respect it is analogous to the "General Scholium" of Newton's *Principia*, the locus for Newton's famous remark that he was content to specify how to calculate gravitational forces but would not form speculative "hypotheses" about the hidden causes of gravity. It comes as no surprise to find Ampère claiming to emulate Newton; we could hardly expect him to do otherwise in his situation. For example, in an analogy to Newton's reliance upon Kepler's laws for planetary motion to establish the law of gravitation, Ampère alluded to his equilibrium demonstrations as the experimental basis of his force law.[29]

337

First, to observe the facts while varying the circumstances as much as possible, then to accompany this initial work with precise measurements in order to deduce from them general laws based solely upon experience, and to deduce from these laws, independent from any hypothesis on the nature of the forces that produce the phenomena, the mathematical value of these forces, that is, the formula that represents them, such is the procedure Newton followed. In general, it has been adopted in France by those savants to whom physics owes the immense progress it has made in recent times, and it is what has served me as a guide in all my research on electrodynamic phenomena. I have consulted only experience in order to establish the laws of these phenomena, and I have deduced from them the formula which alone can represent the forces to which they are due. I have done no research on the cause itself that might be assigned to these forces, being quite convinced that all research of this type should be preceded by purely experimental knowledge of laws and by the determination, deduced solely from these laws, of the value of the elementary forces, the direction of which is necessarily that of the straight line drawn through the material points between which they act.

An admonition to determine phenomenological laws prior to speculations about causes was standard rhetorical fare in Ampère's environment. On the other hand, we know Ampère expected an ether theory to eventually provide an explanatory basis for his force law. He alluded to this subject several times in the *Théorie*, particularly when discussing the ambiguity of the phrase "elementary forces." For example, he made the following comments about his phenomenological law for the force between current elements.[30]

I have only said that it should be considered *elementary* in the sense in which chemists rank in the class of simple bodies all those which they have not been able to decompose, whatever might be the additional expectations based on analogy that might incline one to believe that they really are composites; and also because after one has deduced its value from experiments and calculations presented in this memoir, it was by starting from this single value that one could calculate those of all the forces which are manifested in the most complicated cases.

But all the same, whether it might be due to the reaction of a fluid of which the rarity does not permit the supposition that it reacts in virtue of its mass, or to the combination of forces proper to the two electric fluids, it follows no less that the action would be opposed to the reaction along the same straight line; for, as has been seen in the considerations that have just been read, this circumstance is necessarily encountered in any complex action when it takes place for truly elementary forces from which the complex action is composed.

Ampère thus qualified his earlier ambiguous comments about the advantages of his force law being a central force. As usual, his arguments against the rival programs of Biot and Poisson were advanced on two levels. He pointed out that Biot relied upon a "fundamental" force between a current element and a magnetic molecule, a force that was not directed along the line that joined them. This was misleading on Ampère's part, since he knew that Biot did not consider his transverse force to be truly *fundamental*. But Ampère also responded to Biot's vague allusions to a possible future understanding of electric currents as a *static* distribution of magnetic fluid particles; how could a distribution of this type produce rotary effects? On the other hand, not only was Ampère's phenomenological law a central force, but there was no difficulty in thinking that *elementary forces* between electric fluid particles *in motion* were also central. An electric current is significantly different from a static distribution of charges.[31]

> When, on the contrary, one supposes that they are put in motion in conducting wires by the action of the pile, that they continually change position there and combine at each instant into neutral fluid, separate again, and immediately combine with other molecules of the opposite fluid, it is not contradictory to suppose that from actions in inverse square of distances exerted by each molecule there might result between two elements of conducting wire a force which depends not only upon their distance, but also upon the directions of the two elements along which the electric molecules move, combine with molecules of the opposite type, and separate the following instant to unite with others.

Ampère had discussed the conduction process in his 1822 letter to van Beek and in much more detail in his 1824 lectures at the Collège de France. In conjunction with those lectures he circulated a manuscript that he initially intended to become the sixth and concluding *paragraphe* of the *Précis*. The text was preserved by one of his students, the Abbé Moigno, who published it as part of his 1849 *Traité de Télégraphie électrique*.[32] It is clear from this text that Ampère did not think of conduction as literally a "current" of electricity flowing as a liquid might from one part of a circuit to another. Rather, he began with the assumption that each molecule of a conductor is positively charged and surrounded by a negatively charged atmosphere of electric fluid. The action of a voltaic pile induces a temporary polarization of the two electric fluids in molecular regions adjacent to the pile. This state is immediately neutralized by the atmosphere of the neighboring molecules and this results in the polarization of those molecules. This process continues over the entire path of the circuit; Ampère sometimes referred to a current as a "chain" of these temporary separations and reunions of the two electric fluids. With this model in mind, he considered two possible explanatory theories for how currents produce electrodynamic forces. One possibility would be to rely upon forces acting directly between electric fluid particles; the alternative would involve the transmission of effects through an ether and thus would require mechanisms similar to those proposed in Fresnel's optics. We have noted that Ampère thought of the ether as made up of superpositions of the two electric fluids. In his Collège de France lectures he proposed that light propagation might be due to the transmission of a transverse polarization and oscillation of the two fluids about their initial state of equilibrium. Furthermore, the radiation of heat and light from powerful electric currents might be traced to the chain of polarizations within the conductor. As we have seen, Ampère also held an electric theory of chemical forces. In his most optimistic moments he felt that he was on the verge of discovering the electrical foundation for chemical, optical, thermal, and electrodynamic phenomena. But an understanding of the transmission of wave motions through an electrical ether required a prior explication of the forces between moving particles of the electric fluids. In the *Théorie* Ampère referred to the two alternative explanations and mentioned that he would prefer to see electrodynamic forces traced to electric forces.[33]

> If it was possible, starting from this consideration, to find that the mutual action between two elements is in fact proportional to the formula by which I have represented it, this explication of the fundamental fact of the entire theory of electrodynamic phenomena evidently should be preferred to any other; but it would require research with which I have had not had the time to occupy myself, nor with the even more difficult research to which one would have to commit oneself to see if the alternative explication, where one attributes the electrodynamic phenomena to motions impressed on the ether by electric currents, might lead to the same formula.

Since Ampère had not been able to advance either of these two explanatory possibilities, he concentrated on the accuracy of his phenomenological force law and its experimental foundation; he insisted that, regardless of what explanatory foundation might be discovered in the future, his law would always be recognized as an accurate empirical law. We have followed Ampère's gradual discovery and implementation of the equilibrium demonstration technique; he naturally highlighted his direct analysis of the force law in the *Théorie*. Although Ampère never completely stopped revising his equilibrium demonstrations, the set that appeared in the *Théorie* became the most widely known and cited. As Ampère himself admitted, he had not constructed the apparatus he described for the fourth demonstration of this particular set. It is unlikely that he ever did so. By 1827 the choice between alternative sets of demonstrations was motivated more by aesthetics than by experimental feasibility. It was obviously still important to Ampère that he present his force law as a derivation from "phenomena" regardless of which particular set of demonstrations served that purpose. This derivation was the basis of his claim to have followed Newton's lead, and he was duly heralded as the "Newton of electricity" by Maxwell. On the other hand, Ampère also realized that he had to show that his force law could account for magnetic and electromagnetic phenomena just as effectively as the alternative theories of Poisson and Biot. He thus devoted a major portion of the *Théorie* to the mathematical relation of his theory to those of his rivals.

Ampère, Poisson, and "Ampère's Theorem"

Although prior to 1826 Ampère had constructed elaborate mathematical connections between his electrodynamics, the Biot-Savart law, and Poisson's theory of magnetism, most of his results had only been verbally summarized at Académie meetings and had not been published in journals that circulated widely in France. In the *Théorie* he made an effort to marshal his calculations into a compelling argument. He devoted most of his attention to cumulative effects rather than translation at the level of individual magnetic elements and tiny electric circuits. Unfortunately, Ampère's meager organizational skills once again failed him. He posed the problem in more than one form and interrupted the relevant calculations with long digressions on other subjects; nor did these calculations include some of the most pertinent results Ampère had derived in the Brussels memoir.[34] It could not have been easy for a casual reader of the *Théorie* to determine precisely what Ampère had or had not accomplished on this subject. The following discussion is thus far more accessible than the original.

Throughout the *Théorie*, Ampère applied his electrodynamic force law using the *directrice* formalism he had developed in his 1824 *Précis*. Before discussing the relation of his program to Poisson's, he introduced slightly different terminology to give a new derivation of his earlier results.[35] Given a closed circuit containing a current element ds, this circuit is imagined to act on an arbitrary circuit element ds' specified

341

by direction angles (λ,μ,ν). The two circuit elements are separated by a distance r and the line between them makes an angle θ' with respect to ds'. Ampère applied the force law in the form:

$$ii'ds'r^k \frac{d(r^n \cos \theta' ds)}{ds}$$

He then let R represent the magnitude of the resulting force exerted by the closed circuit on the circuit element ds', and he calculated its three components (X,Y,Z):

$$X = \tfrac{1}{2}ii'ds'(C \cos \mu - B \cos \nu)$$
$$Y = \tfrac{1}{2}ii'ds'(A \cos \nu - C \cos \lambda)$$
$$Z = \tfrac{1}{2}ii'ds'(B \cos \lambda - A \cos \mu)$$

The quantities (A,B,C) are defined as follows with all integrations taken over the entire circuit:

$$A = \int \frac{ydz - zdy}{r^{n+1}}$$

$$B = \int \frac{zdx - xdz}{r^{n+1}}$$

$$C = \int \frac{xdy - ydx}{r^{n+1}}$$

As had been the case in the *Précis*, the quantities (A,B,C) can be thought of as components of a directed line segment A'E, of magnitude D, which Ampère called the *directrice* of the closed circuit with respect to the circuit element ds'. He also restated the intriguing area property he had discovered about the *directrice* components (A,B,C), namely, the expressions that appear in the numerators of these components are proportional to the projections on the three coordinate planes of the area corresponding to a triangle determined by the circuit element ds and the midpoint of the element ds'.

Since the force R and the *directrice* D are easily shown to be perpendicular to each other, R is in a plane normal to the *directrice*, a plane Ampère called the *plan directeur*. Furthermore, letting ε be the angle between ds' and the *directrice*, the magnitude of the force becomes:

$$R = \tfrac{1}{2}Dii'ds' \sin \varepsilon$$

The magnitude of the *directrice* D thus can be used to calculate the force R directly without calculating the components (X,Y,Z). As he had done in the *Précis*, Ampère also showed how to calculate the magnitude of the force in any given plane; if a line normal to the plane in question has direction angles (ξ, η, ζ), the force in that plane is $\frac{1}{2} U ii' ds'$ where

$$U = A \cos \xi + B \cos \eta + C \cos \zeta$$

With these preliminaries in place, Ampère could start a series of complicated applications. First, he repeated almost verbatim the section of the *Précis* in which he had calculated the force exerted on a current element by a circular current. For a circuit parallel to the xy plane and located at a specified distance l from the circuit element, Ampère found a value for U that is proportional to the area of the circle. U also depends on the distance l, the location of the circuit, and the orientation of the plane of interest. The integrations required to calculate the quantities (A,B,C) turned out to be fairly complicated and Ampère simplified them by assuming that the radius of the circular circuit is small in comparison to other distances.

Second, Ampère considered the action on a circuit element due to a small planar circuit of arbitrary shape and an orientation given by three direction angles (ξ, η, ζ). Relying upon the area relation he had derived in the *Précis*, Ampère carried out the integrations needed to calculate the quantities (A,B,C) and introduced some approximations for the case of a very small circuit. He wrote his results in terms of the circuit area, λ, the distance l from the closed circuit to the circuit element, the perpendicular distance q from the circuit element to the plane of the small circuit, and the location of the circuit "center of mass" (x_1, y_1, z_1).

$$A = \lambda \left[\frac{\cos \xi}{l^3} - \frac{3qx_1}{l^5} \right]$$

$$B = \lambda \left[\frac{\cos \eta}{l^3} - \frac{3qy_1}{l^5} \right]$$

$$C = \lambda \left[\frac{\cos \zeta}{l^3} - \frac{3qz_1}{l^5} \right]$$

It might seem surprising that at this point Ampère made no comment about these intriguing quantities. Only at the very end of his monograph did he mention that they are exactly analogous to the three force components Poisson had calculated for the action between a magnetic element and a single particle of magnetic fluid. If Ampère's circuit is imagined to encircle the *line of magnetization* of Poisson's magnetic element in a plane normal to that line, Poisson's angles (a,b,c) become Ampère's angles (ξ,η,ζ); Poisson's angle i has a cosine which in Ampère's terminology is q/l; Poisson's other direction angles (l', l'', l''') have cosines in Ampère's terminology given by (x_1/l, y_1/l, z_1/l). The two sets of expressions thus are interchangeable when Ampère's factor λ replaces Poisson's $h^3\delta$. Ampère withheld comment on these relations until the conclusion of his memoir, and even at that point he did not make them explicit. Nor did he point out that the magnitude D calculated using the components (A,B,C) is proportional to the magnitude of Poisson's force.

From one point of view Ampère's silence is not hard to understand. Lacking a detailed ether theory for the motion of individual electric fluid particles, he could not draw a connection between Poisson's theory and his own at the level of fluid particles. The two sets of force components did indicate analogies between two pairs of entities, that is, between Poisson's magnetic element and Ampère's small circuit, and between one of Poisson's magnetic fluid particles and one of Ampère's current elements. Ampère recognized that although he could exploit the analogy between a magnetic element and an electric current, there was no obvious way to draw a detailed connection between a current element and a particle of magnetic fluid. Without an ether theory that specified how circuit elements are the cumulative effect of the motion of electric fluid particle motions, there was no point in Ampère speculating about what the exact analogue of a magnetic fluid particle was in his own theory.

Ampère's predominant strategy was thus to compare calculations for the cumulative effects of appropriately distributed tiny circuits and equivalent magnetic elements. He began by following Savary's example to calculate the electrodynamic forces exerted by solenoids of small diameter. Recalling the result he had achieved for the action of a single small closed circuit on a current element, Ampère applied it to an indefinitely long solenoid with cross section λ and spacing between its coils g. He found that the resultant force is:

$$\frac{\lambda i i' ds' \sin \varepsilon}{2gl'^2}$$

where ε is the angle between the direction of the current element ds' and the line drawn from the element to the end of the solenoid.[36] Although Ampère did not say so, this expression is of course analogous to the Biot-Savart law. Similarly, Ampère calculated the interaction

344

between two solenoids and found an inverse square relation analogous to Coulomb's law for small bar magnets.

With these preliminaries in place, Ampère could concentrate on the effects of large-scale distributions of magnetic elements and solenoids; he developed two relevant arguments. First, temporarily adopting the point of view of the magnetic fluid theory, Ampère calculated the three components of the force exerted between a dipole magnetic layer and a magnetic area element located on a second surface. He showed that this effect is the same as that which would be produced by forces acting between the elementary magnetic area and each differential segment of the contour spanning the surface; the magnitude of these forces is given by the Biot-Savart law. Based on his earlier demonstration that the action of a solenoid on a circuit element is also given by the Biot-Savart law, Ampère concluded that any of Poisson's elementary magnetic surface areas or Biot's "magnetic molecules" could be replaced by a tiny solenoid with its endpoint located at the position of the imaginary magnetic surface area.[37] In the interest of keeping assumptions about unobservable entities to a minimum, there was no need to introduce the concept of a magnetic molecule; Biot's and Poisson's references to magnetic molecules and magnetic poles could be interpreted simply as their way of talking about what were actually tiny electrodynamic solenoids.

Second, Ampère carried out a second analysis to discuss what from Poisson's point of view were purely magnetic phenomena. To do so he once again calculated the force components produced by the action between two closed circuits and compared them to forces that would be generated between two magnetic surfaces bounded by these circuits. He showed that the total action of one of the magnetic layers could be analyzed into a collection of forces exerted on the differential elements of the contour of the second surface. Furthermore, these forces are equivalent, or at least proportional, to Ampère's electrodynamic forces acting between a contour circuit spanning the first magnetic surface and each current element of the contour circuit around the second surface.[38] Although Ampère compared his results to those of Poisson, he did not include the explicit calculations he had performed for the Brussels memoir and he never did explicitly calculate forces between two magnetic elements. As we have already noted, he concluded by simply noting the analogy between Poisson's force components for a magnetic element acting on a molecule of magnetic fluid and his own *directrice* components for a small circuit acting on a current element.[39] Both of Ampère's arguments deserve attention.

To address the Biot-Savart law, Ampère considered two surfaces, σ and σ', with magnetic fluid thicknesses ε and ε' respectively; the magnetic force between surface elements $d^2\sigma$ and $d^2\sigma'$ separated by a distance r is given by

$$\frac{\mu\varepsilon\varepsilon' d^2\sigma \, d^2\sigma'}{r^2}$$

where μ is a scaling constant. Ampère then constructed another surface separated from the surface σ by an infinitesimal distance h and with a distribution of the magnetic fluid opposite to the type on σ. Ampère thus represented a magnetic layer in a slightly different manner than Poisson had. Poisson used a single surface of polarized magnetic elements; Ampère used two surfaces of oppositely charged magnetic fluids, what we might call a dipole surface. Letting g represent the product $h\varepsilon$, which Ampère assumed to be constant, he calculated the following expression for the x component of the total force between a surface element $d^2\sigma'$ and the entire dipole layer made up of the surface σ and its neighboring surface of opposite fluid type:

$$\mu g \varepsilon' \, d^2\sigma' \int \frac{u^2 d\phi}{r^3}$$

where the integration is around the contour of the surface σ and similar expressions result for the y and z components.[40] These are the same expressions Ampère had found in the Brussels memoir but with the magnetic "intensity" now represented in terms of the thickness of the magnetic fluid layer. Similarly, he followed the procedure of that memoir to show that these forces have the same effect as forces applied to all the contour elements ds spanning the dipole layer when these forces have magnitude

$$\mu g \varepsilon' \, d^2\sigma' \, \frac{ds \sin\theta}{r^2}$$

and are all directed perpendicular to the plane of the sector formed by the contour element ds and the location of the surface element $d^2\sigma'$.[41] This expression is, of course, a version of the Biot-Savart force and Ampère noted that he had already shown that it is equivalent to the action of a solenoid with an endpoint at the location of the magnetic element $d^2\sigma'$. Ampère had thus drawn one connection between his program and those of his rivals. The action between an elementary magnetic surface area and all the elementary magnetic surface areas on a second arbitrarily shaped magnetic surface can be translated into forces acting between the first elementary area and all the elementary segments of the contour of the surface; the magnitude of these equivalent forces is given by the Biot-Savart law. Biot's interpretation of the Biot-Savart law in terms of a "magnetic molecule" or magnetic pole could thus be replaced by a purely electrodynamic interaction between a tiny solenoid and an electric current element. In his more extended treatment of the Biot-Savart law in the Brussels memoir Ampère noted that the law is actually an "approximation" of the cumulative effect of electrodynamic forces.

Second, after a long discussion of force directions in his own program and that of Biot, Ampère concluded the *Théorie* with a final summary of his electrodynamic interpretation of magnetism.[42] A particularly perplexing aspect of Ampère's *Théorie* is his propensity to reformulate his approach whenever he addressed a slightly different application. Rather than build upon results he had derived early in the monograph, Ampère decided that in order "to leave nothing to be desired on this subject" he should reformulate the force law "in a more general and symmetric form" and give a new derivation of the components (X,Y,Z) for the force exerted by a closed circuit on a current element.

So, starting anew, Ampère considered two circuit elements on different circuits and separated by a distance r; the set of lengths (u,v,w) are the projections of r on the three coordinate planes and these projections make angles (ϕ,χ,ψ) with the coordinate axes. By applying the force law and integrating over one of the circuits, Ampère calculated new expressions for the set of components (X,Y,Z) for the force exerted by a closed circuit on one current element of the other circuit; the integrations are around the contour of the acting circuit.[43]

$$X = \tfrac{1}{2} ii' \left(dz \int \frac{v^2 d'\chi}{r^3} - dy \int \frac{w^2 d'\psi}{r^3} \right)$$

$$Y = \tfrac{1}{2} ii' \left(dx \int \frac{w^2 d'\psi}{r^3} - dz \int \frac{u^2 d'\phi}{r^3} \right)$$

$$Z = \tfrac{1}{2} ii' \left(dy \int \frac{u^2 d'\phi}{r^3} - dx \int \frac{v^2 d'\chi}{r^3} \right)$$

Second, Ampère explored some consequences of Poisson's approach. Ampère had already calculated expressions for the action a magnetic element $d^2\sigma'$ exerts on a magnetic layer bounded by a contour s; he now extended these results to two magnetic layers so that he could compare them to the interaction between two closed circuits following contours that bound the surfaces. By constructing another integration similar to the one he had carried out over the first layer, Ampère once again transformed this surface integral into a contour integral and showed that the force components for the action between the two magnetic layers are equivalent to those produced by forces exerted on each contour element, forces with a mathematical structure exactly the same as the expressions he had calculated using his own force law. For example, the x component of the force between the two magnetic layers is equivalent to the following force exerted by one of the surfaces to each element of the contour of the second surface:

$$-\mu g g'\left(dz\int\frac{v^2 d'\chi}{r^3} - dy\int\frac{w^2 d'\psi}{r^3}\right)$$

Aside from the fact that the constant factor $\mu g g'$ replaces $\frac{1}{2}ii'$, this expression is the same as the one generated by Ampère's program for the force X exerted by a closed circuit on a circuit element. Ampère thus concluded that[44]

> and so it is completely demonstrated that the mutual action of two solid and closed circuits traversed by electric currents can be replaced by that of two assemblages each composed of surfaces having for contours these two circuits and on which would be fixed molecules of the austral and boreal fluid, attracting and repelling each other along the lines that join them in proportion to the inverse square of distances.

This derivation provided a different perspective from the earlier analysis involving the Biot-Savart law. Ampère had now shown that the action between a magnetic surface and a second surface could be translated into a resultant of forces applied to each differential segment of the second surface's contour. Since these forces had expressions proportional to the ones Ampère had calculated for the action of a contour circuit acting on each contour element of the second surface, it was reasonable for him to conclude that magnetic surfaces have the same effect as closed circuits following their contours.

Although Ampère also briefly noted the analogy between Poisson's force components for the action of a magnetic element on a single particle of magnetic fluid and Ampère's own components for the action of a small planar circuit on a current element, he did not pause to make this identity explicit. He merely made an obscure allusion to the "theorem" he had derived in his Brussels memoir, a theorem from which he claimed that "one sees immediately"[45]

1° That the action of an electrodynamic solenoid, calculated according to my formula, is in all cases the same as that of a series of magnetic elements of the same intensity distributed uniformly along the straight line or curve which captures all the little solenoid currents and by giving to the axes of the elements at each of these points the same direction as this line;

2° That the action of a solid and closed voltaic current, calculated once again according to my formula, is precisely that which magnetic elements of the same intensity would produce when distributed uniformly on an arbitrary surface terminated by this circuit when the axes of the magnetic elements are everywhere normal to this surface.

Although the argument had been tortuously disjointed, readers who persevered must have been impressed. Until the end of the nineteenth century, the equivalence of a closed electric circuit and a layer of magnetic elements was generally referred to as "Ampère's theorem."[46] Ampère's own achievements should not be confused with a quite different law that is misleadingly named after him. Sometimes referred to as "Ampère's circuital law" or more simply as "Ampère's law," this law depends upon field theoretic concepts and is often stated in the form:

$$\oint \vec{B} \cdot d\vec{l} = \mu \Sigma I$$

Here the line integral of a magnetic field is taken with respect to a closed contour and the result is said to be proportional to the sum of the currents that pass through the space spanned by the contour. Maxwell discussed this law in his *Treatise on Electricity and Magnetism* and he correctly did not attribute it to Ampère.[47] He eventually developed it into one of what are now known as "Maxwell's equations," the equations that relate spatial flux and temporal variation of electric and magnetic fields and are the basis of the wave equation for electromagnetic radiation. The contrast between Ampère's own theorems and "Ampère's law" reflects the conceptual gap that separates him from the field theorists of Maxwell's generation. For Ampère the material basis for electrodynamic and magnetic effects was simply electric fluid particles in motion. All references to "equivalent" magnetic dipoles or magnetic layers were only alternative ways of describing electric circuits. A magnetic "pole," for example, was for Ampère only a convenient way to refer to the cumulative effects of electric circuits within magnets. For Maxwell, on the other hand, magnetic fields assumed a reality as robust as that of electric fields. It is interesting that in the notes Ampère prepared with Joseph Liouville for lectures at the Collège de France, he carried out calculations very similar to those of Maxwell for the work exerted on a magnetic pole as it circulates a linear current. In spite of Ampère's long-standing familiarity with rotary effects, for him the concept of a magnetic pole was too illusory to give these calculations any significance.[48]

A full discussion of the story that links Ampère to Maxwell and William Thomson cannot be attempted here. Nor can we examine the development of Ampère's ideas by German physicists such as Weber. For Ampère himself, the *Théorie* was a culmination of his debates with Biot and Poisson rather than a prelude to future developments. He did publish a few more short electrodynamics memoirs, chiefly in connection with some new rotary effects such as "Arago's wheel." He also took close note of Faraday's discovery of induction in 1831. Although initially tempted to claim priority based on his 1822 experiment, Ampère quickly conceded that he had misunderstood the phenomenon and credited Faraday with the full discovery. By this point Ampère's attention had shifted to other topics.

Part IV

Closing Years

10

The Final Synopsis

During the last decade of his life Ampère gradually lost interest in electrodynamics. Although he did complete a few short memoirs and kept abreast of Faraday's discovery of induction, Ampère preferred to spend his time revising his encyclopedic classification of academic disciplines. Unfortunately, he was not granted the luxury of a peaceful old age in which to explore these ideas at leisure. Financial problems became a daily concern. His sister ran up large debts maintaining his household, while his son Jean-Jacques used his inheritance from Ampère's mother to enjoy leisurely journeys abroad. Ampère's domestic life deserves some attention as the context in which he attempted to bring his intellectual life to closure.

The relationship between Ampère and his son was uneven at best. Both men were temperamental and subject to long periods of brooding followed by explosive outbursts of anger. Ampère's home simply was not expansive enough to house both of them for any extended period of time. Jean-Jacques met the famous Madame Récamier in 1820. Although she was twice his age, he immediately fell under her spell and joined the entourage that included Ballanche and, more sporadically, Chateaubriand.[1] Nourished by this literary circle and with his father's encouragement and editing, Jean-Jacques completed his first drama, *Rosemonde*, during September of 1823. His father dutifully organized preliminary readings and tried to make the required arrangements to bring the production to the stage. But at this point

Madame Récamier decided to spend a year in Italy; Jean-Jacques promptly lost all interest in his father's plans and followed her to Rome and Naples. By the time he returned to Paris in December of 1824 he and his father had drifted even farther apart. The elder Ampère had hopes that his son might marry Georges Cuvier's daughter Clémentine. Jean-Jacques soon found the situation intolerable and escaped to Germany early in 1826. He stopped writing dramatic verse and gradually found his calling as a literary historian. He never married and he maintained a strained and unsatisfying relationship with Madame Récamier until her death in 1850.

Ampère's health deteriorated badly, beginning in 1829. He had suffered from severe bouts of bronchitis as early as 1821. He sometimes had recourse to mesmerists and claimed to experience some relief, at least temporarily. Other physicians prescribed the application of leeches, a fairly standard procedure to which Ampère submitted with resignation. Laryngitis and pneumonia forced him to spend the winter of 1829 in the South of France, where the climate was less aggravating to his throat and chest. His isolation that winter was made tolerable to some extent by the companionship of his son. Although Jean-Jacques described the ailing man's mood as sour and carping, he did provide some solace to his father by gradually applying himself to an academic career. He taught a literature course at the Marseille Athénée in the spring of 1830, which his father attended approvingly. By 1833 he was teaching foreign literature at the Collège de France.

Meanwhile, Ampère's daughter Albine became a source of constant concern. While Ampère's second marriage to Jenny Potot had been an unmitigated disaster, Albine's 1827 marriage to Gabriel Ride had even more disruptive consequences. Ride was a military man with powerful inclinations for strong drink, violence, and gambling. He also had a dangerous habit of trying to avoid payment of his gambling debts. Absorbed in his own problems as usual, Ampère could never recall the exact circumstances of his introduction to the man who became his son-in-law. The affair began in October of 1827, when Ride was 32 years old. He had been one of Napoleon's lieutenants and had suffered a head wound at Leipzig in 1813. Following his convalescence, he had returned to active duty and had been decorated as a *chevalier* of the Legion of Honor in 1823. True to form, Albine's mother, Madame Potot, protested that Ride was not sufficiently well-to-do; she was fairly easily persuaded to withdraw her objection. Ironically, in this case she should have been paid more heed. The wedding took place in November of 1827, immediately following the return of Jean-Jacques from abroad, but the marriage quickly deteriorated. Ride combined drunkenness with a bizarre delight in brandishing weapons. Albine later described some of his behavior in her 1831 suit for separation.[2]

> Toward the end of our first year of marriage, Mr. Ride began to put loaded pistols to my forehead and daggers to my heart to make me brave. One day in the month of November 1828, upon returning in a cab after dining near the Champ-de-Mars, he held me for several minutes with a loaded pistol at my forehead, while telling me that he wanted to make me as brave as a soldier.

Episodes of this type were frequent; in one case Ride had to be restrained by two physicians who subdued the drunkard only by bleeding him. Beginning early in 1830, he was repeatedly placed in sanitariums for treatment of his alcoholism. He always emerged recalcitrant and immediately relapsed into his old habits. During October of 1830 Albine fled to her father's home, where Ride confronted her eight days later. Much to Jean-Jacques' dismay, Ampère allowed the couple to live in his house under the proviso that Ride would not have access to weapons or drink. As might be expected, these restrictions were not honored and new episodes took place. Although the couple left Paris temporarily during 1831, they returned later that year; Jean-Jacques promptly withdrew in disgust and took lodgings with a friend in Madame Récamier's neighborhood. His father rented out his son's room to Antoine-Frédéric Ozanam, a young law student from Lyon who later pursued a literary career and helped to form the Saint Vincent de Paul Society. Ozanam happened to lodge in the house during a peaceful period; he was duly impressed by Ampère's erudition and the sparse living conditions.

In 1832 a legal separation was granted to Albine; Ride was packed off to a relative in Guadaloupe. Matters did not rest there; Ride returned in 1834 and once again lived with Albine in Ampère's home. Ampère described one of the subsequent events in a letter to Jean-Jacques.[3]

> All my life I will remember the danger I ran shortly before my departure from Paris when he came into my room one night with his sword – a danger from which I escaped only by indulging his mania, appearing to agree with his ideas and persuading him to have the porter come to our defense. He then opened the door to the street where he ran in his nightshirt with his bare sword ever in his hand.

Police disarmed the crazed swordsman and placed him in custody. In November of 1835 he was sent off to his two brothers in New Orleans, where he remained for the last few months of Ampère's life.

Ampère remained remarkably poised and compassionate throughout these sad developments. He carried out a lengthy correspondence trying to rectify his son-in-law's tangled finances. He always referred to Ride's old military wound as the source of the man's unfortunate condition. While one might question Ampère's judgment in encouraging the couple to stay together, he certainly displayed remarkable patience and sympathy. Ampère's death in 1836 granted him a reprieve from witnessing the final years of Albine's suffering. As had been the case with her father, she was always willing to accept her spouse for yet another attempt to make their marriage work. Ride returned from the United States shortly after Ampère's death; in 1838 he and Albine had a daughter, Marie, who survived only ten months. Ride's old behavior re-established itself and he was institutionalized until 1841. Thereafter Albine became increasingly deranged; she claimed to experience terrible encounters with the devil that made her delirious with fear of hell. She died in 1842 and was followed by Ride the next year.

While Ampère's domestic life was obviously not as serene as he might have wished, he did manage to organize a major publication during the last few years of his life. During 1832, with the danger of cholera in Paris, Ampère wintered in Claremont, where his host Gonod took enthusiastic interest in Ampère's classification project. With Gonod's help he compiled the manuscript for his *Essai sur la Philosophie des Sciences*, the major intellectual achievement of his twilight years. Partially motivated by his teaching at the Collége de France, his *Essai* functioned as a final synthesis of both his experiences as a scientist and his philosophical reflections about the relations that link academic disciplines. Nevertheless, there is an anticlimactic tone to this period, a realization that his years of scientific creativity had come to an end.

Physics, Zoology, and the Essai sur la Philosophie des Sciences

Ampère's last major composition was published in two volumes, the first by Ampère himself in 1834, and the second posthumously in 1843 under the supervision of his son. The *Essai* represents the culmination of Ampère's abiding conviction about the holistic nature of human knowledge. It is unique in that it represents the only case in which Ampère managed to organize his philosophical views into a publishable format. His title is apt to induce inappropriate expectations in modern readers. During the twentieth century, philosophy of science became a highly specialized discipline with its own esoteric debates about the nature of scientific explanation and the epistemological status of the unobservable micro-entities that arise in scientific discourse. Although these themes lurk below the surface of Ampère's text, they are not explicitly addressed. While this may come as a disappointment to modern philosophers of science, it tells us quite a bit about Ampère. He obviously inherited a passion for classification as part of his Enlightenment heritage; he also held a less typical conviction that ultimately there is a "natural" manner in which to depict the vast array of scholarly disciplines. Although Ampère's ostensive subject matter of classification is thus of little concern to many modern philosophers, a closer reading reveals some interesting cultural themes. For example, in contrast to thinkers with either a Comptian or Darwinist orientation, Ampère did not take a historical approach to classification. Just as his earlier classification of the chemical elements did not include a discussion of their origin, Ampère's *Essai* was intended to depict a "natural" classification scheme quite independent of how various disciplines might have evolved historically.

Two developments in Ampère's scientific career contributed to his frame of mind as he composed the *Essai*. First, he pursued to the fullest extent possible a unified ether theory for the propagation of heat and light. This was a causal theory and as such it went beyond the phenomenological laws that had preoccupied Ampère during most of his career as a physicist. Although incomplete and nonquantitative, it represented what for him was a crucial final stage in the pursuit of scientific knowledge. Second, he participated in the widely publicized natural history debate between Georges Cuvier and Etienne Geoffroy Saint-Hilaire. This controversy called attention to "natural" classification schemes and homologies, that is, similarities in structure among organisms placed in different locations within a taxonomy.

Taking the physics case first, we find that during the 1830s Ampère published two short memoirs on a unified theory of heat and light.[4] He began by reviewing his earlier atomic theory, using the more modern terminology of particles, molecules, and atoms. He still considered atoms to be material points that act as the sources of repulsive and attractive forces and are located, at least for the most part, at the apexes of polyhedral-shaped molecules. He attributed sound to molecular vibrations, while heat and light were accounted for by atomic vibrations within molecules and "a propagation in the ether." According to Ampère, radiant heat[5] cannot be distinguished from light, for light is only radiant heat made capable of traversing the humors of the eye because the frequency and intensity of the vibrations which constitute it are in that case sufficiently large for these vibrations to be able to be transmitted through these humors.

In the case of radiant heat, the ether served as the medium of propagation and transmitted the state of atomic vibration from one molecule to another. On the other hand, Ampère felt that the conduction of heat through a solid might involve transmission of vibration simply in terms of the forces between molecules. If so, the ether would be superfluous for this phenomenon. As had been the case in his earlier thinking about the transmission of electrodynamic forces, he declared the issue unresolved because of "calculations which he has not made," and he even noted that the caloric theory could not be ruled out. Although tentative and nonquantitative, his proposal that heat propagation could be attributed to causes similar to those responsible for light was a final example of Ampère's conviction that science should go beyond phenomenological laws. This explanatory stage in scientific research was on Ampère's mind during the composition of the *Essai*, and we will see that he eventually included it as one of the four "points of view" that gave structure to his classification procedure.

More significantly, however, Ampère drew motivation from his longstanding interest in botany and zoology. During the 1820s and early 1830s he participated in a vigorous debate between France's two leading naturalists: Georges Cuvier and Etienne Geoffroy Saint-Hilaire. Although Ampère did not explicitly refer to this controversy in the *Essai*, it contributed to the context in which he organized his final thoughts about classification. The debate has received extended treatment in an excellent book by Toby Appel, so we will only very briefly note Ampère's contribution to it.[6] Ampère was, of course, well acquainted with the Cuvier family, owing to the unsuccessful marriage arrangements he had attempted for Jean-Jacques and the philosophical discussions he had enjoyed with Cuvier and his brother Frédéric. Widely recognized as the greatest naturalist of his time, Cuvier had achieved a synthesis of paleontology, taxonomy, and comparative anatomy through his insistence that the key to these disciplines is an understanding of how the anatomical functions of an animal's organs were designed by "the Creator" with an integrity appropriate to thrive within the "conditions of existence" it encounters. This integrity or organic harmony places severe restrictions on how organs can be combined within a living organism. In 1812 he introduced a "natural" classification scheme for animals based upon the four *embranchements* of nervous-system structure: vertebrate, mollusk, articulate, and radiate.

Geoffroy, on the other hand, espoused a "philosophical anatomy" according to which morphology was of primary importance rather than function. For example, he claimed that a single design was responsible for the structure of all vertebrates; homologies in bone structure thus became one of his primary research topics. Beginning in 1820, he argued that homologies could also be found between the vertebrates and the articulates, the *embranchement* that includes insects. Cuvier would not tolerate attempts to find a common structural design for two different *embranchements*; his objections to Geoffroy's general methodology and theoretical position became increasingly public throughout the 1820s.

It is quite a testament to the scope of Ampère's intellect that in 1824, during the height of his work in electrodynamics, he found time to publish a detailed memoir in support of Geoffroy's thesis.[7] He also used his classes at the Collège de France as a platform to support Geoffroy when the debate with Cuvier became a public display at the weekly meetings of the Académie in 1830. Ampère's article is primarily empirical and does not address the more general theoretical dispute between Geoffroy and Cuvier. In unpublished notes, Ampère claimed that his own position was simply based upon recognition of empirical facts and was not an example of the "absurdity of a natural history submitted to metaphysical views."[8] Nevertheless, he also pointed out that homologies are to be expected because "when the creator wished the same organs for a group of living beings, he added new parts for a new goal, but while leaving the old ones, for he had seen *that they were good*."[9] Ampère was thus willing to use the rhetoric of natural theology while at the same time insisting that only observation could decide the issue. During the 1820s Geoffroy and Ampère were among a small minority of naturalists who were willing to cross swords with the powerful Cuvier. On the other hand, Ampère consistently cited Cuvier's classification scheme as a shining example of a "natural" taxonomy for zoology. It is characteristic of pre-Darwinian biology that, without the unifying theme of evolution, taxonomy and homologies could be viewed as independent issues. The Geoffroy-Cuvier debate became a notorious public display at Académie meetings during March, 1830. Ampère's defense of Geoffroy was thus fresh in his mind during the composition of the *Essai*. Although he cited Cuvier as an exemplar for a "natural" system, we can see the analogue to Geoffroy's homologies in the tightly symmetrical structure Ampère built into his scheme.

Ampère mentioned in his preface that his *Essai* had been generated by his thinking about disciplinary demarcations while teaching his physics courses at the Collège de France. Initially he had made a rather conventional distinction between *physique générale élémentaire* and *physique mathématique*; the former of these he thought of as concerned with observation and measurement, while the latter was an attempt to determine phenomenological laws and causal explanations. Upon further reflection, he divided each of these categories into two subdivisions. *Physique générale élémentaire* included both *physique expérimentale*, concerned with relatively direct observation, and *chimie*, the analysis of materials into their elementary components. There is a resonance here with Ampère's inaugural lecture in Bourg so many years earlier. Similarly, *physique mathématique* could be divided into *stéréonomie*, the search for laws, and *atomologie*, the search for causal explanations, that is, the discovery of causes which for Ampère "reduce, in the final analysis, to forces of attraction or repulsion which take place between the molecules of bodies and between the atoms of which these molecules are composed."[10] Physics could thus be approached from four distinct "points of view," and Ampère ultimately proposed that the same fourfold process of discrimination could be applied to all other sciences. Ampère also pointed out that his final scheme was deliberately modeled on the taxonomic efforts of Cuvier and the botanist Bernard de Jussieu, that is, he took as his goal the stipulation of a "natural" classification scheme analogous to those he admired in botany and zoology. Twenty years earlier he had proposed a scheme of this type for the chemical elements, a scheme that elicited little response from working chemists. As we have noted, his strategy in chemistry was to arrange the elements in a cyclic pattern so that the transition from one element to another was "natural" in the sense that neighboring elements shared a significant number of properties. Similarly, his classification of the sciences was intended to discover "a natural series which makes evident the more or less intimate relations they have with each other."[11] In 1832 he published a preliminary classification of 32 "sciences of the first order."[12] Shortly thereafter he noticed that his results agreed with those produced by relying upon an analysis in terms of the four "points of view," an agreement he accounted for by claiming that these points of view constitute a "principle in the very nature of our intelligence."[13]

Ampère relied upon this discovery as part of his effort to convince his readers that he had indeed discovered a "natural" classification scheme. As usual, Ampère mentioned Linnaeus' systems as the type he wanted to avoid. The very fact that Ampère had repeatedly revised his scheme struck him as evidence that it was not an artificial set of categories based on an algorithmic application of an arbitrary principle. In particular, his revisions were typically based upon new perceptions of "analogies" between disciplines. More important, he emphasized the fact that he had arrived at his categories from two independent procedures, a consilience he interpreted as strong evidence in his favor. Nevertheless, since he had noticed this agreement late in the composition of the *Essai*, he did not emphasize it within the main body of his text; instead, after each major division he included "observations" printed in smaller type, in which he commented on the significance of the four "points of view" for the discipline in question.

Ampère's introduction to the 1834 volume amplified this theme of natural classification. To provide historical context, he noted the organizational scheme of the *Encyclopédie*, the text that had been his own initial encounter with organized knowledge. D'Alembert's introductory *Système figuré des connaissances* classified disciplines according to the faculties of memory, reason, and imagination. Ampère was particularly offended by the fact that this scheme grouped together sciences that lacked any significant analogy with each other. For example, the historical sciences were all thrown together, regardless of whether the subject matter was minerals, plants, animals, elements, or societies. As we have seen in the case of his classification of the chemical elements, Ampère placed great importance on how a classification scheme presents a rational sequence. Each transition from one science to another should be suggested by the characteristics they share and the sequence should progress along a continuum of increasing complexity. Perhaps Ampère's conviction had its origin in his frustrating efforts to read the alphabetically arranged entries in the *Encyclopédie*. For example, he remarked that a natural classification scheme would make it possible for[14]

> a man who wished pass through the entire series to find them arranged in a sequence in such a manner that, by following them in this order, for the study of a science, he would never need to have recourse to any information than that which he would have acquired by studying the preceding sciences.

Ampère thus chose mathematics as his starting point, a discipline based upon the smallest possible number of simple ideas.

Ampère also emphasized that his approach began by grouping specific sciences according to their common characteristics; only thereafter did he determine larger organizational categories in which they

361

could be grouped. An artificial system could in principle operate in the reverse direction by establishing group definitions first. Ultimately, the subject matter of the discipline and the "point of view" of the researcher became Ampère's organizational characteristics. But he did not make these criteria explicit except in the *observations* that followed each major subdivision of the *Essai*.

As another motivation for his *Essai*, Ampère cited his recognition that his principle of classification, the four points of view, correspond to the four *époques* he assigned to individual psychological development. He had been intrigued by developmental psychology even prior to his arrival in Paris in 1804; he now considered himself in a position to link this topic to his classification of the sciences. His manner of distinguishing three aspects of his project is revealing.[15]

> It is one thing to class the objects of our knowledge and something else to class our knowledges themselves; finally, it is something else to class the faculties through which we acquire them. In the first case, one should pay attention only to the characteristics which depend upon the nature of objects; in the second, one must combine these characteristics with those which pertain to the nature of our understanding; in the third, the latter alone should be taken into consideration, and one must only take into account the former insofar as it influences the intellectual operations required by the study of the objects with which one is occupied.

Ampère had always been outspoken about his aversion to what he called the "idealism" of Kantian philosophy. In more modern terminology, what he also objected to was a relativist doctrine that it is meaningless to talk about "nature" as an objective world which we gradually come to know through our interactions with it. From this relativist point of view, our knowledge is always constructed out of subjective experiences and social conventions; the truths we discover owe their status to these conventions rather than to a correspondence with an objective reality. In sharp contrast to this view, Ampère adamantly drew a distinction between the world we wish to know and the knowledge we have of that world. In this sense he was a philosophical "realist." Although this realist presupposition contributed significantly to the composition of Ampère's *Essai*, it did so in a subtle manner. Ampère pointed out that he considered the sciences he intended to classify to be jointly produced by both the objective world and human faculties, that is, the *Essai* did not include any explicit attempt to clarify the distinction between objective reality and the subjective reality of human perceptions; Ampère simply claimed that the fourfold set of "points of view" placed rigid constraints on the possible types of human knowledge. He summarized this position in one of his *observations*.[16]

> In a word, to observe what is patent; to discover what is hidden; to establish the laws which result from the comparison of the observed facts and all the modifications which they experience according to time and place; finally, to proceed to the research of a still more hidden unknown than that of which we have just spoken, that is, to work back to the causes of the known effects, or to predict the effects to come, according to the knowledge of causes; this is what we do in succession, and the only things we can do in the study of any object, according to the nature of our understanding [*intelligence*].

Regardless of the extent to which we might disagree with Ampère's generalization, his account certainly corresponds to the stages he passed through in his own research in electrodynamics. His initial experimental observations, his more careful equilibrium demonstrations, his discovery of phenomenological laws, and his search for a causal ether dynamics correspond rather closely to the four points of view he held to be characteristic of all scientific investigation.

As we might expect, Ampère's choice of terminology was an esoteric product of his knowledge of Greek. In analogy to the two kingdoms of plants and animals, he distinguished two *règnes*: the *sciences noologiques* pertain to human thought and all its social manifestations; the *sciences cosmologiques* address "all the material entities of which the universe is composed." He divided each *règne* into two *sous-règnes*; in the case of the *sciences cosmologiques* these were the divisions of inanimate and animate matter. Each *sous-règne* in turn was divided into two *embranchements*, each *embranchement* into two *sous-embranchements*, and each *sous-embranchement* into two "first-order sciences" corresponding to the category of "classes" used by botanists and zoologists. Finally, each first-order science was divided into two second-order sciences and each second-order science was subdivided into four third-order sciences determined by the four possible "points of view." Each of the two *règnes* thus includes 64 sciences, resulting in a grand total of 128.

Ampère's entire 1834 volume, the only one published during his lifetime, was dedicated to the *règne* or "kingdom" of the cosmological sciences. Some examples are worth brief discussion. Ampère divided mathematics into two sciences of the first order: *arithmologie* and *géométrie*. Application of the four "points of view" resulted in eight sciences of the third order. In the case of *arithmologie* the result was *arithmographie, analyse mathématique, théorie des fonctions,* and *théorie des probabilités.* The terminology was slightly idiosyncratic in that Ampère followed Lagrange by including calculus within the *théorie des fonctions;* that is, from the third "point of view" this discipline is the study of mathematical laws that permit the solution of a two-fold problem.[17]

> Knowing the relations through which simultaneously varying quantities are connected, to find those which result between these very quantities and the *limits* of the ratios of their respective increments, and when one knows, on the contrary, these latter relations, to work back to those of the primitive variables.

On the other hand, the theory of probability is generated by the fourth "point of view" and concerns either research into causes or the prediction of future events. Ampère was not concerned to draw a sharp line between "pure mathematics" and applied mathematics. After all, his primary thesis was that his classification scheme was "natural," due to the way each science suggested another that followed it in the scheme. Composition of this section of the *Essai* must have evoked memories of his own mathematical education at Poleymieux, the process that generated his first mathematics publication on probability theory.

Many of Ampère's categories reflect other stages in his scientific career; indeed, much of the *Essai* can be read autobiographically. For example, application of the fourth point of view to geometry results in a science with the striking name of *géométrie moléculaire.* He chose this name as a more accurate replacement for the customary name of crystallography. According to Ampère,[18]

> This science, which has for an object the determination of what are called primitive forms in materials susceptible of crystalization, according to the secondary forms given by observation, or, reciprocally, to explain the existence of secondary forms when the primitives are known, is known under the name of crystallography.

For Ampère it was natural to think that mathematics places restraints on the possible forms of molecules; after all, this assumption had been the basis of the highly abstract chemical atomism he had constructed 20 years earlier.

We have already noted Ampère's application of the four "points of view" to his first-order science of *physique générale.* The four resulting third-order sciences of *physique expérimentale, chimie, stéréonomie,* and *mécanique moléculaire* represent the practices of direct observation, more detailed analysis, phenomenological laws, and causal explanation in terms of inter-atomic forces. Although Ampère claimed that his sequence would invigorate scientific education, it is hard to imagine how it could ever be put into practice. Students would presumably have to digest a survey of highly empirical science prior to any exposure to phenomenological relationships or theoretical ideas. While this might be possible within a localized domain such as electricity or optics, a more general application would hardly be practical. Ampère did admit that he expected some flexibility to prevail.

Within the *sous-règne* of the animate world, Ampère of course included discussion of zoology and his old hobby of botany. Although he was intrigued by the ample evidence that many species have become extinct, no trace of evolutionary theory appeared in his text. Lamarck had proposed an elaborate evolutionary account in his 1809 *Philosophie zoologique,* but it had not been well received. Ampère does not appear to have had any interaction with Lamarck. In some respects this is surprising in the light of Lamarck's efforts to play the role of "naturalist-philosopher."[19] But Lamarck was a generation older and he had fallen into disrepute with most of the Parisian scientific community by the time of Ampère's arrival. Nor did the Geoffroy-Cuvier debate bring evolutionary theory to center stage.

Based upon the manuscripts Ampère had amassed by the time of his death, the second volume of the *Essai* was published by Jean-Jacques Ampère in 1843. Dedicated to the human sciences, this volume allegedly involved a resolute application of the same principle of classification that had structured volume 1. Details are not important enough to merit attention here. Ampère began with the philosophy of individual thought and gradually worked his way into the complexities of social organization. He saw society as a collection of individuals rather than thinking of an individual as a product of society. Natural theology found its place early in his "natural" scheme; for Ampère it was obvious that religious and spiritual issues are of fundamental importance. Although he made some effort to convince his readers of the legitimacy of his sequence, he begged the question all too often. The following passage is typical.[20]

In the natural order, the study of the state of human societies, of the changes or revolutions they have experienced, of the religious beliefs which direct them, should be succeeded by that of the means by which they conserve and ameliorate themselves.

Although Ampère persevered through all 64 of his *sciences noologiques*, it is easy to imagine alternatives to this "natural order."

What are we to make of Ampère's exertions? From one perspective, the entire enterprise was a misguided and quixotic pursuit of a chimera. Ampère's "natural" classification may well strike the modern eye as a thoroughly artificial and contrived sequence of disciplines reflective only of the academic categories operative in Ampère's own milieu. Nor did Ampère devote any attention to the processes through which knowledge is legitimated and promulgated. On the other hand, his efforts surely reveal a great deal about his own unique mentality. Most professional scientists reach a stage quite early in their careers when they realize that they cannot master more than a tiny subset of all possible research avenues. Ampère never fully adjusted to this realization. His classification efforts were a way of coping with his inability to stay abreast of developments in more than a few highly specialized disciplines. Although during the 1830s the *Essai* was routinely praised as the last achievement of a great *savant*, it generated little serious interest.

Ampère's health became increasingly worrisome throughout the period in which he composed the *Essai*. The end came on 10 June, 1836, while he was on an inspection tour in Marseille. He was stricken by a fever, together with a combination of his chronic ailments, and had to be put to bed at the home of Deschamps, the director of the Marseille college. He succumbed quickly before any of his friends or relatives could arrive. Buried in a simple fashion in Marseille, his passing was not marked by elaborate public display. Ampère had limited his circle of activity to small groups of scientists and scholars. Among the general public he was relatively unknown except as an eccentric professor at the Collège de France. Ballanche and Ozanam wrote short eulogies and Arago presented a much more detailed one to the Académie. According to Arago, Ampère's last words were uttered when Deschamps began reading from the *Imitation of Christ* and Ampère assured him that he knew the text by heart. Arago also took the occasion of his eulogy to bemoan the fact that Ampère's life had been cut short by the exertions required by his educational inspection duties. Arago was not alone in considering it a national disgrace that a man of Ampère's talent and accomplishments was forced to eke out an income travelling about the country listening to students recite their lessons. With appropriate government support, his time might have been better spent. And it is of course ironic that Ampère should have met his death on an inspection tour of institutions he had had no contact with as a student.

In 1881 the Paris Congress of Electricians stipulated that the unit of electric current should be known as the ampere, thereby establishing the primary means by which his name is still known to electricians and the general public. Initially an ampere was defined electrochemically as a current that would deposit 0.001118 grams of silver per minute within a specified type of voltmeter; this unit was subsequently referred to as the "international ampere." As electrodynamic technology became more sophisticated, the International Committee of Weights and Measures decided in 1928 to adopt an "absolute" system of electrical units. During the 1930s and 1940s, increasingly accurate measurements of the attractive forces between currents and coils made it possible to define the ampere in "absolute" units as follows, effective from 1948:[21]

> One ampere is the current in each of two long parallel wires placed one meter apart in vacuum when the force per unit length acting on each wire is 2×10^{-7} newton per meter.

In this way Ampère's initial discovery of the forces between current-bearing wires is symbolically commemorated in the definition of the unit of electric current.

In 1869 Ampère's remains were transferred to the Montmartre Cemetery in Paris, where he was eventually joined by his son. To this day their joint grave can be found within those densely occupied grounds. Ampère might well have preferred burial in Poleymieux, perhaps high on the slope overlooking the verdant valley below, the only place he had experienced unalloyed happiness. Characteristically, he left no arrangements for mundane details of that sort. To the last, he was consumed only by his "thirst for certitude"; in death as in life, provisions for his physical body took last priority. Although his birthdate is incorrectly listed, the engraving on his stone has a succinct grace he might have appreciated.

André Marie
AMPÉRE
Born in Lyon 21 June 1775
Died in Marseille 10 June 1836
Member of the Académie des Sciences
He contributed to human knowledge
in the mathematical, physical,
metaphysical, and moral sciences
He created electrodynamic theory
He wrote the Essai sur
la Philosophie des Sciences
Truly Christian, he loved Man
He was Simple Good and Great

367

Notes

References to the Ampère archives at the Académie des Sciences in Paris are abbreviated to AS, followed by the relevant *chemise* or carton number.

References to Louis de Launay's three-volume edition of Ampère's correspondence are abbreviated to CORR, followed by the relevant volume and page numbers. Although this publication is incomplete and contains some erroneous dates, it is much more carefully edited than the selections assembled and published with her own commentary by Hortense Chevreux in 1873 and 1875.

CHAPTER 1   IDYLLIC YOUTH

1   Jardin and Tudesq, 1983, p. 273.
2   Ibid., p. 271.
3   Chateaubriand is an excellent source of insight into the Catholic romanticism Ampère shared with the circle of acquaintances he developed in Lyon. *René* initially appeared in 1802; the quotation is from p. 89 of the Irving Putter translation (Berkeley: University of California Press, 1960).
4   Vernay, 1936, p. 158.
5   Académie des Sciences Ampère archives: *chemise* 326. Hereafter references to these archives will be abbreviated AS, followed by the relevant *chemise* or carton number.
6   Ampère, *Correspondence*, ed. Launay, 1936–1943, vol. 1, p. 279. Hereafter, Launay's three-volume edition of Ampère's correspondence will be cited as CORR, followed by the volume and page numbers. Although this edition is incomplete and includes some erroneous dates, it is much more

carefully edited than the volumes assembled by Hortense Cheuvreux in 1873 and 1875.

7   Rousseau, *Emile*, 1979 Allan Bloom translation (New York: Basic Books), p. 328.
8   Arago, 1854a, p. 5.
9   J. J. Ampère's introduction to Saint-Hilaire, 1866, p. 67.
10 Quoted in Maurois, 1938, p. 307.
11 Thomas, 1765; reprinted in 1819: *Oeuvres de Thomas*, vol. 1, pp. 476–551.
12 Ibid., p. 480.
13 Ibid., p. 496.
14 Ibid., p. 538.
15 Ibid., pp. 513–14.
16 AS, chemise 326.
17 See Launay's comments in CORR, 1: 1–2 and 3: v–vi and 827–30; the short memoir is published in CORR, 3: 829–30.
18 AS, chemise 326.
19 Arago, 1854a, pp. 6–7.
20 AS, *chemise* 326.
21 A report of the events in Poleymieux can be found in Taine, 1899, vol. 4, pp. 196–8. Vernay, 1936 provided a summary drawn from the Poleymieux archives.
22 Flach, Costes, and Boucrot, 1986, p. 65.
23 Quoted in Launay, 1925, p. 25.
24 AS, *chemise* 326.

CHAPTER 2   MARRIAGE AND PROVINCIAL LIFE (1800–1804)

1   Arago, 1854a, p. 12.
2   CORR, 1: 7 and 15.
3   In 1986 L. Pearce Williams discovered a set of 19 of Ampère's letters to Couppier during 1795 and 1796; they are now preserved at Cornell University and copies are in the Académie des Sciences Ampère archives.
4   CORR, 1: 7.
5   AS, *chemise* 330.
6   CORR, 1: 27.
7   CORR, 1: 9.
8   CORR, 1: 30.
9   CORR, 1: 31.
10 CORR, 1: 31–2.
11 CORR, 1: 33.
12 CORR, 1: 33.
13 CORR, 1: 13. The ruins of the mill survived into the twentieth century and a commemorative plaque was affixed to them.
14 CORR, 1: 37.
15 Cheuvreux, 1873, pp. 124–9. Ballanche is also cited as a witness by Sainte-Beuve, 1956 edition, p. 956.

16 See Williams, 1953, p. 314.
17 CORR, 1: 111.
18 CORR, 1: 259.
19 CORR, 1: 268.
20 Fox, 1974.
21 This terminology has been relied upon by Nancy Cartwright in other contexts (Cartwright, 1989). Although not used by the Laplacians themselves, it is particularly useful for a discussion of their methodology.
22 Some excerpts from this treatise have been published in Potier's edition of Coulomb's memoirs; Potier, 1884, pp. 1–62.
23 Quoted in Gillmor, 1971, p. 181.
24 Coulomb, 1789, in Potier, 1884, p. 303.
25 Coulomb, 1785, in Potier, 1884, p. 138.
26 Coulomb, 1801, pp. 180–1.
27 Coulomb, 1789, in Potier, 1884, p. 297.
28 CORR, 1: 108–9.
29 CORR, 1: 109.
30 Quoted in CORR, 1: 109.
31 AS, *chemise* 203.
32 Ibid.
33 The remainder of this passage is a scribe's copy of Ampère's original draft.
34 Ibid.
35 Lagrange, 1811, vol. 1, p. 172.
36 AS, *chemise* 203.
37 Lavoisier, 1789, pp. 22–5.
38 AS, *chemise* 203.
39 Ibid.
40 Ibid.
41 AS, *chemise* 205$_{bis}$.
42 AS, *chemise* 215.
43 CORR, 1: 182.
44 Quoted in Kline, 1972, vol. 2, p. 465.
45 Ibid., p. 465.
46 CORR, 1: 135.
47 Dhombres and Dhombres, 1989, pp. 262–5.
48 Ampère, 1806b.
49 Lagrange, 1762a and 1762b.
50 See Kline, 1972, vol. 2, pp. 582–7.
51 Lagrange, 1806, *Oeuvres*, vol. 10, p. 386.
52 Lacroix, 1806, p. 486.
53 Grattan-Guinness, 1990b.
54 Lagrange, 1788, pp. v–vi.
55 Lagrange, 1788, p. 46; 1811, second edition, vol. 1, p. 70.
56 Lagrange, 1788, p. 51; 1811, second edition, vol. 1, p. 75.

57 Lagrange, 1788, p. 54; 1811, second edition, vol. 1, p. 77. Lagrange uses the notation S to represent integration over the entire mass of the body.
58 Lagrange, 1788, pp. 89–92; 1811, second edition, vol. 1, pp. 128–31.
59 Ampère, 1806b, p. 496.
60 Ibid., p. 496.
61 Ibid., p. 499.
62 Ibid., pp. 503–4.
63 ibid., p. 504.
64 Ibid., pp. 505–10.
65 Ibid., p. 502.
66 Lagrange, 1762a and 1762b.
67 Lagrange, 1806, in his *Oeuvres*, vol. 10, pp. 364–451.
68 Ibid., pp. 407–11.
69 Lagrange, 1811, vol. 1, pp. 88–104.
70 Ibid, p. 91.
71 See Bos, 1974.
72 See Fraser, 1985, p. 167.
73 Lagrange, 1811, volume 1, p. 95.
74 Ampère, 1825b.
75 Ibid., p. 134.
76 Ibid., pp. 146–7.
77 Ibid., p. 148.
78 See Fraser, 1985, p. 170.
79 CORR, 1: 249.
80 Ampère, 1808. See CORR, 2: 257, 262–3, and 270 for relevant correspondence between Ampère and Delambre. Mallez, 1936, pp. 65–8 provides excerpts from the Bourg Société report on the memoir.
81 Darnton, 1968, p. 177.
82 Ibid., pp. 62–4.
83 Joly, 1938, pp. 220–2.
84 Ibid., p. 229.
85 Ibid., p. 306.
86 Cheuvreux, 1873, p. 238.
87 Petetin, 1787.
88 Ibid., p. 106.
89 Ibid., 1787, p. 35.
90 Ibid., pp. 53–5.
91 Petetin, 1808, p. 156.
92 Ibid., pp. 264–9.
93 Degérando, 1804.
94 CORR, 3: 845.
95 Frainnet, 1903, p. 101.
96 Valson, 1886.
97 Viatte, 1928, p. 22.
98 Valson, 1886, p. 408.

99 Ibid., p. 411.
100 Valson, 1886, pp. 409–10.
101 Frainnet, 1903, p. 174.

CHAPTER 3 LAPLACIAN PHYSICS

1 For extensive discussion of these institutions see Hahn, 1971 and Crosland, 1978.
2 Crosland, 1978, p. 75. Also see Grattan-Guinness, 1990a, vol. 1, pp. 78–84.
3 For examples see Heilbron, 1982, p. 4.
4 See Heilbron, 1982, pp. 4–5, and Hahn, 1971, pp. 98–100.
5 See Hahn, 1971, pp. 224–5.
6 Ibid., p. 304.
7 Williams, 1956, pp. 378–9.
8 Bradley, 1976, pp. 434–5.
9 Quoted by Arnold, 1983, p. 249.
10 For a detailed listing of teaching positions and their occupants between 1794 and 1825, see Dhombres and Dhombres, 1989, pp. 839–43.
11 Dhombres and Dhombres, 1989, p. 572, and Grattan-Guinness, 1990a, vol. 2, p. 527.
12 Dhombres and Dhombres, 1989, pp. 496–504.
13 See Frankel, 1978.
14 Haüy, 1787, pp. viii–ix.
15 Ibid., p. xvii.
16 See Frankel, 1978, and Crosland, 1967.
17 Biot, 1804; see Brown, 1969.
18 Biot, 1809, p. 112.
19 Ibid., p. 113.
20 See Arago, 1854b, p. 599 and Libri, 1840, p. 415.
21 Libri, 1840, p. 432.
22 Arago, 1854b, p. 662.
23 Quoted and translated in Home, 1983, p. 251 from the archives of the Académie des Sciences, *chemise* for the *séance* of 6 January, 1812.
24 Poisson, 1812, p. 3.
25 Ibid., p. 16.
26 See Grattan-Guinness, 1990a, vol. 1, pp. 496–513 for discussion of Poisson's derivations.
27 Poisson, 1814c, p. 170.
28 Home, 1983, p. 254.
29 Gardini's essay is discussed in Home, 1983, pp. 254–8.
30 Biot, 1816b, pp. 84–5.
31 Biot, 1816a, p. xxiii.
32 Fox, 1974, pp. 127–30.
33 Biot, 1821c, in Fresnel's *Oeuvres complètes*, vol. i, p. 588.
34 Biot, 1816a, p. xv.

35 Laplace, 1809, p. 306.
36 Ampère, 1822j.
37 Biot, 1821a, p. 235.
38 Ibid., p. 233.

CHAPTER 4 AMPÈRE IN PARIS

1 Jardin and Tudesq, 1983, p. 372.
2 Crosland, 1967, p. xii.
3 CORR, 1: 294–5.
4 CORR, 1: 307.
5 CORR, 1: 325.
6 CORR, 1: 337.
7 Launay, 1925, p. 120.
8 CORR, 1: 313.
9 Bredin, *Lettres inédites*, p. 33.
10 CORR, 1: 322.
11 Bredin, *Lettres inédites*, p. 34.
12 CORR, 2: 461.
13 CORR, 2: 465.
14 CORR, 2: 466.
15 CORR, 2: 483.
16 CORR, 2: 523.
17 CORR, 2: 533.
18 CORR, 2: 540.
19 CORR, 2: 541.
20 CORR, 2: 537.
21 AS, *chemise* 97.

CHAPTER 5 METAPHYSICS: AMPÈRE, KANT, AND MAINE DE BIRAN

1 The most accessible overview of Ampère's philosophy can be found in his correspondence with Maine de Biran, as published by Tisserand in volume 7 of Maine de Biran's *Oeuvres*; my own discussion is supplemented by a study of Ampère's manuscripts and a collection of his letters to Maine de Biran at the Bibliothèque Nationale (n.a. fr 14605).
2 Arago, 1854a, pp. 34–5.
3 Villers, 1801, p. 205.
4 Kinker, 1801 and Villers, 1801. Although a Latin edition of Kant's most important works was published by Born between 1796 and 1798, few readers managed to penetrate the obscurity of this translation; neither Ampère nor Maine de Biran refer to it in their correspondence. See Vallois, 1924, pp. 46–7.
5 Degérando, 1804, pp. 528 and 537, and 1810, p. 397.

6 Ampère never completed his essay, but some drafts were collected by J. J. Ampère and published by Saint-Hilaire; Saint-Hilaire, 1866, pp. 333–461.
7 AS, *chemise* 261.
8 See CORR, 2: 488–9 and 529–30, and Gouhier, 1948, pp. 226–8.
9 AS, *chemise* 287.
10 CORR, 1: 286.
11 Ampère, BN n.a. fr 14605, f. 79.
12 *Oeuvres*, ed. Tisserand, vol. 7, p. 460.
13 AS, *chemise* 281.
14 Maine de Biran, *Oeuvres*, ed. Tisserand, vol. 7, p. 464.
15 AS, *chemise* 261.
16 Saint-Hilaire, 1866, pp. 12–13.
17 Published in Maine de Biran, *Oeuvres*, ed. Tisserand, vol. 7, pp. 453–4.
18 AS, *chemise* 261.
19 Ampère, BN n.a. fr 14606, ff. 35–6.
20 AS, *chemise* 261.
21 AS, *chemise* 275_2.
22 AS, *chemise* 261.
23 Ibid.
24 Ibid.
25 CORR, 2: 466.
26 AS, carton 27.
27 Ibid.
28 See Giard, 1972, for a discussion of Gergonne as a scholar and teacher of logic.
29 Gergonne, 1817. See Dahan-Dalmedico, 1986, and Grattan-Guinness, 1990a, pp. 135–7.
30 Gergonne, 1817, p. 348.
31 Ibid., p. 349.
32 Ibid.
33 AS, *chemise* 261. J. J. Ampère published a summary of some of his father's most interesting lectures in Saint-Hilaire, 1866, pp. 132–6.
34 AS, *chemise* 261.
35 Ibid.
36 Ibid.
37 Ibid.
38 Ibid.
39 Ibid.
40 Ibid.
41 Ibid.
42 AS, *chemise* 205_{bis}.
43 Ampère, 1814, p. 47.
44 AS, *chemise* 269.
45 For further discussion of Ampère's views about the importance of "novel predictions" see Hofmann, 1988.

CHAPTER 6   MATHEMATICS, CHEMISTRY, AND PHYSICS (1804–1820)

1   CORR, 1: 281–2.
2   Ampère, 1806a.
3   Grabiner, 1990, pp. 189–201.
4   Ampère, 1806a, p. 149.
5   For discussion of this proof see Grabiner, 1990, pp. 192–6.
6   Ampère, 1806a, p. 156.
7   This procedure was incorporated into textbook presentations. See, for example, Lacroix, 1810, vol. 1, p. 240.
8   Ampère, 1806a, pp. 162–3.
9   Ibid., pp. 169–78.
10 Ibid., pp. 178–81. There are notes in AS, *chemise* 88$_{bis}$.
11 Ampère, 1809.
12 Ampère, 1809, pp. 275–6.
13 Lacroix, 1810, vol. 1, pp. 241–2.
14 See Grattan-Guinness, 1990a, vol. 2, pp. 709–15 and 800–3.
15 Ampère, 1824a. See Grattan-Guinness, 1990a, vol. 2, pp. 781–4.
16 Ampère, 1824a, pp. 11–12.
17 Kline, 1972, vol. 2, p. 671.
18 Ampère, 1815b and Ampère, 1820a. For discussion of Ampère's techniques and results see Grattan-Guinness, 1990a, vol. 2, pp. 700–5; Forsyth, 1906, vol. 6, chapters 12 and 17; Goursat, 1896, chapter 2.
19 See Grattan-Guinness, 1990a, vol. 1, pp. 158–60.
20 Ampère, 1815b, p. 550.
21 Ibid., pp. 552–4.
22 Ibid., pp. 571–3.
23 Ibid., pp. 583–6.
24 Ibid., pp. 590–7. The technique is discussed in Forsyth, 1906, vol. 6, pp. 17–19 and 267–9.
25 Ampère, 1815b, pp. 601–2.
26 Ampère, 1815b, pp. 608–9. This example is discussed by Forsyth, 1906, vol. 6, pp. 280–1.
27 Ampère, 1820a, pp. 46–64; Forsyth, 1906, vol. 6, 272–6.
28 Ampère, 1820a, p. 128.
29 Poisson, 1814a.
30 Ibid., p. 109.
31 Forsyth, 1906, vol. 6, p. 266.
32 Lavoisier, 1789, p. 130.
33 Ampère, 1816a and 1816b.
34 See Gough, 1981, pp. 27–32.
35 See Siegfried, 1982.
36 Ibid.
37 Berthollet, 1809, p. 163.
38 See Rocke, 1984.

39 See Mauskopf, 1970, and Emerton, 1984, pp. 278–84.
40 CORR, 1: 356–7.
41 CORR, 2: 430–1.
42 See Partington, 1964, vol. 4, pp. 58–9.
43 See Crosland, 1978, pp. 80–7, Fullmer, 1975, and Knight, 1992, pp. 96–101.
44 Ampère, 1815a, pp. 145–6.
45 Ibid., p. 152.
46 CORR, 2: 458–9.
47 CORR, 3: 885–6.
48 CORR, 2: 463.
49 CORR, 2: 467.
50 Ampère, 1814, p. 47.
51 CORR, 2: 492.
52 Dumas, 1839, p. 351.
53 Ibid., pp. 352–3. Also cited and discussed in Fisher, 1982, p. 91.
54 See Mauskopf, 1969, and Cole, 1975.
55 CORR, 2: 482.
56 CORR, 3: 887.
57 CORR, 2: 513.
58 See Weeks, 1960.
59 Ampère, 1816a, p. 297.
60 Ibid., p. 380.
61 Ibid., p. 394.
62 Ampère, 1816b, pp. 20–2.
63 Quoted in Lemay and Oesper, 1948, p. 176, from a 5 August, 1816, letter in the private collection of Pierre Lemay.
64 AS, *chemise* 114.
65 Ibid. The details of these techniques were recorded in a set of notes taken by Joseph Liouville for a "Cours de physique mathématique" given by Ampère at the Collège de France during 1826 and 1827. Ampère corrected these notes himself and they have been cited in Hamamdjian, 1978; see Lützen, 1990, pp. 8–9.
66 Cited from Alfred Tulk's translation of Oken's *Elements of Physiophilosophy* in Williams and Steffens, 1978, vol. 3, pp. 143–4.
67 Caneva, 1980.
68 Oersted, 1806 and 1813.
69 Oersted, 1813, pp. 109–10.
70 Ibid., p. 130.
71 Ampère, 1822a, p. 449.
72 For detailed discussions of Laplacian optics see Buchwald (1989), Frankel (1974), and Grattan-Guinness (1990a), vol. 1, pp. 470–88.
73 See Buchwald, 1989, pp. 311–24.
74 Ampère, 1816c, p. 235.
75 Ibid., p. 247.
76 *Bibliothèque Universelle*, anonymous report, 1816, vol. 1, p. 319.
77 Ibid., pp. 320–1.

78 Fresnel, *Oeuvres*, vol. 2, p. 835.
79 Ibid., pp. 841–2.
80 See Buchwald, 1989, pp. 205–14.
81 Fresnel, *Oeuvres*, vol. 1, pp. 629–30.
82 Ibid., p. 630.
83 Ibid., p. 631.
84 AS, *chemise* 261.

CHAPTER 7 AMPÈRE'S RESPONSE TO THE DISCOVERY OF
ELECTROMAGNETISM (1820)

1   Ampère, 1826f; revised and reprinted as Ampère, 1827d.
2   See, for example, Duhem, 1974.
3   Maxwell, 1954 reprint, vol. 2, pp. 175–6.
4   Blondel, 1985, and Williams, 1983 and 1985a.
5   Blondel, 1982, and Grattan-Guinness, 1990a and 1991.
6   See Hofmann, 1987a.
7   Translated from the French publication of Oersted's memoir: *Bibliothèque Universelle*, 1820, vol. 14, pp. 279–80.
8   Biot, 1821a, p. 228.
9   Biot and Savart made reports to the Académie on 30 October, 1820, and 18 December, 1820. These reports were never published but a short résumé of the first one appeared in a "Note" published in both the *Annales de Chimie et de Physique* and the *Journal de Physique* late in 1820 (Biot and Savart, 1820). Biot gave a more detailed retrospective account of both sets of measurements at the public session of the Académie for 2 April, 1821; his discourse was published soon thereafter (Biot, 1821a). The title and much of the content of this publication were incorporated into the second edition of Biot's *Précis* in 1821 (Biot, 1821b, pp. 117–28), and a much longer version appeared in the 1824 third edition (Biot, 1824). The quotation cited here is from Biot and Savart, 1820, p. 223 of the *Annales de Chimie* publication.
10 Biot, 1824, vol. 2, pp. 742–6.
11 Biot, 1821b, vol. 2, p. 123.
12 Biot, 1824, vol. 2, p. 773.
13 Caneva, 1980 and 1981.
14 See, in particular, Williams, 1983.
15 Ampère, 1820i and 1820j.
16 Ampère, 1820b.
17 See Ampère, 1820c–1820h.
18 For evidence on this point see Williams, 1983.
19 This important manuscript appears to be a rejected draft for the concluding part of the memoir Ampère eventually presented to the Académie on 26 December, 1820 (Ampère, 1820j). The manuscript is preserved in AS, *chemise* 158. It has been partially edited and published in Blondel, 1978.

When Ampère taught physics at the Collège de France in November 1826, he and Joseph Liouville composed a long manuscript that includes a symmetry proof by Liouville in support of Ampère's long-standing assumption that the electrodynamic force acts along the line joining any two current elements; AS, *chemise* 208$_{bis}$.

20 AS, *chemise* 158; edited by Blondel, 1978, p. 64.

21 Ibid.

22 Ampère and Babinet, 1822.

23 Ampère mentions a completion date in a 23 January 1822 letter to Faraday edited in Ross, 1965, pp. 247–248. Although Babinet composed most of this *Exposé*, Ampère made important additions and corrections. In the *Académie* Ampère archives, carton 11, *chemise* 208, there is a valuable set of proof sheets for the first nineteen sections; these provide considerable insight into Ampère's thinking during this period, particularly due to passages that Ampère let stand at this time but which he later saw in need of correction.

24 Ampère and Babinet, 1822, p. 19.

25 Ampère, 1820b, p. 240 and Ampère, 1820j, p. 172.

26 AS, *chemise* 158; edited in Blondel, 1978, p. 182.

27 Ampère, 1820j, p. 182.

28 Ibid., p. 173.

29 Ampère, 1820c, p. 168. Much of the chronology in this publication was repeated in an unsigned report by Laumont (Laumont, 1820). Ampère identified the author in CORR, 2: 567.

30 L. Pearce Williams has argued for a 25 September date but he cites no textual evidence. Why would Ampère say that his anomalous discovery suddenly delayed his plans to test his force law hypothesis if he had made that discovery prior to the detection of electrodynamic forces between linear currents? It is, of course, possible that Ampère simply confused the order of his own discoveries. For further discussion of this issue see Williams, 1983 and 1985a, and Blondel, 1985.

31 Ampère, 1820j, p. 176.

32 To my knowledge, no immediate record was made of what Ampère actually said at the 6 November meeting. AS, *chemise* 164 includes a page of Ampère's notes for his presentation. His earliest recollections are probably those published in Ampère, 1820c, p. 168.

33 AS, *chemise* 158. For one of the fullest published discussions of this issue, see Ampère, 1827d, pp. 199 and 296–302.

34 Ampère, 1820j, p. 174.

35 In his most detailed description of these demonstrations, Ampère refers to an instrument of the type shown in plate 6; Ampère, 1820c, pp. 168–9.

36 Ampère, 1820h.

37 Ibid., pp. 226–7. A more detailed argument is provided by a manuscript draft for the 4 December memoir: AS, *chemise* 162.

38 Ampère, 1820h, p. 229. Ampère adopted the notation of n and k for his

two parameters during 1821. He originally simply used r^{-2} for the assumed distance dependency and wrote the parameter k as the ratio n/m to express the fact that it indicates the relative strength of two forces.

39 The primary sources for these arguments are the manuscripts Ampère composed for his lectures to the Académie during December, 1820, and January, 1821: AS, *chemises* 162, 163, and 166.

40 Ampère, 1820j, p. 174.

41 Laumont, 1820, p. 544.

42 Ampère, 1821a, p. 161. The same passage appears in Ampère, 1820g, p. 554.

43 AS, *chemise* 164.

44 Ampère, 1821a, p. 162. To my knowledge, Ampère used the engraving depicted in plate 9 for the first time when he reprinted Ampère, 1820g in his anthology Ampère 1823b, pp. 87–92.

45 Ampère, 1821a, p. 162.

46 Ampère, 1821b, pp. 311 and 318.

47 Ibid., p. 318.

48 AS, carton 32 (*non-classée*).

49 Ampère, 1826f and 1827d.

50 Gooding, 1986.

51 Ampère, 1820h and AS, *chemises* 162, 163, and 189.

52 AS, *chemise* 162. A similar passage appears in Ampère, 1820h, p. 229.

53 AS, *chemise* 163.

54 Ampère, 1822d, p. 407.

55 AS, *chemise* 189.

56 Ampère and Babinet, 1822, pp. 76–78.

57 Ampère's discussion appears in a draft of an undated letter that he may never have sent. It was probably written very early in 1821; AS, *chemise* 182.

58 AS, *chemise* 163.

59 Savary, 1823.

60 Ampère, 1823c, pp. 221–2.

61 Ibid., p. 222.

CHAPTER 8   AMPÈRE'S ELECTRODYNAMICS (1821–1822)

1 Ampère's first extensive discussion of electric current was in his 1822a. He provided a much more sophisticated version in the notes for his physics course at the Collège de France; these have been edited in Blondel, 1982, pp. 177–86.

2 Ampère, 1823b, p. 87; he inserted some recollections in this reprint of his 1820g for the 1823 *Recueil*.

3 As usual, Ampère's memory is not entirely trustworthy. His claim that 4 December is when Biot verbally communicated the results of a second set

of measurements is included in a draft found in AS, *chemise* 179; this claim is repeated in another draft held in the manuscript collection of the Bibliothèque de la Ville de Lyon (Recueil A.1, ff. 91–114); it also appears less forcefully in the edited version of Ampère 1820g included in his *Recueil*: Ampère, 1823b, p. 87. For some reason, Ampère later claimed that he only became aware of Biot's conclusions from the second edition of his *Précis* in 1822; Ampère, 1827d, pp. 281–2.

4 AS, *chemise* 163.
5 AS, *chemise* 166.
6 CORR, 2: 570.
7 CORR, 2: 569.
8 CORR, 2: 569.
9 CORR, 3: 922.
10 CORR, 2: 566.
11 CORR, 3: 925.
12 CORR, 3: 907.
13 Ampère, 1821g, p. 102.
14 Ampère, 1821b, and Delambre, 1824, pp. cxxxix–cxl.
15 Ampère, 1821f, p. 128.
16 CORR, 3: 916–17.
17 Prechtl, 1821a and 1821b.
18 AS, *chemise* 182.
19 Biot, 1821a, p. 231.
20 Ampère, 1821d, p. 17. This is the published version of Ampère's lecture.
21 Ibid., p. 20.
22 AS, *chemise* 206$_{bis}$.
23 Biot, 1824, vol. 2, pp. 704–74.
24 Ibid., p. 773.
25 Ibid., p. 768.
26 Biot, 1821a, p. 233.
27 Ampère, 1822d, p. 399–400.
28 Ampère, 1820b, p. 249.
29 Ibid., p. 250.
30 Fresnel, 1820.
31 Fresnel, 1821a, p. 141.
32 Ampère, 1820b, pp. 252–4 and Ampère, 1820j, p. 190.
33 Ampère, 1821c, p. 189.
34 G. de La Rive, 1821a.
35 Ampère, 1822a, p. 448.
36 Ibid., 447–8.
37 Ampère, 1822a, p. 449 and in a letter to Faraday dated 10 July, 1822: CORR, 2: 587.
38 A. de La Rive, 1822, p. 47.
39 A. van Beek, 1821.
40 Faraday, 1821a; for a discussion of Faraday's criticism, see Williams, 1985b and 1986.

41 Fresnel, 1821b, p. 146 and Ampère, 1822f. Also see Ampère's 12 June, 1822, letter to Gaspard de La Rive; CORR, 2: 580–2.

42 Savary, 1821, p. 370.

43 See Ampère's letters to Erman on 20 April, 1821, and Auguste de La Rive during 11–31 October, 1821; CORR, 3: 917 and 2: 618.

44 Savary, 1821, pp. 377–8.

45 CORR, 2: 576–7.

46 Savary, 1821, p. 372.

47 Ampère, 1822a, p. 467.

48 See Ampère's 23 January, 1822, letter to Faraday edited in Ross, 1965, p. 217.

49 According to Ampère's own reports, which of course are not entirely reliable, he designed the instrument by the spring of 1822 (Ampère, 1823c, p. 234). The apparatus was apparently not constructed until one year later, approximately at the time of the publication of the second edition of the *Recueil*; at any rate, Ampère wrote a short note describing the instrument for the 16 June 1823, *Procès-verbaux* of the Académie and said that it was "presented" to the Académie on that date.

50 Ampère, 1822b, p. 66.

51 Ibid.

52 AS, *chemise* 206_{bis}.

53 Ibid.

54 AS, *chemise* 173.

55 Ampère, 1823c, p. 236.

56 Prechtl, 1821a, pp. 263–4; see AS, *chemise* 192.

57 Ampère, 1822d, pp. 407–13.

58 For a detailed summary of Ampère's derivation, see Grattan-Guinness, 1990a, vol. 2, pp. 930–3.

59 Ampère, 1822d, p. 418.

60 Ibid., p. 400.

61 Ampère, 1823d, pp. 300–2.

62 AS, *chemise* 179.

63 *Procès-verbaux*, vol. 7, p. 512.

CHAPTER 9   DEFENSE AND ELABORATION OF THE THEORY

1 Mendoza, 1985.

2 These notes were published by Joubert in 1885; see Ampère, 1885.

3 Ampère, 1885, p. 334.

4 For a detailed account see Hofmann 1987a, pp. 65–7.

5 CORR, 2: 765.

6 Auguste de La Rive reported that the direction of the rotation of the copper loop reversed "following the direction of the current in the surrounding conductors"; Auguste de La Rive, 1822, p. 48. This gloss was so inaccurate that some readers interpreted it to mean that the induced current

direction was the same as that of the primary current. Jean-Baptiste Demonferrand passed along this erroneous claim in his widely read text (Demonferrand, 1823, pp. 173–4). Ampère sent a copy of Demonferrand's book to Faraday, and this contributed to Faraday's inaccurate understanding of Ampère's experiment.

7   Ampère, 1885, p. 332.
8   AS, *chemise* 182; also see Pouillet, 1822.
9   Ampère, 1822h, 1822i, 1823a.
10 For example, see Mascart and Joubert, 1896, vol. 1, pp. 531–2.
11 Ampère, 1823b, p. 360.
12 Savary, 1823.
13 CORR, 2: 624.
14 Ampère, 1824b, 1824d, 1824e, 1825e, 1826d, and 1826e.
15 Poisson, 1826a and 1826b.
16 Poisson, 1826a, pp. 249–50.
17 Ampère 1824b and 1824d. The former of these was published as a monograph and included three long "Notes" not included in the 1824d publication in the *Annales de Chimie et de Physique*. Another summary was published as Ampère, 1824c. Joubert published a manuscript version of the memoir in Joubert, 1885, vol. 2, pp. 395–410.
18 Ampère, 1826e. Joubert published a manuscript version in Joubert, 1887, vol. 3, pp. 194–202.
19 Ampère, 1826e, p. 43.
20 Ampère, 1827d, pp. 229–231.
21 Relevant calculations are in the manuscript version of the November, 1825, memoir; Joubert, 1887, vol. 3, pp. 198–202.
22 Ampère, 1826e, p. 45. Some of the calculations are included in the manuscript version published in: Joubert, 1887, vol. 3, pp. 194–202.
23 Ampère, 1826e, pp. 46–7.
24 Ampère, 1827b.
25 Ibid., p. 19.
26 Ibid., p. 21.
27 Ampère, 1826f.
28 Ampère, 1827d.
29 Ibid., pp. 176–7.
30 Ibid., pp. 294–5.
31 Ibid., p. 299.
32 Moigno, 1849, pp. 222–40 and, in the 1852 second edition, pp. 302–17. Ampère's text is published under the title "Sur le mode de transmission des courants électriques et la théorie électro-chimique." Blondel also published it in Blondel, 1982, pp. 177–86.
33 Ampère, 1827d, p. 301.
34 Ampère, 1827b.
35 Ampère, 1827d, pp. 212–17.
36 Ibid., p. 270.
37 Ibid., pp. 304–20.

38 Ibid., pp. 360–5.
39 Ibid., pp. 366–73.
40 Ibid., p. 318.
41 Ibid., p. 320.
42 Ibid., pp. 360–73.
43 Ibid., p. 311.
44 Ibid., p. 365.
45 Ibid., pp. 366–7.
46 For example, Mascart and Joubert, 1896, p. 488.
47 Maxwell, 1891, vol. 2, 1954 reprint, p. 155.
48 This contrast between Ampère and Maxwell has been explored in an interesting article by Pierre-Gérard Hamamdjian, 1978.

CHAPTER 10   THE FINAL SYNOPSIS

1 Louis de Launay has drawn from Jean-Jacques' journals to assemble a chronicle of this relationship: Launay, 1927.
2 CORR, 2: 741.
3 CORR, 2: 791.
4 Ampère, 1832b and 1835. See Brush, 1970.
5 Ampère, 1832b, p. 228.
6 Appel, 1987.
7 Ampère, 1824g.
8 AS, *chemise* 255.
9 AS, *chemise* 254.
10 Ampère, 1834, p. x.
11 Ampère, 1834, p. xiii.
12 Ampère, 1832c.
13 Ampère, 1834, p. xviii.
14 Ibid., pp. 13–14.
15 Ibid., pp. xx–xxi.
16 Ibid., pp. 42–3.
17 Ibid., p. 39.
18 Ibid., p. 48.
19 See Burkhardt, 1977.
20 Ampère, 1843, p. 121.
21 Smith, 1948, p. 3.

Bibliography

The following bibliography includes most of Ampère's publications; it does not include multiple references to memoirs published in more than one format with only slight variations.

ANDRÉ-MARIE AMPÈRE

1802–1817

Ampère, A.-M. (1802): *Considerations sur la Théorie mathématique du jeu* (Lyon and Paris: Frères Perisse).

—— (1806a): "Recherches sur quelques points de la théorie des fonctions dérivées qui conduisent à une nouvelle démonstration de la série de Taylor, et à l'expression finie des termes qu'on néglige lorsqu'on arrête cette serie à un terme quelconque," *École Polytechnique Journal*, 6 (cahier 13): 148–81.

—— (1806b): "Recherches sur l'application des formules générales du calcul des variations aux problèmes de la mécanique," *Mémoires présentés à l'Institut des Sciences, Lettres et Arts par divers savants et lus dans ses assemblées. Sciences mathématiques et physiques*, 1: 493–523.

—— (1806c): "Démonstration générale du principe des Vitesses Virtuelles, dégagée de la considération des infiniment petits," *École Polytechnique Journal*, 6 (cahier 13): 247–69.

—— (1808): "Sur les avantages qu'on peut retirer, dans la théorie des Courbes, de la considération des Paraboles osculatrices, avec des Réflexions sur les fonctions différentielles dont la valeur ne change pas lors de la transformation des axes," *École Polytechnique Journal*, 7 (cahier 14): 159–81.

—— (1809): "Mémoire sur la fonction dérivée, ou coefficient différenciel du

384

premier ordre, lu par M. Binet, professeur de mathématiques transcendantes au Lycée de Rennes", *Bulletin de la Société Philomathique de Paris*, 1 (1807–9): 275–8.

—— (1814): "Lettre de M. Ampère à M. le comte Berthollet, sur la détermination des proportions dans lesquelles les corps se combinent d'après le nombre et la disposition respective des molécules dont leurs particules intégrantes sont composées," *Annales de Chimie*, 90: 43–86.

—— (1815a): "Démonstration de la relation découverte par Mariotte, entre les volumes des gaz et les pressions qu'ils supportent à une même température," *Annales de Chimie*, 94: 145–60.

—— (1815b): "Considérations générales sur les intégrales des équations aux différentielles partielles," *Journal de l'École Royale Polytechnique*, 10 (cahier 17, January): 549–611.

—— (1816a): "Essai d'une Classification naturelle pour les Corps simples," *Annales de Chimie et de Physique*, 1: 295–308 and 373–94.

—— (1816b): "Suite d'une Classification naturelle pour les Corps simples," *Annales de Chimie et de Physique*, 2: 5–32 and 105–25.

—— (1816c): "Démonstration d'un Théorême d'où l'on peut déduire toutes les lois de la Réfraction ordinaire et extraordinaire", *Mémoires de la Classe des Sciences Mathématiques et Physiques de l'Institut de France*, 14 (1813–15): 235–48.

—— (1816d): "Mémoire sur la réfraction de la lumière" (anonymous), *Correspondance sur l'École Polytechnique*, 3 (1814–16): 238–42.

—— (1817): "Problème des quadratures. Rapport à l'académie royale des sciences, *sur le mémoire de M. Bérard*, inséré à la page 110 du VII° volume de ce recueil," *Annales de Mathématiques Pures et Appliquées*, 8 (numéro 4): 117–24.

1820

Ampère, A.-M. (1820a): "Mémoire Contenant l'Application de la Théorie exposée dans le XVII° Cahier du *Journal de l'École polytechnique*, à l'Intégration des Équations aux Différentielles partielles du premier et du second ordre," *Journal de L'École Royale Polytechnique*, 11 (cahier 18, January): 1–188.

—— (1820b): "Analyse des mémoires lus par M. Ampère à l'Académie des Sciences, dans les séances des 18 et 25 septembre, des 9 et 30 octobre 1820", *Annales Générales des Sciences Physiques*, 6: 238–57.

—— (1820c): "Notes de M. Ampère sur les lectures qu'il a faites à l'Académie des Sciences," *Journal de Physique, de Chimie, d'Histoire Naturelle et des Arts*, 91 (September): 166–9.

—— (1820d): "Conclusions d'un Mémoire sur l'Action mutuelle de deux courans électriques, sur celle qui existe entre un courant électrique et un aimant, et celle de deux aimans l'un sur l'autre; lu à l'Académie royale des Sciences, le 25 septembre 1820," *Journal de Physique, de Chimie, d'Histoire Naturelle et des Arts*, 91: 76–8.

—— (1820e): "Suite de l'analyse des mémoires de M. Ampère," *Annales Générales des Sciences Physiques*, 7: 252–8.

—— (1820f): "Mémoire sur l'expression analytique des attractions et répulsions des courans électriques. Lu le 4 décembre 1820," *Annales des Mines*, 5: 546–53.

—— (1820g): "Exposition du moyen par lequel il est facile de s'assurer directement, et par des expériences précises, de l'exactitude de la loi des attractions et répulsions des courans électriques, suivie de quelques observations sur cette loi. Mémoire lu le 26 décembre 1820," *Annales des Mines*, 5: 553–8.

—— (1820h): "Note sur un Mémoire lu à l'Académie royale des Sciences, dans la séance du 4 décembre 1820," *Journal de Physique, de Chimie, d'Histoire Naturelle et des Arts*, 91: 226–30.

—— (1820i): "Mémoire présenté à l'Académie royale des Sciences, le 2 octobre 1820, où se trouve compris le résumé de ce qui avait été lu à la même Académie les 18 et 25 septembre 1820, sur les effets des courans électriques," *Annales de Chimie et de Physique*, 15: 59–76.

—— (1820j): "Suite du Mémoire sur l'Action mutuelle entre deux courans électriques, entre un courant électrique et un aimant ou le globe terrestre, et entre deux aimans," *Annales de Chimie et de Physique*, 15: 170–218.

1821

Ampère, A.-M. (1821a): "Sur deux Mémoires lus par M. Ampère à l'Académie royale des Sciences, le premier dans la séance du 26 décembre 1820; le second dans les séances des 8 et 15 janvier 1821," *Journal de Physique, de Chimie, d'Histoire Naturelle et des Arts*, 92: 160–5.

—— (1821b): "Exposé sommaire des divers Mémoires lus par Mr. Ampère à l'Académie Royale des Sciences de Paris, sur l'action mutuelle de deux courans électriques, et sur celle qui existe entre un courant électrique et le globe terrestre ou un aimant," *Bibliothèque Universelle des Sciences, Belles-lettres et Arts*, 16: 309–19.

—— (1821c): "Lettre de M. Ampère à M. Erman, Secrétaire perpétuel de l'Académie royale," *Journal de Physique, de Chimie, d'Histoire Naturelle et des Arts*, 92: 304–9 and *Bibliothèque Universelle des Sciences, Belles-lettres et Arts*, 17: 183–91.

—— (1821d): "Notice sur les expériences électro-magnétiques de MM. Ampère et Arago, par Mr. Ampère; lue à la séance publique de l'Académie Royale des sciences de Paris le 2 avril 1821," *Bibliothèque Universelle des Sciences, Belles-lettres et Arts*, 17: 16–20.

—— (1821e): "Extrait d'une lettre de Mr. Ampère au Prof. De La Rive," *Bibliothèque Universelle des Sciences, Belles-lettres et des Arts*, 17: 192–4.

—— (1821f): "Lettre de M. Ampère à M. Arago," *Annales de Chimie et de Physique*, 16: 119–29.

—— (1821g): "Note sur un appareil à l'aide duquel on peut vérifier toutes les propriétés des conducteurs de l'électricité voltaïque, découvertes par M. Ampère," *Annales de Chimie et de Physique*, 18: 88–106.

—— (1821h): "Suite de la Note sur un Appareil à l'aide duquel on peut vérifier toutes les propriétés des conducteurs de l'électricité voltaïque, découvertes par M. Ampère," *Annales de Chimie et de Physique*, 18: 313–33.

1822

Ampère, A.-M. (1822a): "Réponse de M. Ampère à la lettre de M. Van Beck [sic], sur une nouvelle Expérience électro-magnétique," *Journal de Physique, de Chimie, d'Histoire Naturelle, et des Arts*, 93: 447–67.

—— (1822b): "Notice sur les nouvelles Expériences electro-magnétiques faites par différens Physiciens, depuis le mois de mars 1821, lue dans la séance publique de l'Académie royale des Sciences, le 8 avril 1822," *Journal de Physique, de Chimie, d'Histoire Naturelle et des Arts*, 94: 61–6.

—— (1822c): "Expériences relatives à de nouveaux phénomènes électro-dynamiques," *Annales de Chimie et de Physique*, 20: 60–74.

—— (1822d): "Mémoire sur la Détermination de la formule qui représente l'action mutuelle de deux portions infiniment petites de conducteurs voltaïques," *Annales de Chimie et de Physique*, 20: 398–421.

—— (1822e): "Extrait d'un Mémoire lu à l'Académie royale des Sciences, dans la séance du 16 septembre 1822; par M. Ampère," *Bulletin de la Société Philomathique* (October): 145–7.

—— (1822f): Untitled Note inserted on pp. 245–6 in Delambre (1822): "Notice sur de nouvelles recherches de Mr. Ampère relatives aux phénomènes électro-magnétiques," *Bibliothèque Universelle des Sciences, Belles-lettres et Arts*, 19: 244–7.

—— (1822g): Untitled note appended to Auguste de La Rive (1822): "Sur l'Action qu'exerce le globe terrestre sur une portion mobile du circuit voltaïque," *Annales de Chimie et de Physique*, 21: 24–48, on pp. 48–53.

—— (1822h): "Exposé méthodique des phénomènes électro-dynamiques et des lois de ces phénomènes," *Journal de Physique, de Chimie, d'Histoire Naturelle et des Arts*, 95: 248–57.

—— (1822i): "Exposé méthodique des phénomènes électro-dynamiques et des lois de ces phénomènes," *Bulletin de la Société Philomathique* (November): 177–83.

—— (1822j): "Nouvelles expériences électro-magnétiques de MM. Faraday, Ampère, H. Davy, et De La Rive," *Bulletin de la Société Philomathique* (1822): 21–3.

—— (1822k): "Extrait d'une Lettre de Mr. Ampère au Prof. De La Rive sur des expériences électro-magnétiques," *Bibliothèque Universelle des Sciences, Belles-lettres et Arts*, 20: 185–92.

—— (1822l): Untitled anonymous note appended to Van der Heyden (1822): "Lettre à M. Ampère," *Journal de Physique, de Chimie, d'Histoire Naturelle et des Arts*, 95: 70–1.

Ampère, A.-M. and Babinet, J. (1822): *Exposé des nouvelles découvertes sur l'électricité et le magnétism de MM. Oersted, Arago, Ampère, H. Davy, Biot, Erman, Schweiger, de La Rive, etc.* (Paris).

1823

Ampère, A.-M. (1823a): *Exposé méthodique des phénomènes électro-dynamiques, et des lois de ces phénomènes* (Paris: Bachelier).

—— (1823b): *Recueil d'observations électro-dynamiques, contenant divers mémoires, notices, extraits de lettres ou d'ouvrages périodiques sur les sciences, relatifs à l'action mutuelle de deux courans électriques, à celle qui existe entre un courant électrique et un aimant ou le globe terrestre, et à celle de deux aimans l'un sur l'autre* (Paris: Crochard, with publication date 1822).

—— (1823c): "Notes sur cet exposé des nouvelles Expériences relatives aux Phénomènes produits par l'action électro-dynamique, faites depuis le mois de mars 1821," in *Recueil* (Ampère, 1823b), pp. 207–36.

—— (1823d): "Second mémoire sur la Détermination de la formule qui représente l'action mutuelle de deux portions infiniment petites de conducteurs voltaïques," in *Recueil* (Ampère, 1823b), pp. 293–318.

—— (1823e): "Extrait d'une Lettre de M. Ampère à M. Faraday," *Annales de Chimie et de Physique*, 22: 389–400.

—— (1823f): "Notice de deux Mémoires lus à l'Académie des sciences dans la séance du 3 février 1823, par MM. Savary et de Monferrand," *Bibliothèque Universelle des Sciences, Belles-lettres et Arts*, 22: 259–64.

—— (1823g): "Extrait des Mémoires de M. Savary et de M. de Monferrand sur des applications du calcul à la théorie des phénomènes électro-dynamiques," *Bulletin de la Société Philomathique* (1823): 61–3.

—— (1823h): "Observations sur le Mémoire de Mr. Savary dont on a rendu compte dans le cahier de janvier 1823 de ce Recueil," *Bibliothèque Universelle des Sciences, Belles-lettres et Arts*, 24: 109–15.

Ampère, A.-M. and Roche, J. P. L. A. (1823a): "Mémoire sur l'application du calcul aux phénomènes électro-dynamiques," *Bulletin Général et Universel des Annonces et des Nouvelles Scientifiques*, 3: 20–2.

—— (1823b): "Détermination de l'action électro-dynamique d'un fil d'acier aimanté curviligne faite par M. Savary, et communiq. à l'Acad. des sc., le 28 juillet 1823," *Bulletin Général et Universel des Annonces et des Nouvelles Scientifiques*, 3 (numéro 2): 219–20.

1824

Ampère, A. (1824a): *Précis de Calcul différentiel et de calcul intégral.* (Incomplete text privately printed in Paris.)

Ampère, A.-M. (1824b): *Précis de la Théorie des phénomènes électro-dynamiques pour servir de supplément au "Recueil d'observations électro-dynamiques" et au "Manuel d'électricité dynamique" de M. de Monferrand* (Paris: Crochard and Bachelier).

—— (1824c): "Extrait d'un Mémoir sur les phénomènes électro-dynamiques; par M. Ampère: présentè à l'Académie des sciences, dans sa séance du 22 décembre 1823," *Bulletin de la Société Philomathique* (June): 79–85.

—— (1824d): "Extrait d'un Mémoir sur les Phénomènes électro-dynamiques," *Annales de Chimie et de Physique*, 26: 134–62 and 246–58.

—— (1824e): "Description d'un Appareil électro-dynamique," *Annales de Chimie et de Physique*, 26: 390–411.

—— (1824f): "Note sur une Expérience relative à la nature du courant électrique, faite par MM. Ampère et Becquerel," *Annales de Chimie et de Physique*, tome 27: 29–31.

—— (1824g): "Considérations philosophiques sur la détermination du système solide et du système nerveux des animaux articulés," *Annales des Sciences Naturelles*, 2: 295–310 and 3: 199–203 and 453–6.

Ampère, A.-M. and Dulong, P. L. (1824): "Rapport sur un Mémoire de M. Rousseau relatif à un nouveau moyen de mesurer la conductibilité des corps pour l'électricité", *Annales de Chimie et de Physique*, 25: 373–379.

1825

Ampère, A.-M. (1825a): "Analogie entre les facultés numériques et les puissances; Démonstration générale de la formule du Binôme de Newton; Développement des fonctions exponentielles et circulaires," *Annales de Mathématiques Pures et Appliquées*, 15 (numéro 12, June): 369–87.

—— (1825b): "Exposition des principes du calcul des variations," *Annales de Mathématiques Pures et Appliquées*, 16 (numéro 5, November): 133–67 .

—— (1825c): "Extrait d'un Rapport fait à l'Académie, par M. Ampère, sur les Piles sèches de M. Zamboni," *Annales de Chimie et de Physique*, 29: 198–200.

—— (1825d): "Lettre de M. Ampère à M. Gerhardi sur divers phénomènes électro-dynamiques", *Annales de Chimie et de Physique*, 29: 373–81.

—— (1825e): "Mémoire sur une nouvelle Expérience électro-dynamique, sur son application à la formule qui représente l'action mutuelle de deux élémens de conducteurs voltaïques, et sur de nouvelles conséquences déduites de cette formule", *Annales de Chimie et de Physique*, 29: 381–404 and 30: 29–41.

—— (1825f): "Note sur une nouvelle expérience électro-dynamique et sur son application à la formule qui représente l'action mutuelle de deux élémens de conducteurs voltaïques; par M. Ampère, de l'Institut de France, etc.," *Correspondance Mathématique et Physique*, 1 (numéro 5): 276–80.

1826

Ampère, A.-M. (1826a): "Essai sur un nouveau mode d'exposition des principes du calcul différentiel, du calcul aux différences et de l'interpolation des suites, considérées comme dérivant d'une source commune," *Annales de Mathématiques Pures et Appliquées*, 16 (numéro 11, May): 329–49.

—— (1826b): "Nouvelle démonstration du principe des vitesses virtuelles, par M. Ampere, de l'Institut de France," *Correspondance Mathématique et Physique*, 2: 276–81.

—— (1826c): "Mémoire sur quelques nouvelles propriétés des Axes permanens de rotation des corps et des Plans directeurs de ces axes," *Mémoires de l'Académie Royale des Sciences de l'Institut de France*, 5 (1821-2): 86–152.

—— (1826d): "Note sur une nouvelle expérience électro-dynamique de M. Ampère, qui constate l'action d'un disque métallique en mouvement sur une portion de conducteur voltaïque pliée en hélice ou en spirale," *Bulletin de la Société Philomathique* (1826): 134.

—— (1826e): "Extrait d'un Mémoire sur l'action exercée par un circuit électro-dynamique, formant une courbe plane dont les dimensions sont considerées comme infiniment petites; sur la manière d'y ramener celle d'un circuit fermé, quelles qu'en soient la forme et le grandeur; sur deux nouveaux instrumens destinés à des expériences propres à rendre plus directe et à vérifier la détermination de la valeur de l'action mutuelle de deux élémens de conducteurs; sur l'identité des forces produites par des circuits infiniment petits, et par des particules d'aimant; enfin sur un nouveau théorème relatif à l'action de ces particules, lu à l'Académie royale des sciences dans sa séance du 21 novembre 1825," *Correspondance Mathématique et Physique*, 2: 35–47.

—— (1826f): *Théorie des Phénomènes électro-dynamiques, uniquement déduite de l'expérience* (Paris: Méquignon-Marvis).

—— (1826g): "Note sur une nouvelle expérience électro-dynamique, qui constate l'action d'un disque métallique en mouvement sur une portion de conducteur voltaïque, plié en hélice ou en spirale," *Bulletin Universel des Sciences et de l'industrie*, 6: 211–214.

—— (1826h): "Note sur quelques Phénomènes électro-magnétiques," *Annales de Chimie et de Physique*, 32: 432–43 (unsigned).

1827

Ampère, A. (1827a): "Démonstration du théorème de Taylor, pour les fonctions d'un nombre quelconque de variables indépendantes, avec la détermination de l'erreur que l'on commet lorsqu'on arrête la série donnée par ce théorème à l'un quelconque de ses termes," *Annales de Mathématiques Pures et Appliquées*, 17 (numéro 11, May): 317–29.

—— (1827b): "Mémoire sur l'action mutuelle d'un conducteur voltaique et d'un aimant," *Nouveaux Mémoires de l'Académie Royale des Sciences et Belles-Lettres de Bruxelles*, 4: 1–70.

—— (1827c): "Supplément. Lettre à M. Le Docteur Gherardi," *Nouveaux Mémoires de l'Académie Royale des Sciences et Belles-Lettres de Bruxelles*, 4: 71–88.

Ampère, A.-M. (1827d): "Mémoire sur la théorie mathématique des phénomènes électro-dynamiques uniquement déduite de l'expérience, dans lequel se trouvent réunis les Mémoires que M. Ampère a communiqués à la Académie royale des Sciences, dans les séances des 4 et 26 décembre 1820, 10 juin 1822, 22 décembre 1823, 12 septembre et 21 novembre 1825," *Mémoires de l'Académie Royale des Sciences de l'Institut de France*, 6 (1823): 175–387.

1828

Ampère, A.-M. (1828a): "Note sur l'Action mutuelle d'un Aimant et d'un Conducteur voltaïque," *Annales de Chimie et de Physique*, 37: 113–39.

—— (1828b): "Mémoire sur la Détermination de la surface courbe des ondes lumineuses dans un milieu dont l'élasticité est différente suivant les trois directions principales, c'est-à-dire celles où la force produite par l'élasticité a lieu dans la direction même du déplacement des molécules de ce milieu," *Annales de Chimie et de Physique*, 39: 113–45.

1829

Ampère, A. (1829a): "Solution d'un problème de dynamique, suivie de considérations sur le problème général des forces centrales," *Annales de Mathématiques Pures et Appliquées*, 20 (numéro 2, August): 37–58.

—— (1829b): "Démonstration élémentaire du principe de la gravitation universelle," *Annales de Mathématiques Pures et Appliquées*, 20 (numéro 4, October): 89–96.

1831–1836

Ampère, A.-M. (1831a): "Sur la théorie des forces centrales," *Annales de Mathématiques Pures et Appliquées*, 22 (July): 1–30.

—— (1831b): "Faits proposés comme objet de recherche par M. Ampère à M. Becquerel, et qu'ils ont obtenus ensemble," *Annales de Chimie et de Physique*, 48: 404–5.

—— (1831c): "Expériences sur les Courans électriques produits par l'influence d'un autre courant," *Annales de Chimie et de Physique*, 48: 405–12.

—— (1832a): "Note de M. Ampère sur une Expérience de M. Hippolyte Pixii, relative au Courant produit par la Rotation d'un aimant, à l'aide d'un appareil imaginé par M. Hippolyte Pixii," *Annales de Chimie et de Physique*, 51: 76–9.

—— (1832b): "Idées de Mr Ampère sur la chaleur et sur la lumière," *Bibliothèque Universelle des Sciences, Belles-lettres et Arts*, 49: 225–35.

—— (1832c): "Classification des connaissances humaines," *Revue Encyclopédique*, 54: 223–9.

—— (1834): *Essai sur la Philosophie des sciences* (Paris: Bachelier).

Ampère, A. (1835): "Note de M. Ampère sur la Chaleur et sur la Lumière considérées comme résultant de mouvemens vibratoires," *Annales de Chimie et de Physique*, 58: 432–44 and *Bibliothèque Universelle des Sciences, Belles-lettres et Arts*, 60: 26–37.

Ampère, A.-M. (1836): "Mémoire sur les Équations générales du Mouvement," *Journal de Mathématiques Pures et Appliquées*, 1: 211–28.

Posthumous

Ampère, A.-M. (1837): "Recherches mathématiques inédites de M. Ampère," *Correspondance Mathématique et Physique*, 9: 144–8.

—— (1838): "Nouvelle discussion de l'équation générale des courbes du second degré," *Correspondance Mathématique et Physique*, 10: 90–103.

—— (1843): *Essai sur la Philosophie des sciences, seconde partie* (Paris: Bachelier).

—— (1845): "Théorie du Calcul Élémentaire," *Nouvelle Annales de Mathématiques*, 4: 105–9, 161–4, 209–13, and 278–85.

—— (1849): "Sur le mode de transmission des courants électriques et la théorie électro-chimique," in F. Moigno, *Traité de Télégraphie électrique* (Paris: A. Franck), pp. 224–40; second edition, 1852, pp. 302–17.

—— (1885): "Notice sur quelques expériences nouvelles relatives à l'action mutuelle de deux portions de circuit voltaique et à la production des courants électriques par influence, et sur les circonstances dans lesquelles l'action électrodynamique doit, d'après la théorie, produire dans un conducteur mobile autour d'un axe fixe un mouvement de rotation continu, ou donner à ce conducteur une direction fixe," in J. Joubert (ed.), *Memoires sur l'Electrodynamique*, 1885, vol. 2, pp. 329–37.

—— (1936–43): *Correspondance du Grand Ampère*, edited by L. de Launay, 3 vols (Paris: Gauthier Villars).

References

Appel, T. A. (1987): *The Cuvier-Geoffroy Debate: French Biology in the Decades before Darwin* (Oxford: Oxford University Press).

Arago, F. (1854a): "Ampère" in *Oeuvres complètes*, edited by J.-A. Barral vol. 2, pp. 1–116.

—— (1854b): "Poisson" in *Oeuvres complètes*, vol. 2, pp. 593–671.

Arnold, D. H. (1983): "The *Mécanique Physique* of Siméon Denis Poisson: the evolution and isolation in France of his approach to physical theory (1800–1840)," *Archive for History of Exact Sciences*, 28: 243–367; 29: 37–51; and (1984), 29: 287–307.

Beek, A. van (1821): "Sur les procédés électriques par lesquels l'acier reçoit la vertu magnétique," *Bibiliothèque Universelle des Sciences, Belles-lettres, et Arts*, 18: 184–94.

Belhoste, B. (1981): *Augustin-Louis Cauchy: A Biography*, trans. F. Ragland (New York: Springer-Verlag).

Berthollet, C. L. (1809): "Introduction" to *Système de Chimie* by Thomas Thomson, translated by J. Riffault (Paris: Bernard).

Biot, J. B. (1804): "Sur les variations du magnétisme terrestre à différentes latitudes," *Journal de Physique, de Chimie et d'Histoire Naturelle*, 59: 429–50.

—— (1809): "Sur l'esprit du système", *Mercure de France*, 36: 247–252; reprinted in Biot, 1858, vol. 2, pp. 109–16.

—— (1816a): *Traité de Physique expérimentale et mathématique* (Paris: Deterville).

—— (1816b): "Lettre au Prof. Pictet sur le *Traité de physique*", *Bibiliothèque Universelle des Sciences, Belles-lettres, et Arts*, 2: 81–6.

—— (1821a): "Sur l'aimantation imprimée aux métaux par l'électricité en mouvement," *Journal des Savans*, April: 221–35.

—— (1821b): *Précis élémentaire de Physique expérimentale*, second edition (Paris: Deterville).

—— (1821c): "Remarques de M. Biot, sur un rapport lu, le 4 juin 1821, à l'Académie des Sciences par MM. Arago et Ampère," *Annales de Chimie et de Physique*, 17: 225–58.

—— (1824): *Précis élémentaire de Physique expérimentale*, third edition (Paris: Deterville).

Biot, J. B. and Savart, F. (1820): "Note sur le magnétisme de la pile de Volta," *Annales de Chimie et de Physique*, 15: 222–3.

Blondel, C. (1978): "Sur les premières recherches de formule électrodynamique par Ampère (octobre 1820)," *Revue d'Histoire des Sciences et de leures Applications*, 31: 53–65.

—— (1982): *A.-M. Ampère et la création de l'électrodynamique (1820–1827)* (Paris: Bibliothèque Nationale).

—— (1985): "Ampère and the programming of research," *Isis*, 76: 559–61.

—— (1989): "Vision physique «éthérienne», mathématisation «laplacienne»: l'électrodynamique d'Ampère," *Revue d'Histoire des Sciences et de leures Applications*, 42: 123–37.

Bos, H. J. M. (1974): "Differentials, higher-order differentials and the derivative in the Leibnizian calculus," *Archive for History of Exact Sciences*, 14: 1–90.

Bradley, Margaret (1976): "The facilities for practical instruction in science during the early years of the École Polytechnique," *Annals of Science*, 33: 425–46.

Bredin, C.-J. (1936): *Lettres inédites de Claude-Julien Bredin à André-Marie Ampère, au Pasteur A. Touchon et à Mme Touchon*, edited by Louis de Launay (Lyon: Académie des Sciences, Belles-Lettres et Arts de Lyon).

Brown, T. M. (1969): "The elctroic current in early nineteenth-centiry French physics," *Historical Studies in the Physical Sciences*, 1: 61–104.

Brush, S. G. (1970): "The wave theory of heat: a forgotten stage in the transition from caloric theory to thermodynamics," *British Journal for the History of Science*, 5: 145–67.

Buchwald, J. Z., (1989): *The Rise of the Wave Theory of Light: Optical Theory and Experiment in the Early Nineteenth Century* (Chicago: University of Chicago Press).

Burkhardt, R. W. (1977): *The Spirit of System: Lamarck and Evolutionary Biology* (Cambridge: Harvard University Press).

Caneva, K. (1980): "Ampère, the Etherians, and the Oersted connection," *British Journal for the History of Science*, 13: 121–38.

—— (1981): "What should we do with the monster? Electromagnetism and the psychosociology of knowledge," in E. Mendelsohn and Y. Elkana (eds), *Sciences and Cultures* (Dordrecht: Reidel), pp. 101–31.

Cartwright, N. (1989): *How the Laws of Physics Lie* (New York: Oxford University Press).

Cheuvreux, H. (1873): *Journal et correspondance d'André-Marie Ampère* (Paris: J. Hetzel).

Cheuvreux, H., (1875): *André-Marie et Jean-Jacques Ampère: Correspondance et souvenirs (de 1805 à 1864) recuellis par Madame H.C.*, 2 vols (Paris: J. Hetzel).

Cole, T. M. (1975): "Early atomic speculations of Marc Antoine Gaudin: Avagadro's hypothesis and the periodic system," *Isis*, 66: 334–60.

Coulomb, C. (1801): "Détermination théorique et expérimentale des forces qui ramènent différentes aiguilles aimantées à saturation à leur méridien magnétique," *Mémoires de la Classe des Sciences Mathématiques et Physiques de l'Institut de France*, 3: 176–97.

Crosland, M. (1967): *The Society of Arcueil: A view of French Science at the time of Napoleon I* (Cambridge: Harvard University Press).

—— (1978): "The French Academy of Sciences in the Nineteenth Century," *Minerva*, 16: 73–102.

Dahan-Dalmedico, A. (1986): "Un texte de philosophie mathématique de Gergonne," *Revue d'Histoire des Sciences et de leurs Applications*, 39: 98–126.

Darnton, R. (1968): *Mesmerism and the End of the Enlightenment in France* (Cambridge: Harvard University Press).

Degérando, J. M. (1804): *Histoire comparée des systèmes de philosophie* (Paris: Henrichs).

—— (1810): "Rapport historique sur les progrès de la philosophie depuis 1789 et sur son état actuel," reprinted in the 1847 second edition of *Histoire comparée*, pp. 385–456.

Delambre, J. B. J. (1824): "Analyse des travaux de l'Académie royale des Sciences, pendant l'année 1820, partie mathématique," *Mémoires de l'Académie Royale des Sciences Mathématiques et Physiques de l'Institut de France* (History section), 4: cxxvii–ccii.

Demonferrand, J. B. F. (1823): *Manuel de l'Électricité dynamique, ou Traité sur l'Action mutuelle des conducteurs électriques et des aimans, et sur la nouvelle théorie du magnétisme* (Paris: Bachelier).

Devons, S. (1978): "The Search for Electromagnetic Induction," *Physics Teacher*, 16: 625–31.

Dhombres, N. and Dhombres, J. (1989): *Naissance d'un nouveau pouvoir: sciences et savants en France 1793–1824* (Paris: Editions Payot).

Duhem, P. (1974): *The Aim and Structure of Physical Theory*, trans. P. P. Wiener (New York: Atheneum).

Dumas, J. B. (1839): *Leçons sur la Philosophie chimique* (Brussels: Société Belge de Libraire; reprinted 1972, Catholic University of Louvain).

Emerton, N. E. (1984): *The Scientific Reinterpretation of Form* (Ithaca and London: Cornell University Press).

Faraday, M. (1821a): "On some new Electromagnetical Motions, and on the Theory of Magnetism," *Quarterly Journal of Science*, 12: 74–96; reprinted in *Experimental Researches in Electricity*, 2: 127–47; translated as "Sur les Mouvemens électro-magnétiques et la théorie du magnétisme," *Annales de Chimie et de Physique*, 18: 337–70.

Fisher, N. (1982): "Avagadro, the Chemists, and Historians of Chemistry: Part I," *History of Science*, 20: 77–102.

Flach, L., Costes, C., Boucrot, F. (1986): *Les Moments poétiques d'André-Marie Ampère*, (Lyon: Société des Amis d'Ampère et d'Électricite de France).

Forsyth, A. (1906): *Theory of Differential Equations*, 6 vols (Cambridge: Cambridge University Press; reprint, (New York: Dover, 1959).

Fox, R. (1974): "The Rise and Fall of Laplacian Physics," *Historical Studies in the Physical Sciences*, 4: 89–136.

Frainnet, G. (1903): *Essai sur la Philosophie de Pierre-Simon Ballanche* (Paris: Alphonse Picard et Fils).

Frankel, E. (1974): "The search for a corpuscular theory fo double refraction: Malus, Laplace and the prize competition of 1808," *Centaurus*, 18: 223–46.

—— (1978): "Career-making in post-Revolutionary France: the case of Jean-Baptiste Biot," *British Journal for the History of Science*, 11: 37–48.

Fraser, C. G. (1985): "J. L. Lagrange's Changing Approach to the Foundations of the Calculus of Variation," *Archive for History of Exact Sciences*, 32: 151–91.

394

Fresnel, A. (1820): "Note sur des essais ayant pour but de décomposer l'eau avec un aimant," *Annales de Chimie et de Physique*, 15: 219–22.

—— (1821a): "Comparaison de la supposition des courants autour de l'axe avec celle des courants autour chaque molécule," in J. Joubert (ed.), vol. 2, 1885, pp. 141–3.

—— (1821b): "Deuxième note sur l'hypothèse des courants particulaires," in J. Joubert (ed.), vol. 2, 1885, pp. 144–7.

Fresnel, A. J. (1866–70): *Oeuvres complètes*, edited by H. de Senarmont, E. Verdet, and L. Fresnel, 3 vols (Paris: Imprimerie Impériale).

Fullmer, J. Z. (1975): "Davy's priority in the iodine dispute: further documentary evidence," *Ambix*, 22: 39–51.

Gergonne, J. D. (1817): "De l'analise et de la synthèse, dans les sciences mathématiques," *Annales de Mathématiques Pures et Appliquées*, 7: 345–72.

Giard, L. (1972): "La «dialectique rationalle» de Gergonne," *Revue d'Histoire des Sciences et de leurs Applications*, 25: 97–124.

Gillmor, C. S. (1971): *Coulomb and the Evolution of Physics and Engineering in Eighteenth-Century France* (Princeton: Princeton University Press).

Gooding, D. (1986): "How do scientists reach agreement about novel observations?" *Studies in History and Philosophy of Science*, 17: 205–30.

Gough, J. B. (1981): "The origins of Laboisier's theory of the gaseous state," in H. Woolf (ed.), *The Analytic Spirit: Essays in the History of Science in Honor of Henry Guerlac* (Ithaca, New York, and London: Cornell University Press), pp. 15–39.

Gouhier, H. (1948): *Les Conversions de Maine de Biran* (Paris: J. Vrin).

Goursat, E. (1896): *Leçons sur l'Integration des équations aux dérivées partielles du second ordre à deux variables indépendantes*, 2 vols (Paris: Libraire Scientifique A. Hermann).

Grabiner, Judith (1981): *The Origins of Cauchy's Rigorous Calculus* (Cambridge: MIT Press).

—— (1990): *The Calculus as Algebra: J.-L. Lagrange, 1736–1813* (New York and London: Garland).

Grattan-Guinness, I. (1988): "*Grandes écoles, petitie université*: some puzzled remarks on higher education in mathematics in France, 1795–1840," *History of Universities*, 7: 197–225.

—— (1990a): *Convolutions in French Mathematics, 1800–1840*, 3 vols (Boston: Birkhauser Verlag).

—— (1990b): "The varieties of mechanics by 1800," *Historia Mathematica*, 17: 313–38.

—— (1991): "Lines of mathematical thought in the electrodynamics of Ampère," *Physis*, 28: 115–29.

Hahn, R. (1971): *The Anatomy of a Scientific Institution: The Paris Academy of Sciences, 1666–1803* (Berkeley, University of California Press).

Hamamdjian, P. G. (1978): "Contribution d'Ampère au «théorème d'Ampère»," *Revue d'Histoire des Sciences et de leurs Applications*, 31: 249–68.

Haüy, R. J. (1787): *Exposition raisonné de la théorie de l'électricité et du magnétisme, d'après les principes de M. Aepinus* (Paris: Desaint).

—— (1803): *Traité élémentaire de physique* (Paris: Delance; second edition, 1806).

Heilbron, J. L. (1982): *Elements of Early Modern Physics* (Berkeley and Los Angeles: University of Claifornia Press).

Hofmann, J. (1987a): "Ampère, electrodynamics, and experimental evidence," *Osiris*, 3: 45–76.

—— (1987b): "Ampère's invention of equilibrium apparatus: a response to experimental anomaly," *British Journal for the History of Science*, 20: 309–41.

—— (1988): "Ampère's electrodynamics and the Acceptability of Guiding Assumptions," in A. Donovan, L. Laudan, and R. Laudan (eds), *Scrutinizing Science: Empirical Studies of Scientific Change* (Dordrecht: Kluwer), pp. 201–17.

Home, R. W. (1983): "Poisson's memoirs on electricity: academic politics and a new style in physics," *British Journal for the History of Science*, 16: 239–59.

Jardin, A. and Tudesq, A.-J. (1983): *Restoration and Reaction, 1815–1848*, trans. E. Forster (Cambridge: Cambridge University Press).

Joly, A. (1938): *Un Mystique lyonnais et les secrets de la franc-maçonnerie 1730–1824* (Macon: Protat Frères).

Joubert, J. (ed., 1885–7): *Mémoires sur l'Électrodynamique*, vols II and III of *Collection de Mémoires relatifs à la physique*, 5 vols (Paris: Société Française de Physique).

Kinker, J. (1801): *Essai d'une Exposition succincte de la critique de la raison-pure*, trans. J. Févre (Amsterdam: Changuion & den Hengst).

Kline, M. (1972): *Mathematical Thought from Ancient to Modern Times*, 3 vols (New York and Oxford: Oxford University Press).

Knight, D. (1992): *Humphry Davy: Science and Power* (Oxford: Blackwell Publishers).

Lacroix, S. L. (1806): *Traité élémentaire du calcul différentiel et du calcul intégral*, second edition (Paris: Courcier).

—— (1810): *Traité du Calcul différentiel et du calcul intégral*, vol. 1, second edition (Paris: Courcier).

Lagrange, J. L. (1762a): "Essai d'une nouvelle méthode pour déterminer les maxima et les minima des formules intégrales indéfinies," in *Oeuvres*, vol. 1, 334–62.

—— (1762b): "Application de la méthode exposée dans le mémoire précédent à la solution de différentes problèmes de dynamique," in *Oeuvres*, vol. 1, pp. 365–468.

—— (1788): *Méchanique analitique* (Paris: Desaint).

—— (1806): *Leçons sur le Calcul des fonctions*, in *Oeuvres*, vol. 10, pp. 1–455.

—— (1811): *Mécanique analytique*, vol. 1 (Paris: Courcier).

Laplace, P. S. (1809): "Mémoire sur les movemens de la lumière dans les milieux diaphanes," *Mémoires de la Classe des Sciences Mathématiques et Physiques de l'Institut de France*, 10: 300–42.

La Rive, A. de (1822): "Mémoire sur l'Action qu'exerce le globe terrestre sur une portion mobile de circuit voltaïque," *Annales de Chimie et de Physique*, 21: 24–48.

La Rive, G. de (1821): "Notice sur quelques expériences électro-magnétiques," *Bibliothèque Universelle des Sciences, Belles-lettres, et Arts*, 16: 201–4.

Laumont, G. de (1820): "Note sur les expériences électro-magnétiques de MM. Oersted, Ampère et Arago, relatives à l'identité de l'aimant avec l'électricité," *Annales des Mines*, 5: 535–46.

Launay, L. de (1925): *Le Grand Ampère d'après des documents inédits* (Paris: Perrin).

Launay, L. de (1927): *Un Amoureux de Madame Récamier; le journal de J.-J. Ampère* (Paris: Libraire Ancienne Honoré Champion).

Lavoisier, A. (1789): *Traité élémentaire de Chimie, présenté dans un ordre nouveau et d'après les découvertes modernes*, translated as *Elements of Chemistry, in a new systematic order, containing all the modern discoveries*, by Robert Kerr (Edinburgh, 1790; New York reprint, Dover 1965).

Lemay, P. and Oesper, R. E. (1948): "Pierre Louis Dulong, his life and work," *Chymia*, 1: 171–90.

Lewandowski, M. (1936): *André-Marie Ampère: La Science et La Foi* (Paris: Bernard Grasset).

Libri, G. (1840): "Lettres à un Americain sur l'état des sciences en France; iii: M. Poisson," *Revue des Deux Mondes*, 23: 410–37.

Lützen, J. (1990): *Joseph Liouville 1809–1882: Master of Pure and Applied Mathematics* (New York: Springer-Verlag).

Maine de Biran, F. P. (1920–39): *Oeuvres*, edited by P. Tisserand, 14 vols (Paris: Alcan).

—— (1942): *Oeuvres choisis de Maine de Biran*, edited by H. Gouhier (Paris: Aubier).

Mallez, L. (1936): *A.-M. Ampère, Professeur à Bourg, Membre de la Société d'Émulation de l'Ain, d'après des documents inédits* (Lyon: Marc Camus).

Mascart, E. and Joubert, J. (1896): *Leçons sur l'Électricité et le Magnétisme*, second edition, 2 vols (Paris: Masson).

Mathieu, Jean-Paul, (1990): "Sur le théorème d'Ampère," *Revue d'Histoire de les Sciences et de leurs Applications*, 43: 333–8.

Maurois, A. (1938): *Chateaubriand: Poet Statesman Lover* (New York: Harper).

Mauskopf, S. (1969): "The atomic structural theories of Ampère and Gaudin: molecular speculation and Avagadro's hypothesis," *Isis*, 60: 61–74.

—— (1970): "Haüy's model of chemical equivalence: Daltonian doubts exhumed," *Ambix*, 17: 182–91.

—— (1988): "Molecular geometry in 19th century France: shifts in guiding assumptions," in A. Donovan, L. Laudan and R. Laudan (eds), *Scrutinizing Science: Empirical Studies of Scientific Change* (Dordrecht: Kluwer), pp. 125–44.

Maxwell, J. C. (1891): *A Treatise on Electricity and Magnetism*, 2nd edn, 2 vols (London: Clarendon Press: New York reprint, Dover, 1954).

Mendoza, E. (1985): "Ampère's experimental proof of his law of induction," *European Journal of Physics*, 6: 281–6.

Moigno, F. (1849): *Traité de Télégraphie électrique*, second edition, 1852 (Paris: A. Franck).

Moore, F. C. T. (1970): *The Psychology of Maine de Biran* (Oxford: Clarendon Press).

Oersted, H. C. (1806): "Sur la propagation de l'électricité," *Journal de Physique, de Chimie d'Histoire Naturelle et des Arts*, 62: 369–75.

—— (1813): *Recherches sur l'identité des forces chimiques et électriques*, French translation by Marcel de Serres (Paris: J.G. Dentu).

—— (1820): "Experimenta circa Effectum, etc. Expériences sur l'effet du conflict électrique sur l'aiguille aimantée, par Mr. J. Chr. Oersted, Prof. de physique dans l'université de Copenhague," French translation, *Bibliothèque Universelle des Sciences, Belles-lettres, et Arts*, 14: 274–84 and *Annales de Chimie et de Physique*, 14: 417–25.

Partington, J. R. (1961–1964): *A History of Chemistry*, 4 vols (London: Macmillan).

Petetin, J. H. D. (1787): *Mémoire sur la Découverte des phénomenes que présentent la catalepsie et le somnambulisme* (Lyon; Liechtenstein: Kraus Reprint, 1978).

—— (1808): *Électricité animale* (Paris: Brunot-Labbe).

Poisson, S. D. (1812): "Mémoire sur la Distribution de l'Électricité à la surface des Corps conducteurs," *Mémoires de la Classe des Sciences Mathématiques et Physiques de l'Institut Impérial de France* (1811, part 1): 1–92.

—— (1814a): "Mémoire sur l'Intégration des Equations aux differentielles partielles; par M. Ampère," *Bullétin de la Société Philomathique de Paris*, 1: 107–9.

—— (1814b): "Mémoire sur les équations aux différences partielles," *Bullétin de la Société Philomathique de Paris*, 1: 163–5.

—— (1814c): "Second Mémoire sur la Distribution de l'Électricité à la surface des corps conducteurs," *Mémoires de la Classe des Sciences Mathématiques et Physiques de l'Institut Impérial de France* (1811, part 2): 163–274.

—— (1826a): "Mémoire sur la théorie du magnétisme," *Mémoires de l'Académie Royale des Sciences Mathématiques et Physiques de l'Institut de France*, 5 (1821–2): 247–338.

—— (1826b): "Second Mémoire sur la théorie du magnétisme," *Mémoires de l'Académie Royale des Sciences Mathématiques et Physiques de l'Institut de France*, 5 (1821–2): 488–533.

Potier, A. (ed., 1884): *Mémoires de Coulomb*, vol. I of *Collection de Mémoires relatifs à la Physique*, 5 vols (Paris: Société Française de Physique).

Pouillet, C. (1822): "Sur les phénomènes électromagnétiques," *Annales de Chimie et de Physique*, 21: 77–9.

Prechtl, J. J. (1821a): "Ueber die wahre Beschaffenheit des magnetischen Zustandes des Schliessungs-Drahtes in der Voltäischen Saule," *Annalen der Physik*, 67: 259–76.

—— (1821b): "Zur Theorie des Magneten," *Annalen der Physik*, 68: 187–202.

Rocke, A. (1984): *Chemical Atomism in the Nineteenth Century: From Dalton to Cannizzaro* (Columbus: Ohio State University Press).

Ross, S. (1965): "The search for electromagnetic induction," *Notes and Records of the Royal Society*, 20: 184–219.

Sainte-Beuve, C. A. (1956): "M. Ampère" in *Oeuvres* (Paris: Editions Gallimard), vol. 1, 943–78.

Saint-Hilaire, J. B. (1866): *Philosophie des deux Ampère publiée par J. Barthélemy Saint-Hilaire* (Paris: Didier).

Savary, F. (1821): "Notes relatives au mémoire de M. Faraday," *Annales de Chimie et de Physique*, 18: 370–9.

—— (1823): "Mémoire sur l'Application du Calcul aux Phénomènes électro-dynamiques," *Journal de Physique, de Chimie, d'Histoire Naturelle et des Arts*, 96 (February): 1–25.

Siegfried, R. (1982): "Lavoisier's table of simple substances: its origin and interpretation," *Ambix*, 29: 29–48.

Smith, A. W. (1948): *Electrical Measurements in Theory and Application* (New York: McGraw-Hill).

Taine, Hippolyte (1899): *Les Origines de la France contemporaine*, 22nd ed. (Paris: Hachette).

Thomas, A. L. (1765): "Éloge de René Descartes", in *Oeuvres*, 1819 (Paris: Belin), vol. 1, part I, pp. 476–551.

Tisserand, P. (1916): "Quatre nouveaux manuscrits inédits de Maine de Biran," *Revue de Metaphysique et de Morale*, 23: 295–330.

Vallois, M. (1924): *La Formation de l'influence Kantienne en France* (Paris: F. Alcan).

Valson, C. A. (1886): *La Vie et les Travaux d'André-Marie Ampère*, second edition 1897, (Lyon: Libraire Générale Catholique & Classique).

Vernay, A. (1936): "Histoire du village de Poleymieux pendant les années où Ampère y vécut (1782–1797)," *Bulletin de la Société des Amis d'André-Marie Ampère*, no. 6: 155–81.

Viatte, A. (1928): *Un Ami de Ballanche: Claude-Julien Bredin (1776–1854): Correspondance philosophique et littéraire avec Ballanche* (Paris: Boccard).

Villers, C. (1801): *Philosophie de Kant* (Metz).

Weeks, M. E. (1960): *Discovery of the Elements*, 6th edn (Easton: Journal of Chemical Education).

Williams, L. P. (1953): "Science, education and the French Revolution," *Isis*, 44: 311–30.

—— (1956): "Science, education and Napoleon I," *Isis*, 47: 369–82.

—— (1983): "What were Ampère's earliest discoveries in electrodynamics?" *Isis*, 74: 492–508.

—— (1985a): "Reply," *Isis*, 76: 561.

—— (1985b): "Faraday and Ampère: a critical dialogue," in D. Gooding and F. James (eds) *Faraday Rediscovered: Essays on the Life and Work of Michael Faraday 1791–1867* (New York: Stockton Press), pp. 83–104.

—— (1986): "Why Ampère did not discover electromagnetic induction," *American Journal of Physics*, 54: 306–11.

Williams, L. P. and Steffens, H. J. (eds) (1978): *The History of Science in Western Civilization*, 3 vols (Washington DC: University Press of America).

Index

Académie des Sciences
  and Ampère's reports on
    electrodynamics 236–8
  and the French Revolution 100–2
  Old Regime organization 99–100
  1785 reform 100
  see also Institut National des Sciences
    et des Arts
action at a distance 45
  Ampère's objection 52, 154, 214, 221
Alembert, Jean Le Rond d' 17, 58, 61,
    361
Ampère, André-Marie
  and astronomy 26
  and botany and zoology 13, 18, 24,
    187, 358–61, 365
  and chemistry 194–212
    atomic theory 151, 199–205, 357, 364
    Avagadro's hypothesis 192, 197, 200
    classification of elements 206–12
    relations with Davy 194–6, 205, 210
    electric theory of affinities 203–5,
      214, 266
    fluorine and iodine 195–6, 210
    study of Lavoisier 187–8
    Mariotte's law derivation 196–8
  education 10–18
  and electric current 266–7, 339–40

Ampère, André-Marie (*cont'd*)
  and electric current elements 248–50,
    266–7
  and electric fluid 266–7, 275, 290,
    318–19, 339–40, 349–50
  and electrodynamics
    addition law 237, 238, 248–50,
      253–60
    Ampère's research program 265–8,
      290, 314, 318–19
    rivalry with Biot 268–74, 279–82,
      297–9, 316
    explanation of the Biot-Savart law
      333–6, 344–9
    chronology of 1820 research 236–8
    "crucial experiment" of January,
      1821, 271–4, 276, 322–3
    discovery of electrodynamic forces
      228, 236, 238
    experimental apparatus
      for colinear conductors 316–18
      for equilibrium demonstrations
        228, 250, 254–61, 263–4,
        273–4, 276, 297, 302–5, 315–16,
        322
      using helices 242–7, 250, 258
      for detection of induction 284–7,
        311–15

400

Ampère, André-Marie (*cont'd*)
for parallel conductors 236, 239, 245
for rotary effects 292–302
force law for current elements
derivation 249, 250–2, 306
hypothesis 240–2, 244, 252
direct analysis from equilibrium
demonstrations 273–4, 293, 302–8, 315–16, 322, 337–8, 340–1
early experimental evidence 241–2, 277
with parameter k = 0, 237, 261–4, 292–3, 300
with parameter k = $\frac{1}{2}$, 252, 273–4, 303, 310
as phenomenological law 338–9
reformulation using area theorems 328–30
1826 reformulation 341–3
Newtonian methodology 337–41
response to Oersted's discovery 235, 236, 238
rotary effects 288, 291–302, 332–3, 339
symmetry principle 237, 251, 300, 306
electromagnetic induction 282–8, 309, 310–15, 332–3, 350
ether theories 189, 204, 214
for electrodynamics 216, 266–7, 290, 314, 319, 332–3, 338–40, 344
for optics 164, 214–15, 222–3, 340, 357–8
family 7–8
friendships 88–92
and the Institut 130, 133, 137, 165, 174, 183, 198, 205
as *inspecteur* for the Université Impériale 124, 132, 136, 141–2, 366
and magnetism 267–8, 284–5
molecular hypothesis 283–90, 292, 300–2, 311–15
response to Poisson's theory 327, 331–5, 343–5, 347–9
rejection of transverse magnetism 279
marriage, first 32
marriage, second 127–30

Ampère, André-Marie (*cont'd*)
and mathematics 58–62, 133, 165–6, 173
calculus 80–2, 166–73
calculus of variations 66–79, 305
definition of the derivative 141, 168–9, 171
education 16–18
partial differential equations 173–83
probability 62–6
teaching 105, 138–41, 172–3
and mechanics 72–9, 212–14
and optics 164, 212, 214, 216–19, 221–3
and philosophy 126, 133, 146–7, 164, 166, 362–3
analysis and synthesis 158–61
faits primitifs 155
Kant 147, 362
Leibniz 154
of noumena and phenomena 149–51, 153
of space 147, 151–2
see also analysis and synthesis
and religion 13–14, 18, 25, 32, 38–9, 66, 92–5, 126–7, 133–5, 365
and general scientific methodology 363
analysis and synthesis 158–61, 188
fundamental facts 155
hypotheses 160–4, 202
importance of prediction 56, 162–4, 202, 277, 282, 289, 316–18
and teaching in Bourg and Lyon 31, 33, 36–7, 80
inaugural lecture at Bourg 50–1, 360
and teaching in Paris 103, 105, 124
Collège de France 141–2, 205, 223, 320, 339, 340, 349–50, 359, 360
École Normale 152, 156, 158–61, 223, 230
École Polytechnique 137–43, 172–3, 213–14
Ampère, André-Marie: writings
autobiographical notes 10–11, 16–18, 23
on calculus 80–2, 166, 172, 364
on calculus of variations and mechanics 72–9, 212–14
on chemistry 196, 198–200, 206

Ampère, André-Marie: writings (cont'd)
  on Christianity 92–4, 126
  drama (L'Américide) 25
  on electricity and magnetism (1801
    Bourg manuscript) 50–8, 113, 162, 189
  on electrodynamics 238, 259, 261, 275,
    276
    Exposé méthodique 316, 319–20
    Précis 328–31
    Recueil 321
    Théorie des Phénomènes
      électro-dynamiques 321, 336–7
  Essai sur la Philosophie des Sciences 142,
    156, 356–7, 360–6
  mathematics lectures 141, 172
  on optics 218–19
  on partial differential equations 174
  on philosophy 146–7, 149–50, 152–3,
    155
  poetry 21, 25, 30
  on principle of virtual velocities 212–14
  on probability 62–6, 364
Ampère, Anne-Joséphine-Albine
  [daughter] 128, 354–6
Ampère, Antoinette [sister] 8, 21
Ampère, Jean-Jacques [father]
  career 10, 19–23
  dramatic composition 19
  education of André-Marie 10–14
  execution during the Terror 21–3
Ampère, Jean-Jacques [son] 11, 132, 136,
  142, 151–2
  birth 32, 33
  literary career and Madame Récamier
    353–4
Ampère, Jeanne-Antoinette [mother, née
  Desutières-Sarcey] 7, 22, 129–30, 132
Ampère, Joséphine [sister] 10, 129, 132,
  353
Ampère, Julie [wife, née Cathérine-
  Antoinette Carron]
  courtship 26–32
  illness and death 32–3, 38–9, 85–7
Ampère's law 349
Ampère's theorem 309, 349
analysis and synthesis 15, 144
  Ampère's definitions 158–61
  in Ampère's electrodynamics 273–4,
    293, 302–8, 315–16, 322, 337–8,
    340–1

analysis and synthesis (cont'd)
  Gergonne's definitions 156–8
  in Lavoisier's chemistry 188
Arago, Dominique François Jean 13, 107,
  108, 145, 206, 366
  and electrodynamics 231, 236, 268, 283
  and optics 219–22, 281, 323
atomic theory
  Ampère 151, 199–205
  Berthollet 111, 191
  Dalton 191–2
Avagadro, Amedeo 192
  equal numbers in equal volumes
    hypothesis 192–3, 200, 205

Babinet, Jacques 241, 260, 261, 262–3
Ballanche, Pierre-Simon 89–90, 92, 94–5,
  353, 366
  correspondence with Ampère 133, 205
Barruel, Étienne 104–5
Berthollet, Claude Louis 137, 186
  and Ampère's 1814 memoir 199
  and atomism 111, 191–2
  and the Société d'Arcueil 107–8
Berzelius, Jöns Jacob 193, 203, 278
Binet, Paul René 171–2
binomial theorem 60
Biot, Jean-Baptiste 115, 142, 160, 244
  early career 110–12
  reduction of electromagnetism and
    electrodynamics to magnetism 234,
    278, 280–2, 323
  and Laplacian physics 45, 109, 112–13,
    114, 118–22
  reaction to Oersted's discovery 232–4,
    237, 242, 260, 268–9, 271
  and optics 119, 217, 219, 220–2, 281,
    323
  and the Société d'Arcueil 107–8
Biot-Savart law 232, 278, 319, 322, 323
  and Ampère's theory 333–6, 344–9
Blondel, Christine 229
Bourg: École Centrale 36–7
Bredin, Claude-Julien 88–9, 92, 125, 132
  correspondence with Ampère 126, 130,
    131, 132, 198, 199–200, 206, 274, 291
  discussion of religion with Ampère
    126, 133–5
Bredin, Louis 84, 88
Buche, Joseph 88

Buffon, Georges Louis Leclerc, comte de 13, 206, 207

calculus
in the 18th century 58–62
and Ampère 80–2, 166–73
and Cauchy 59, 61, 140, 167
and Lagrange 62, 166–7
and Newton 59–60
calculus of variations
and Ampère 66–7, 72–6, 78–9, 305
and Lagrange 67–72, 76–8
caloric 51, 185–6, 197, 200
in Ampère's 1801 theory 53–5
and Lavoisier 188
Caneva, Kenneth 215, 235
Carron, Cathérine see Ampère, Julie
Carron, Élise 27, 29–32, 130
correspondence with Ampère 125, 166
Cauchy, Augustin-Louis 133, 138, 166, 168, 172, 173, 183
and calculus 59, 61, 140, 167
as teaching colleague of Ampère 138–41, 172
causes
discussed by Ampère 338, 360, 362–3
discussed by Biot 112
Chalier, Marie Joseph 21
Chaptal, Jean Antoine 105, 107
Chateaubriand, René de 9, 14, 25, 90, 353
chemical elements 184–5
and Ampère 206–12
and Lavoisier 185, 189–91
chemistry
affinities 107, 185, 193, 194, 197, 203–5, 215
and nomenclature 185–7
18th-century revolution 183–7
see also atomic theory
Cheuvreux, Hortense 32, 85, 90
classification 187
in Ampère's Essai 357
by Ampère of partial differential equations 174
in botany 187, 207
of chemical elements 186–7, 206–12, 360
debate between Cuvier and Saint-Hilaire 358–60

Comte, Auguste 119
Condillac, Étienne Bonnot de 35, 156, 188
Coulomb, Charles Augustin 41, 109
Ampère's objections to 51–2
theory of electricity 47–8
scientific methodology 47–9, 108
theory of magnetism 45–8, 275
Couppier [friend] 18, 26
Crosland, Maurice 100, 124
cumul 124, 137, 142–3
Cuvier, Frédéric 133, 147
Cuvier, Georges, baron 103, 133, 142, 147, 196, 354
debate with Saint-Hilaire 358–60

D'Alembert, Jean see Alembert, Jean Le Rond d'
Dalton, John 191–2, 199
Davy, Sir Humphry 203, 205, 210, 292
relations with Ampère 194–6, 205, 210
discovery of fluorine and iodine 195
Degérando, Joseph-Marie 91–2, 124, 127, 133, 145, 146, 147
Delambre, Jean-Baptiste-Joseph 80, 103, 128, 133, 223, 277
Demonferrand, Jean-Baptiste 316, 382
Despretz, César Mansuète 273, 276, 322
Descartes, René 14–16
Dulong, Pierre Louis 107, 108, 211
Dumas, Jean-Baptiste 204–5, 211

Écoles Centrales 35–6
École Normale 152, 156, 158, 223
École Polytechnique
curriculum debates 104–5, 138–41
establishment 35, 103
militarization 104, 137
1816 reform 139
Einstein, Albert 273
electric fluid 158
in Ampère's 1801 theory 53–5
in Ampère's electrodynamics 266–7, 275, 283, 290, 318–19, 339–40, 349–50
and Coulomb 47–8
and Poisson 115–17
electromagnetic induction
Ampère's investigation 280–8, 309, 310–15, 332–3, 350
Faraday's discovery 282, 310–11, 350

electromagnetism
  Ampère's initial reaction 235, 236, 238
  Biot's theory of 232–4, 271
  Oersted's discovery 228, 230–1
Encyclopédie, L' 13–14, 188, 361
Erman, Paul 276, 278, 284
ether theories
  in Ampère's electrodynamics 216,
    266–7, 290, 314, 319, 332–3, 338–40,
    344
  for optics 164, 214–15, 222–3, 340,
    357–8
Euler, Leonhard 68, 71, 77
Euler-Lagrange equations 71, 77, 79

faits primitifs see fundamental facts
Faraday, Michael 195, 235, 268, 292, 313,
    320
  criticism of Ampère 289
  discovery of electromagnetic induction
    282, 310–11, 350
  discovery of rotary effects 291
field theory 349–50
Fourier, Joseph 150, 173, 183, 212, 281
Fox, Robert 40, 119
Freemasons 65, 83–5
French Revolution in Lyon and
    Poleymieux 19–23
  and Ballanche 89–90
  and Bredin 88–9
  and Degérando 91–2
Fresnel, Augustin Jean 119, 136, 141–2,
    166, 219–23
  and Ampère's electrodynamics 283–4,
    289
fundamental facts (*faits primitifs*)
  and Ampère 144, 155, 164, 360
  in Ampère's electrodynamics 252,
    280
  and Biot's electromagnetism 260
  and Haüy 109
  in Laplacian physics 121–2, 155
  and Maine de Biran 147, 155
  in religion 94
  see also fundamental laws
fundamental laws 43–4, 112, 156, 160
  in Ampère's electrodynamics 266–7,
    290, 323, 338–40
  in Biot's magnetic theory 278, 323, 339
  in Laplacian physics 120–1

fundamental laws (*cont'd*)
  in Poisson's theory of electricity 115,
    116
  in Poisson's theory of magnetism 324
  see also phenomenological laws

Gardini, François Joseph 117–18
Gaudin, Marc Antoine 205
Gay-Lussac, Joseph Louis 107, 108, 191,
    193, 194, 196, 198, 202, 205, 206, 222,
    244
  electrodynamic experiment with
    Thenard 237, 261, 263, 303
  magnetic experiment with Welter
    263–4, 279, 322
Gergonne, Joseph-Diez 79
  and analysis and synthesis 156–8
Gooding, David 261
Grabiner, Judith 166, 168
Grattan-Guinness, Ivor 69, 229
Guyton de Morveau, Louis-Bernard 139,
    185, 186, 187

Hahn, Roger 102–3
Haüy, René-Just 109
  and crystallography 194
  and Laplacian physics 51, 109–10, 118,
    194
Huygens, Christiaan 121, 217–18
hypotheses
  in Ampère's electrodynamics 240–2,
    244
  in Ampère's metaphysics 150–1, 153
  in Ampère's methodology 160–4, 202
  discussed by Berthollet 191
  discussed by Biot 112
  discussed by Haüy 110
  in optics 220

idéologues 35, 102, 125, 146
imponderable fluids 41, 45, 110, 112, 117,
    118, 119, 154, 217
  see also caloric; electric fluid; magnetic
    fluid
induction *see* electromagnetic induction
Institut National des Sciences et des Arts
  and Ampère 130, 133, 137, 165, 174,
    183, 198, 205
  1795 ratification 102
  1803 reform 102–3

Institut National des Sciences et des Arts (cont'd)
  and Laplace 106
  and Poisson's election 117
  see also Académie des Sciences
isoperimetric problems 68

Kant, Immanuel 146, 147–9, 215
Kinker, Joseph 146, 148–9
Kline, Morris 173–4

Lacroix, Sylvestre-François 31, 68, 111, 115, 137, 138, 139, 140, 156, 172
Lagrange, Joseph Louis de 105, 113, 115, 137
  and Ampère 18, 128
  and calculus 62, 166–7
  mechanics and calculus of variations 67–72, 76–8
  principle of virtual velocities 43–4, 212
Lalande, Joseph-Jérôme Lefrançois 26, 58, 65–6, 80
Lamarck, Jean-Baptiste de 365
Laplace, Pierre Simon
  and Ampère's probability memoir 64
  and Biot 111
  and the Biot-Savart law 233–4, 282
  career and publications 40–1, 106–7, 111–12, 139, 173
  and the École Polytechnique 105, 137–9
  and the Institut 106, 133
  and optics 121, 217–21
  and Poisson 115
  and the Société d'Arcueil 107–8
Laplacian physics 40–1, 44–5, 154
  and Biot 45, 109, 112–13, 114, 118–22
  and Poisson 115–18
  principles of a physical theory 119–20
La Rive, Auguste de 288, 310, 311, 317
La Rive, Gaspard de 268, 285
Laumont, Gillet de 253
Launay, Louis de 1, 7, 16, 127, 198
Lavoisier, Antoine Laurent 83, 100, 102, 161, 162
  on conservation of mass 184
  Elementary Treatise on Chemistry 184, 187–91
  on elements 185, 189–91, 207

Lavoisier, Antoine Laurent (cont'd)
  Method of Chemical Nomenclature 186–7
  methodology 188
  and oxygen 185–7, 189
Leibniz, Gottfried 58, 74, 77, 154
Libri, G. 114
Linnaeus, Carl 24, 187, 207
Liouville, Joseph 142, 172, 321, 349, 378
Lyon 8, 82–5, 92

magnetic fluid
  in Ampère's 1801 theory 53–5
  rejected by Ampère 265
  in Biot's theory 232–3, 260, 271
  in Coulomb 45–9
  in Kant 148
  in Laplacian physics 45
  in Poisson's theory 324–7
magnetism
  Coulomb's theory 45–9
  Poisson's theory 323–7
  see also Ampère and magnetism, Biot
magnetism, animal see mesmerism
Maine de Biran, François-Pierre 133, 145, 146–7, 149, 150, 155, 166
  critic of Ampère's philosophy 144, 150–1, 153
Malus, Étienne Louis 107, 108, 115, 217–19
Maurice, Frédéric 147, 275, 276
Maxwell, James Clerk 229, 341, 349
Mendeleyev, Dmitri 207
Mendoza, Eric 311–12, 314
Mesmer, Franz-Anton 83–4
mesmerism 82–8, 110, 274
metaphysics 126, 144, 166
Moigno, Abbé 339
Monge, Gaspard 105, 137, 139, 182
moulinet électrique 318–19

Napoleon 36, 39, 57, 66, 102, 106, 111, 123, 125, 132, 136
  and the École Polytechnique 104, 137
  and the Institut 102–3
Naturphilosophie 214–15
Newton, Sir Isaac 68, 161, 162, 337
  and calculus 59–60
  and gravitation theory 44, 47, 112, 159
noumena and phenomena
  for Ampère 149–51, 154, 267
  for Kant 147–9

Oersted, Hans Christian 192, 215–17
  and chemical and electrical forces 215
  discovery of electromagnetism 228,
    230–1
  and electric current 216
optics 214–23
  Ampère's generalization of Huygens'
    construction 218–19
  double refraction 217–19
  Fresnel's wave theory 153, 156, 164
  Laplacian particle theory 217–18
  see also ether theories
osculatory curves 81–2
Ozanam, Antoine-Frédéric 136, 355, 366

Paris 124
Périsse, Jean-Marie (Marsil) 28, 31, 84
Petetin, Jacques Henri Désiré 85–8, 274
phenomena see noumena and
  phenomena
phenomenological laws 42–3, 108,
    109–10, 111, 115, 119, 144, 145, 155,
    160, 358
  in Ampère's electrodynamics 223, 228,
    234, 239, 281, 290, 323, 338–9
  in Biot's magnetic theory 232, 242,
    268, 271, 278, 281, 323
  in chemistry 191, 192
  in optics 217, 218
  see also fundamental laws
phlogiston 185
Planta, Sebastian, 150–1
Poinsot, Louis 137, 138–9, 212, 220
Poisson, Siméon-Denis 45, 107, 113–15,
    137, 146, 173, 222
  electricity memoirs 115–18
  and Laplacian physics 119–21
  magnetic theory 323–7
  review of Ampère's memoir on partial
    differential equations 182
Poleymieux 8–10, 135
Popper, Sir Karl 161
positivism 119, 146, 184
Potot, Jeanne-François (Jenny) 127–30, 354
Pouillet, Claude 141, 316, 319
Prechtl, Johann Joseph 278, 303
primitive facts see fundamental facts
principle of virtual velocities 43–4, 69,
    212–14
Prony, Gaspard de 137, 138, 212

Récamier, Juliette 90, 136, 353–4
Restoration 90, 123–4, 138
Ride, Gabriel 354–6
Rocke, Alan 191
rotary effects
  and Ampère 291–302
  and Biot 323
  and Faraday 291
Rousseau, Jean-Jacques
  educational philosophy 11–13
  essays on botany 24, 207
Roux-Bordier, Jaques 90–1, 131, 198, 275

Saint-Hilaire, Étienne Geoffroy 358–9
Savart, Félix 232, 271
Savary, Félix 213, 264, 273, 289, 290, 291,
    292, 307, 321–3
  and Ampère's force law 322
Société d'Arcueil 107–8, 118
Société des Amis d'André-Marie
  Ampère 10
Société Chrétienne 92–4, 133
Société Philomathique 102, 107, 280
Société Philosophique 147
synthesis see analysis and synthesis

Taylor's theorem 170
terrestrial magnetism
  Ampère's 1820 investigation 242,
    244–5, 294, 297
  and Oersted's discovery 238
Thenard, Louis Jacques 107, 108, 194,
    198
  electrodynamic experiment with
    Gay-Lussac 237, 261, 263
Thomas, Antoine Leonard 14–16
transverse magnetism 278–9
  Ampère's counterarguments 279, 297–9

Université Impériale 136
  and Ampère as inspecteur 124, 132,
    136, 141–2, 366

Valson, Claude Alphonse 92
Van Beek, Albert 285–6, 288, 293, 339
Villers, Charles 146, 148–9

Welter, J. J. 263–4, 279, 322
Willermoz, Jean-Baptiste 83, 84
Williams, L. Pearce xiv, 229, 235, 238, 369